RESTORATION OF TROPICAL FOREST ECOSYSTEMS

Tasks for vegetation science 30

Series Editors

Restoration of Tropical Forest Ecosystems

Proceedings of the Symposium held on
October 7–10, 1991

edited by

HELMUT LIETH and MARTINA LOHMANN

KLUWER ACADEMIC PUBLISHERS
DORDRECHT / BOSTON / LONDON

Library of Congress Cataloging-in-Publication Data

Restoration of tropical forest ecosystems : proceedings of the symposium held on October 7-10, 1991 / edited by Helmut Lieth and Martina Lohmann.
p. cm. -- (Tasks for vegetation science ; 30)
"Organized on behalf of ISTE [and] INTECOL ..."
Includes index.
ISBN 0-7923-1945-1 (acid free)
1. Forests and forestry--Tropics--Congresses. 2. Forest ecology--Tropics--Congresses. 3. Rain forest ecology--Congresses. 4. Restoration ecology--Tropics--Congresses. 5. Reforestation--Tropics--Congresses. 6. Sustainable forestry--Tropics--Congresses. 7. Forest conservation--Tropics--Congresses. 8. Rain forest conservation--Congresses. I. Lieth, Helmut. II. Lohmann, Martina. III. International Society for Tropical Ecology. IV. International Association for Ecology. V. Series.
SD247.R47 1993
634.9'56'0913--dc20 92-33607

ISBN-0-7923-1945-1

The Symposium on: *Restoration of Tropical Forest Ecosystems* was organized in association with:
ISTE, Prof. Dr. H. Lieth, president; INTECOL e. V., Prof. Dr. W. Haber, president;
by Prof. Dr. H. Lieth and Prof. Dr. Drs. h. c. A. Miyawaki

Organizational office: AG Systemforschung, University of Osnabrück, D-4500 Osnabrück, Germany

Organizational Secretary: Dipl. Biol. Martina Lohmann, Osnabrück, Germany

Organizer of poster and video session: Dr. Johannes R. Rheker, Bonn, Germany

Published by Kluwer Academic Publishers,
P.O. Box 17, 3300 AA Dordrecht, The Netherlands.

Kluwer Academic Publishers incorporates
the publishing programmes of
D. Reidel, Martinus Nijhoff, Dr. W. Junk and MTP Press.

Sold and distributed in the U.S.A. and Canada
by Kluwer Academic Publishers,
101 Philip Drive, Norwell, MA 02061, U.S.A.

In all other countries, sold and distributed
by Kluwer Academic Publishers Group,
P.O. Box 322, 3300 AH Dordrecht, The Netherlands.

Printed on acid-free paper

Printed in the Netherlands

Table of Contents

Preface

The destruction of the tropical forests proceeds rapidly. We all know that this has global ecological and economical consequences. The problem is of such magnitude that it can only be compared to warfare. The destruction of tropical forests is not only detrimental to the global ecology but also poses a serious threat to the people living in this area. Furthermore the over-utilization of such a valuable resource poses a serious threat to the next generations.

Apart from the problem generated for the people in those regions and on earth in general there is a moral obligation to preserve the vast biological diversity in the tropical forests. We have a commitment to all species on earth as well as man and, therefore, need to preserve as many of them as possible, if not all.

The threats to the tropical forests come from two main directions: First from increasing population in the tropical countries, and second from large improper timber harvesting operations in the tropics.

Mitsubishi Corporation had asked to organize this symposium. The company was attacked as being a member of the second group but was under the impression that they were acting in the best interests of the population when they initiated their logging operations, until such time when environmentalists came and demonstrated against their activities. Therefore they asked the initiators of this symposium to have the problem discussed openly. If they were doing something which was not in the interests of the local population or to the global environment, they wanted to know about it and they wanted to hear the solutions to the problem.

The discussion during this symposium included all aspects of the problem. We had invited people who could discuss the problem from different angles: forestry, ecology and nature conservation. We had a lively discussion but no outright demonstrations or accusations. Nobody at the symposium believed that the tropical forest area would remain untouched. The population explosion takes care of that argument. The two main problem areas before us are first the wise utilization of that portion of the forest which will be used — especially the introduction of planned forestry in such areas, and second, the development of a good plan for nature conservation in the tropics.

The papers presented at the symposium will most certainly not solve all the problems but we hope they contribute to the very much needed, continued discussion of possible solutions which must be implemented in the near future.

The organizers of the symposium thank the officials of Mitsubishi Intl. who sponsored this meeting and especially the officers of its Düsseldorf head office, Mr. Mita, Mr. Hatanaka, Mrs. Linke, Mrs. Littig, Mrs. Shulze and Mr. Serbesow, as well as Mrs. Bruns and Dr. Rheker for their help in organizing the meeting.

As a follow up of the symposium in Bonn, another International Symposium on Rehabilitation of Tropical Rainforest Ecosystems was held in Kuching/Sarawak, Malaysia on Sept. 2nd–7th 1992. During this symposium the problems with integrating the "forest people", the Penan, were presented by Hatta and Jayl who are actively engaged in integrating these people into a modern sociey.

The topic of the Penan was presented by NGOs during the Bonn symposium. We had the government of Malaysia invited to present their views but the invitation came too late.

We are, therefore, glad that the authors of the paper and the organizers of the symposium in Kuching permitted us to add that paper to the publication of the Bonn symposium. The editors of this volume are especially thankful because it shows now the problem of the Penan in Sarawak in a different light.

The inclusion of this paper initiated a further

H. Lieth and M. Lohmann (eds.), Restoration of Tropical Forest Ecosystems, ix–x.

discussion of the paper presented by Wakker and Hüttche who were invited to provide a written version of their verbal presentation and videos. The paper by Hatta and Jayl shows the problems of forest dwelling populations from a different perspective. The juxtaposition of the two papers in the context of this volume was kept intentional because this problem requires careful attention in the future use of tropical forest utilization.

Osnabrück and Yokohama

A. Miyawaki, M. Lohmann, H. Lieth

List of Contributors

1. Written and oral contributions at the symposium

Abas bin Said, Centre of Applied Sciences, University of Agriculture Malaysia
R. S. Ambasht, Department of Botany, Banares Hindu University
Ashari Muktar, Faculty of Forestry, Universiti Pertanian Malaysia
B. Böer, University of Osnabrück
B. J. Brown, Institute for Environmental Studies, University of Wisconsin–Madison
E.-F. Bruenig, Department of Tropical Forestry, University of Hamburg
M. Clüsener Godt, Division of Ecological Sciences, UNESCO
Demel Teketay, Swedish University of Agricultural Sciences, Faculty of Forestry
J. B. S. Ferraz, Institute Nacional de Pesquisas da Amazônia
K. Fujiwara, Yokohama National University
W. Haber, Abteilung für Landschaftsökologie, TU München
M. Hadley, Division of Ecological Sciences, UNESCO
C. Hüttche, Robin Wood
J. Huss, Institute of Silviculture, University of Freiburg
F. W. Kigenyi, Uganda
D. Lamb, Botany Department, University of Queensland
J. Langub, Majlis Adat Istiadat, Sarawak, Kuching, Sarawak, Malaysia
P. Lawrence, Botany Department, University of Queensland
H. Lieth, Systems Research Group, University of Osnabrück
G. Maury-Lechon, CNRS
F. P. Malaisse, U.R. Ecologie, Faculté des Sciences agronomique
T. Mita, Administration and Coordination Division, Mitsubishi Intl.
A. Miyawaki, Institute for Environmental Science and Technology, Yokohama National University
K. Mori, Environmental Affairs Department, Mitsubishi Corp.
P. Müller, Institut für Biogeographie, Universität des Saarlandes
M. Ozawa, Environmental Affairs Department, Mitsubishi Corp.
J. A. Parrotta, Institute of Tropical Forestry, USDA Forest Service
G. J. Ray, Institute for Environmental Studies, University of Madison–Wisconsin
J. R. Rheker, Department of Geography, University of Bonn
H. Solhee, Department of Development, Kuching, Sarawak, Malaysia
M. Sutisna, Institute of Silviculture, University of Freiburg
O. Setzer, Forest Management Institute Brandys
J. Vanclay, Faculty of Forestry, Royal Veterinary and Agricultural University
E. Wakker, Friends of the Earth
J. Warnken, University of Osnabrück
H. J. Weidelt, Institut für Waldbau, Universität Göttingen

2. List of reviewers

R. S. Ambasht, Department of Botany, Banares Hindu University
B.J. Brown, Institute for Environmental Studies, University of Wisconsin–Madison
E.-F. Bruenig, Institute for World Forestry, University of Hamburg
M. Clüsener Godt, Division of Ecological Sciences, UNESCO

S. Elliot, Department of Biology, Chiang Mai University
J. B. S. Ferraz, Instituto Nacional de Pesquisas da Amazônia
D. Lamb, Botany Department, University of Queensland
H. Lieth, Systems Research Group, University of Osnabrück
F. P. Malaisse, U.R. Ecologie, Faculté des Sciences agronomiques
J. A. Parrotta, Institute of Tropical Forestry, USDA Forest Service
G. J. Ray, Institute for Environmental Studies, University of Wisconsin–Madison
J. R. Rheker, Institute of Geography, University of Bonn
J. Vanclay, Faculty of Forestry, Royal Veterinary and Agricultural University

3. List of participants at the discussion

R. S. Ambasht, Department of Botany, Banares Hindu University
Ashari Muktar, Faculty of Forestry, Universiti Pertanian Malaysia
M. Clüsener Godt, Division of Ecological Sciences, UNESCO
J. B. S. Ferraz, Institute Nacional de Pesquisas da Amazônia
W. Haber, Abteilung für Landschaftsökologie, TU München
C. Hüttche, Robin Wood
F. W. Kigenyi, Uganda
D. Lamb, Botany Department, University of Queensland
H. Lieth, Systems Research Group, University of Osnabrück
G. Maury-Lechon, CNRS
T. Mita, Administration and Coordination Division, Mitsubishi Intl.
A. Miyawaki, Institute for Environmental Science and Technology, Yokohama National University
M. Ozawa, Environmental Affairs Department, Mitsubishi Corp.
J. A. Parrotta, Institute of Tropical Forestry, USDA Forest Service
G. J. Ray, Institute for Environmental Studies, University of Madison–Wisconsin
J. Vanclay, Faculty of Forestry, Royal Veterinary and Agricultural University
E. Wakker, Friends of the Earth
A. Ziegler, Robin Wood

PART ONE

Contributions to the problem of restoration and rehabilitation of disturbed areas

Introduction

The symposium was partially organized to obtain a survey of the problems with restoration and rehabilitation of tropical forests from many angles. We wanted to learn how the forest restoration was approached in the humid tropics of South America, Africa and Southeast Asia. The speakers were asked to present their case studies in such a way that ecological and socio-economic problems were covered along with the likely results of their projects or studies.

Dr. Miyawaki presented his philosophy of rebuilding natural ecosystems in the tropics based on phytosociological studies prior to selecting tree species to be used for the plantations. He started his work in Malaysia based on his extensive experience with restoration projects in Japan.

Drs. Clüsener-Godt and Hadley then explained the various projects of UNESCO in helping regeneration and restoration of tropical forests. Dr. Maury-Lechon presented her experimental work in improving the development of valuable tree seedlings, mainly dipterocarp species, from different proveniences in order to maintain the genetic resources of such valuable timber species.

After these more general topics, eight specific projects of forest restoration were presented; three from the Americas, two from Africa and three from Southeast Asia.

The topics of concern were different in each project. They begin with problems in nurseries and extend to successional analyses of various forest stands on their way to climax forest. Reforestation trials on heavily degraded sites, and even mining spoils, were reported as well as the greening of urban areas with suitable species from the nearby tropical forests. Included in the presentations was the first report of the joint venture for a careful ecological analysis of the rebuilding of a tropical rainforest ecosystem near the University Campus in Bintulu, Sarawak, Malaysia.

Sarawak, Malaysia

H. Lieth and M. Lohmann (eds.), Restoration of Tropical Forest Ecosystems, 3.

1. Restoration of native forests from Japan to Malaysia

AKIRA MIYAWAKI
Yokohama National University, Yokohama, Japan

Keywords: Tropical forest restoration, Malaysia, Dipterocarpaceae, succession, Japan

Abstract. The paper describes the method and some applications of accelerating the succession from barren land to tropical forest ecosystems following the "Miyawaki method".

It has been proven that restoration of native forest is possible by collecting seeds and seedlings of climax species in the forest, raising them in a nursery and after adaptation planting the young plants on sites adequately prepared.

The species to be planted were selected from a phytosociological study of comparable sites still covered by natural vegetation.

The paper presents the success of older projects of greenification in Japan and the first steps of an experimental project in Bintulu at the campus of the University of Agriculture of Malaysia. After the success in the earlier steps of the restoration of a tropical forest ecosystem we shall concentrate in the future on rehabilitating and monitoring the development of the field stratum and the soil animal associations of this ecosystem.

I. Introduction

Forests, especially tropical forests, are one of the most important bases for human life and development. Forests of various types once covered about 30% of the land area (Lieth 1973). This area was sharply reduced during the last hundred years; in recent decades mainly due to the destruction of tropical forest types. Forests cover at present about 20% of the land area (Miyawaki 1991a) and are still declining.

Forests contain a substantial portion of the biological resources, richness and diversity. In the tropical forests this is especially relevant, because we know so little about the species diversity and especially the gene pools residing there which we expect to need in the future. The forests hold unlimited possibilities. The concentration of tropical forests is in three big tropical regions. The tropical forests, which are called the treasury of biological resources, have been lost in the last 40 years, due to activities of the developed countries. More than 40% of the tropical forests will be lost by the beginning of the 21st century if they are left as they are (Ishi 1988).

We must regenerate, restore and rehabilitate the forests in order to avoid the loss and devastation of these ecosystems. We must plant, however, not only exotic trees but also native trees in order to restore forest ecosystems. Fortunately, in tropical areas, native trees have also great commercial value. Many exotic trees, such as *Eucalyptus* from Australia and *Pinus* spp. from America were once planted in tropical areas (Miyawaki et al. 1982a). They needed a great deal of management. According to our field survey in Borneo, native trees bear fruit every year (Miyawaki et al. 1982b). These seeds can be planted in pots. It is therefore expected that native trees can be planted in tropical areas. We have to guarantee the tropical forest ecosystem, the life, mind, culture and gene pool, and finally we have to guarantee the wood resources for the future (Miyawaki 1992).

Traditionally, the Japanese created groves around village shrines and temples, which are the typical "native trees in the native land" (Miyawaki 1979, 1984, 1989a). At present, people all over the world are watching the concrete achievement of the village-shrine groves. A comprehensive study of their ecology, their biological resources and environments has been recently published by Miyawaki et al. (1982).

We have succeeded in recreation of environmental protection forests, which are based on exact field surveys and the ecological scenario for these 20 years (Miyawaki 1975, 1981b, 1982a, 1984, 1991b; Miyawaki et al., *Vegetation*

H. Lieth and M. Lohmann (eds.), Restoration of Tropical Forest Ecosystems, 5–24.

of Japan vol. 1∼10, 1980–1989). We planted native trees in more than 280 places from Hokkaido in the north to subtropical Okinawa, including Tokyo Bay, Osaka Bay and Nagoya Bay, under the sponsorship of various enterprises, several government offices and local public bodies which have much foresight and provided substantial cooperation (Fig. 1, Miyawaki et al. 1983; Miyawaki et al. 1979; Miyawaki et al. 1987; Miyawaki et al. 1974).

We have attempted to rehabilitate tropical forests since 1990, under the sponsorship of the Mitsubishi Corporation (Miyawaki 1992). This study is based on the successes at many places in Japan and on some results of field surveys in Southeast Asia, which have been studied since 1978, supported by a Grant-in-aid for overseas Scientific Survey of the Japanese Ministry of Education, Science and Culture (Miyawaki 1982b). It is said that it may be impossible to restore the tropical forests once they have been destroyed. On July 15, 1991, we planted, in a joint study, more than 10,000 young trees of 20 species, including lauans (main native trees such as *Shorea* spp., *Dipterocarpus* spp. and *Hopea* spp., etc. in tropical rain forests), at the 800 ha Bintulu Campus of Universití Pertanian Malaysia in Sarawak (Borneo Island) (Miyawaki 1991b). This may be the first systematic trial based on ecological field studies anywhere in the world. When we started this trial, I introduced it with a speech from which I want to repeat the final paragraph here:

> The period when we accuse opponents is over. We should positively wrestle with recreation of green environments and the tropical forest ecosystems in order to survive on our blue planet. The "Native Trees in the Native Land" campaign, which is based on the biological scenario, must be promoted on a glob al scale. With the passing years, native trees grow without fail. I hope that everybody is sweating to rehabilitate the global environment based on the scientific direction. Everybody is expected to make an effort, based on his or her situation, in order to rehabilitate global environments. Let's try to contribute to sustainable development of the tropical forests, starting with Bintulu, Sarawak, Malaysia.

II. Subject and study field

East Asia is one of the three areas in the world where tropical forests continue for miles. Other areas, such as the Taiga forests (boreal coniferous forests), cold-temperature deciduous forests, warm-temperate evergreen broad leaved forests (Laurel forests), subtropical forests and tropical forests, have many rifts. In the case of Japan, 3,000 km from south to north were once covered by forests (Miyawaki 1989b). But recently the forests disappeared rapidly due to development of industry, urbanization and public transport systems. Especially, the forests in lowlands where almost all of the Japanese have settled, have been forced to disappear (Miyawaki 1976; Miyawaki (Ed.) 1980–1989). In this circumstance, a very important subject in making new cities and industries and traffic facilities is how to use the title role of ecosystems, with living vegetation as a material to cope with iron, cement, petrochemical products and all other kinds of energy.

We have been making an effort to restore the forests since 1973, under the sponsorship of some enterprises, local public bodies and some ministries who have great foresight. Our effort is based on field surveys (Braun-Blanquet 1928, 1951, 1964) which have been performed since 1952 (Miyawaki et al. 1974–1990). The work procedure is explained in Fig. 2. We proceeded with the restoration of green environments actively in these 20 years, from Senpaku Industrial Town in Hokkaido (Miyawaki 1988) to subtropical Okinawa (Miyawaki et al. 1986). The number of places surveyed is 286, including Tokyo Bay, Osaka Bay and Nagoya Bay, where it was said that native forests could not be restored within 300 years. Main study sites are environs of steel works, power stations, factories of automobiles and textiles, roads, schools, sewage disposal places, airports, bays and new towns (Miyawaki, Fujiwara & Box 1987) (Fig. 1, Table 1).

Since 1990 we have started to restore natural ecosystems of tropical rain forests, based on surveys from Japan, Kalimantan in Indonesia (from 1978, Miyawaki 1982b), Sarawak in Malaysia (from 1990, Miyawaki 1992), mangrove forests in Thailand (from 1980, Miyawaki

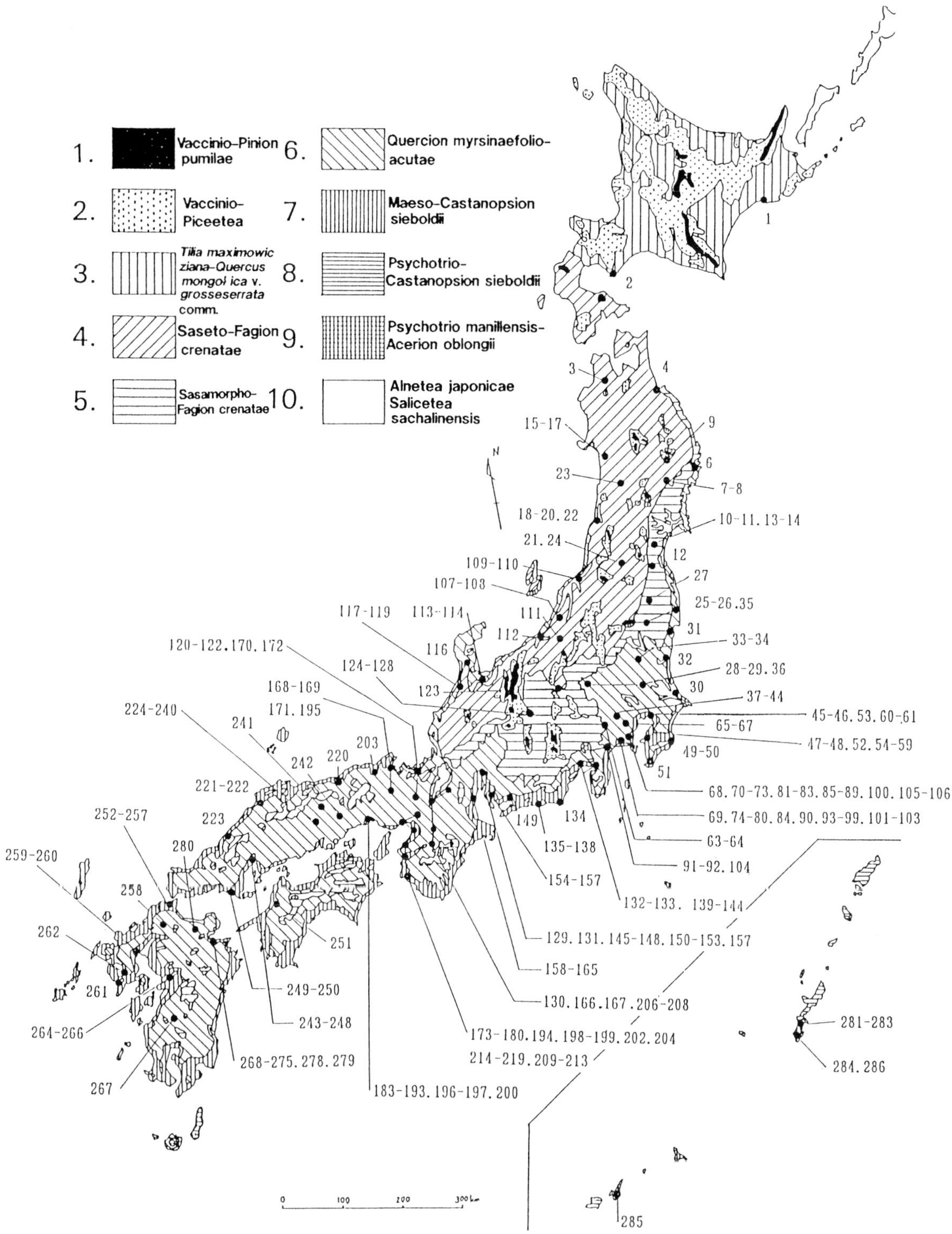

Fig. 1. Locations of recently built "environmental protection forests" in Japan, with regions of potential natural vegetation. The numbers in this figure correspond to their entries in Table 1.

Table 1. List of projects to create green environment and/or restore native forest ecosystems. S.C. = JUSCO Shopping center. TEPCO = Tokyo Electric Power Company. The numbers refer to their entry in Fig. 1

No.	Prefecture	Area
1.	Hokkaido	Senpaku industrial area
2.		Muroran works (Nippon Steel Co.)
3.	Aomori	Kashiwa-mura S.C.
4.		Hatinohe Simoda-cho S.C.
5.	Iwate	East Japan Railway Co. (EJR)
6.		Kamaisi works (Nippon Steel Co.)
7.		Esasi S.C.
8.		Maesawa S.C.
9.		Yahaba S.C.
10.	Miyagi	Tagajo S.C.
11.		Kagitori S.C.
12.		Siraisi factory
13.		Sendai EJR
14.		Sendai Port
15.	Akita	Akita-shi Goshono S.C.
16.		Akita-shintosi S.C.
17.		Takanosu S.C.
18.	Yamagata	Sakata-shi Izumi elementary school
19.		Sakata-shi Daisan junior high school
20.		Sewerage plant
21.		Takahata-cho ground
22.		Sakata-omiya S.C.
23.		Shin-yokote S.C.
24.		Yonezawa S.C.
25.	Fukushima	Hirono power plant (TEPCO)
26.		Tomioka power plant (TEPCO)
27.		Hiwada
28.	Ibaragi	National environment institute
29.		Kasumigaura basin sewage plant
30.		Kashima kyoudo thermal power plant
31.		Kita-ibaragi S.C.
32.		Toukai-mura S.C.
33.	Tochigi	Tochigi proving ground (HONDA)
34.		Mooka parts center (HONDA)
35.	Gunma	Numappara yousui power plant
36.		Takasaki station (EJR)
37.	Saitama	Saitama manufacturing Wakou factory (HONDA)
38.		Saitama manufacturing (HONDA)
39.		Wakou-shi Wakou institute (HONDA)
40.		Wakou-shi Wakou factory (HONDA)
41.		Asagiri-shi Asagiri institute
42.		Moriya new-town (MITUI real estate co.)
43.		Yosikawa S.C.
44.		Omiya-mihashi S.C.
45.	Chiba	Narashino-shi
46.		Honda motorpool
47.		Kimitsu-shi seaside park
48.		Futtsu S.C.
49.		Ohara S.C.
50.		Ohamishirasato-cho S.C.
51.		Tateyama S.C.
52.		Kisarazu S.C.
53.		Kamatori S.C.
54.		Goi thermal power plant (TEPCO)
55.		Anegasaki thermal power plant (TEPCO)
56.		Sodegaura thermal power plant (TEPCO)
57.		Futtsu thermal power plant (TEPCO)
58.		Futtsu park
59.		Kimitsu manufacturing plant (Nippon Steel Co.)
60.		Chiba factory (TORAY)
61.		Ichihara
62.		Kimitsu works (Nippon Steel Co.)
63–64.	Tokyo	Akigawa-shi
65.		Harajuku station (EJR)
66.		Yotsuya station (EJR)
67.		Tamachi station (EJR)
68.	Kanagawa	Isogo thermal power plant
69.		Sagamihara S.C.
70.		Yokohama national university
71.		Yokohama national university attached Tateno Yokohama elementary school
72.		Yokohama national university attached Yamate elementary school
73.		Kawasaki-shi Yurigaoka housing
74–78.		Shyonan kokusai-mura
79.		Yokosuka Kurihama housing
80.		Kurihama high school
81.		Yokohama factory (NISSAN)
82.		Kanazawa factory (MITSUBISHI)
83.		Enoshima women's center
84.		Fujisawa beach
85.		Hokubu sewage plant, Yokohama City
86.		Felice jogakuin high school
87.		Yokohama municipal university
88.		Yokohama municipal university the school of medicine
89.		Foreign countries institute center
90.		Yokosuka-shi Yokosuka factory

Table 1. (Continued)

No.	Prefecture	Area	No.	Prefecture	Area
91.		Hakone sewage plant	137.		Hamamatu City Expressway
92.		Tanzawa mountain road	138.		Hamakita S.C.
93.		Shyonan tsujido beach	139–143.		Fuji City road, factory, ground, reservoir, central park
94.		Shiomidai beach			
95.		Yanagisima beach	144.		Iwamotoyama kindergarten, elementary school
96.		Hiratsuka beach			
97.		Oiso high school	145.	Aichi	Nagoya airport
98.		Sakawaga-nishi repair office	146.		Nagoya-shi Nanyo S.C.
99.		Chigasaki beach sewage plant	147.		Yagoto S.C.
100.		Higasi-Ogishima LNG power plant (TEPCO)	148.		Nagoya-nishi S.C.
			149.		Daigo bureau of harbor construction
101.		Ishinazaka incineration plant			
102.		Oshimizu sewage plant	150.		Nagoya port
103.		Imaizumi Daini junior high school	151.		Aichi office district (TORAY)
104.		Odawara elementary school	152.		Nagoya office district (TORAY)
105.		National highway 16 (Sakonyama)	153.		Tokai office district (TORAY)
			154.		Kariya-shi elementary school, junior high school
106.		National highway 16 Kanazawa junction			
107.	Niigata	Ojiya elementary school	155–157.		Handa-shi elementary school, junior high school
108.		Nagaoka-shi housing lot			
109.		Niigata-higashi S.C.			
110.		Shin-Murakami S.C.	157.		Nagoya works (Nippon Steel Co.)
111.		Muikamachi S.C.	158.	Mie	Suzuka manufacturing plant (HONDA)
112.		Joetsu S.C.			
113.	Toyama	Inariyama park	159.		Shin-Hisai S.C.
114.		Tonami S.C.	160.		Matsuzaka S.C.
115.	Isikawa	Ishikawa factory (TORAY)	161.		Mie S.C.
116.		Noto-Kashima S.C.	162.		Taian S.C.
117.		Kindaigakuen-toshi S.C.	163.		Yokkaichi S.C.
118.		Kaga S.C.	164.		Ueno S.C.
119.		Mattou S.C.	165.		Ohyachi S.C.
120.	Fukui	Daiichi, Daini nuclear power plant (Kansai Electric Power Co., Inc.)	166.	Shiga	Nishi-Otsu S.C.
			167.		Tsuchiyama S.C.
			168.	Kyoto	Miyatsu energy institute (Kansai Electric Power Co., Inc.)
121.		Takahama nuclear power plant (Kansai Electric Power Co., Inc.)			
			169.		Miyatsu gasturbine power plant (Kansai Electric Power Co., Inc.)
122.		Mihama nuclear power plant (Kansai Electric Power Co., Inc.)			
123.	Nagano	Ueda-shi, HOKI Co.	170.		Kisenyama power plant (Kansai Electric Power Co., Inc.)
124.		Enna-tunnel JR			
125.		Chyuo-park reservoir	171.		Shin-Ayabe transformer substation (Kansai Electric Power Co., Inc.)
126.		Hakuba S.C.			
127.		Minami-Matsumoto S.C.			
128.		EJR Chuou-line	172.		Nishi-Kyoto transformer substation (Kansai Electric Power Co., Inc.)
129.	Gifu	Gifu factory (TORAY)			
130.		Yasu factory (IBM)			
131.		Seki S.C.	173.	Osaka	Sakai works (Nippon Steel Co.)
132.	Shizuoka	Mishima factory (TORAY)	174.		Nankou power plant (Kansai Electric Power Co., Inc.)
133.		Numazu factory (TORAY)			
134.		Hamaoka nuclear power plant (Chubu Electric Co.)	175.		Sakaiko power plant (Kansai Electric Power Co., Inc.)
135.		Hamamatsu manufacturing plant (HONDA)	176.		Tanagawa Daini power plant (Kansai Electric Power Co., Inc.)
136.		Industrial area Hamamatu			

Table 1. (Continued)

No.	Prefecture	Area	No.	Prefecture	Area
177.		Minami-Kawachi transformer substation (Kansai Electric Power Co., Inc.)	208.		Nara S.C.
178.		Setunan transformer substation (Kansai Electric Power Co., Inc.)	209–213.	Wakayama	Tanagawa daini transformer substation (Kansai Electric Power Co., Inc.)
179.		Yodogawa transformer substation (Kansai Electric Power Co., Inc.)	214.		Kainan power plant (Kansai Electric Power Co., Inc.)
180.		Takatsuki S.C.	215–219.		Gobo thermal power plant (Kansai Electric Power Co., Inc.)
181.		Takami S.C.	220.	Tottori	Tottori Shin-toshi S.C.
182.		South Port of Osaka Natural Bird Sanctuary	221.	Shimane	Izumo S.C.
183–187.	Hyogo	Hirohata works (Nippon Steel Co.)	222.		Hirata S.C.
188.		Meyaji transformer substation (Kansai Electric Power Co., Inc.)	223.		Hamada S.C.
189.		Aioi thermal power plant (Kansai Electric Power Co., Inc.)	224.	Okayama	Airport road
190.		Akaho transformer substation (Kansai Electric Power Co., Inc.)	225.		Okayama healthy forest
191.		Himeji daiichi power plant (Kansai Electric Power Co., Inc.)	226–233.		Okayama healthy forest road
192.		Himeji daini power plant (Kansai Electric Power Co., Inc.)	234–236.		Healthy forest school
193.		Takasago power plant (Kansai Electric Power Co., Inc.)	237–239.		Kibikogen-toshi
194.		Higashi-Nada gasturbine power plant (Kansai Electric Power Co., Inc.)	240.		National road
195.		Okutataragi power plant (Kansai Electric Power Co., Inc.)	241.		Yatsuka-mura ground
196.		Nishihari transformer substation (Kansai Electric Power Co., Inc.)	242.		Tsuyama S.C.
197.		Himeji transformer substation (Kansai Electric Power Co., Inc.)	243–248.	Hiroshima	Misuzugaoka housing
198.		Hokusetsu transformer substation (Kansai Electric Power Co., Inc.)	249.	Yamaguchi	Hikari works (Nippon Steel Co.)
199.		Inagawa transformer substation (Kansai Electric Power Co., Inc.)	250.		Hikari S.C.
200.		Yamazaki S.C.	251.	Ehime	Nomura dam (Ministry of Construction)
201.		Shikama S.C.	252.	Fukuoka	Yahata works (Nippon Steel Co.)
202.		Miki S.C.	253–257.		Tobata works (Nippon Steel Co.)
203.		Toyooka S.C.	258.		Iizukahonami S.C.
204.		Yashiro S.C.	259.	Saga	Kohoku S.C.
205.	Nara	National highway 1 Ministry of Construction	260.		Saga-minami S.C.
206.		Kashiwara bypass Ministry of Construction	261.		Genkai power plant (Kyushu Electric Co.)
207.		Shin-Ikoma transformer substation (Kansai Electric Power Co., Inc.)	262.	Nagasaki	Nagasaki shipyard
			263.		Ohmura S.C.
			264.	Kumamoto	Kumamoto manufacturing plant (HONDA)
			265.		Kumamoto airport
			266.		Kumamoto municipal park
			267.		Hitoyoshi S.C.
			268–275.	Oita	Oita works (Nippon Steel Co.)
			276.		Aobadai housing (MITUI real estate co.)
			277.		Oita medical college (MITUI real estate co.)
			278.		Takajou S.C.
			279.		Mitsuyoshi S.C.
			280.		Sankoh-mura S.C.
			281.	Okinawa	Tyatan-cho beach
			282–283.		Tyatan-cho elementary school
			284.		Okinawa S.C.
			285.		Ishigaki daini power plant (Okinawa Electric Power Co.)
			286.		Gushikawa power plant (Okinawa Electric Power Co.)

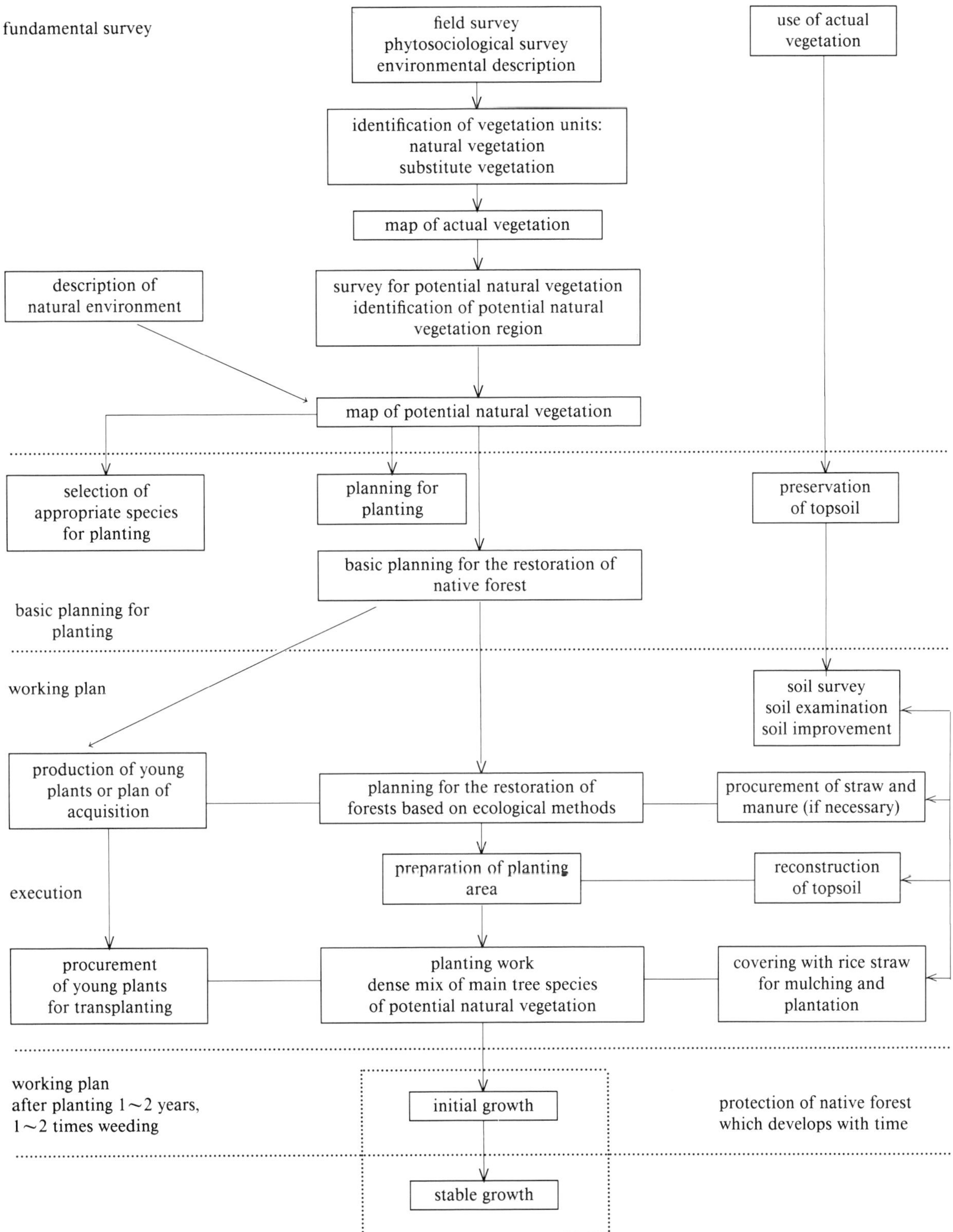

Fig. 2. Flow chart of working procedure for the restoration and creation of native forest (from Miyawaki, 1989 p. 133 partially changed)

1985a) and Dipterocarpus forests at Doi Internon (2,500 m) (Miyawaki 1985b). Main study site, under the sponsorship of Mitsubishi Corporation, is the Bintulu Campus of Malaysia Agriculture University, in Sarawak, which covers 800 ha. This project is a co-study of Yokohama National University (YNU) and Malaysia Agriculture University (UPM). We tried to restore and recover the natural or at least quasi-natural forest ecosystems with native trees, after lumbering and burning for over 50 years.

We are aiming at recovering forest ecosystems on national land, which became bare due to urbanization. For this purpose the recovering of a tropical rain forest in Bintulu (Sarawak), Malaysia is the most important project in restoring such an ecosystem. Presently it is an experiment but in the future this project will be one of the nuclei of the restoration of tropical rain forests in S.E. Asia. So far, afforestation in the tropical area has been aimed at high productivity of wood by early-successional plants, except for the small areas of plantations of Lauan at Kuching in Sarawak and Kepong in the Malay Peninsula. But a forestation based on strict ecological field surveys was never attempted. We have to restore tropical rain forest systematically.

We also made an effort to recover green environments in Malacca and Kuala Lumpur, Malay Peninsula, under the sponsorship of Jasco, a shopping center company, in order to restore the green environments in and surrounding shopping centers, with native trees.

III. Method

Figure 2 shows the scientific vegetation-ecological process for restoration of forest ecosystems from warm-temperate laurel forests to tropical rain forests. First, we select the plant community units to be restored, based on the phytosociologi-

Fig. 3. A new environmental protection forest of recently planted young seedlings of native species on the man-made island of Ohgishima in Tokyo Bay. The plantation is on the site of Ohgishima power station and LNG gas tanks of Tokyo Electric Power Company.

cal method (Braun-Blanquet 1964). Subsequently, we systematize the result for the subject area and draw in spots of actual vegetation and potential natural vegetation (Tüxen 1956; Miyawaki 1975, 1979, 1982a, b; Miyawaki et al. 1980–1989). Further we select the main tree species from the potential natural vegetation and make a prescription for a green environment (Fig. 2).

We restore forest ecosystem based on a new succession concept (Miyawaki 1991b, 1992), and leave it not to the natural process as outlined by Clements (1905, 1916). Main conditions of macro climate of before cutting and after it are almost the same. Where the soil is excavated on construction sites we conserve the topsoil, since topsoil is important for living roots to be 20–30 cm from the earth surface. When topsoil is lost, we bring it from other construction sites and so on. If there was no topsoil to begin with, as on land reclaimed from the sea, we make organic soil mixed with straw and other organic matter (Fig. 3).

Generally speaking, the main tree species in the potential natural vegetation are deep-rooted. Since it is difficult to transplant big trees, we pick up seeds and make pot seedlings. These are transplanted when the plants reach 30–50 cm above the ground, by which time the root system has filled the pot. In Japan, it takes 2 years to transplant. In Southeast Asia, however, it takes only 6 months until transplanting. But in Southeast Asia, root systems do not grow well because the climatic conditions are too favorable. This is a problem to be solved in the future.

In order to form the forest environment in accordance with the natural system, we plant densely mixed pot-seedlings of the main species, as shown in the potential natural vegetation. Ideally though, depending on places, 2–3 pot-seedlings should be planted per square meter.

Fig. 4. Same site as Figure 3, 6 years later (photo in Aug. 1990). It was not hitherto believed that such a new man-made island could develop forests so quickly with native tree species. This was possible only through prudent vegetation–ecological field study (cf. Figure 6).

Fig. 5. An environmental protection forest on a purely man-made island, 500 m from the mainland, in the Pacific Ocean. At first a mound about 5–7 m high was constructed, which was then covered with topsoil. Young seedlings of the main native tree species are used, including *Persea thunbergii*, *Myrica rubra*, *Elaeocarpus sylvestris* var. *ellipticus*, etc. as well as salt-resistant arborescents of the coastal mantle communities, mainly *Pittosporum tobira*, *Quercus phillyraeoides*, and *Litsea japonica*. These were planted densely, 3–5 per m^2, and were covered by rice straw, 4 kg per m^2. (Photo 1984–85, just after planting, at the Gobo Thermal Power Station, Kansai Electric Power Co.)

Fig. 6. Same location as Figure 5, five years after planting. The forest is now 4 m high and has a well-developed crown, despite the harsh environmental conditions. (Photo by Kansai Electric Power Co., Environmental Protection Division.)

Fig. 7. Memorial plantation at Chatan-cho in Okinawa, in the Japanese subtropical region, planted during a satellite international symposium of the IAVS and International Phytogeographical Excursion through Japan. The forest was planted around an elementary school playground with school children (18 Aug. 1984).

Fig. 8. Same place after 5 years, with a 5 m tall forest (photo Apr. 29, 1989).

Fig. 9. New plantation of pot-grown cultivated young seedlings 0.3–0.8 m high on a 40° slope. Main tree species were selected after a field study of potential natural vegetation. Main tree species: *Quercus myrsinaefolia*, *Q. glauca*, *Persea thunbergii*, *Castanopsis cuspidata* var. *sieboldii*, etc. (Main gate of Yokohama National University, planted June 1979.)

Fig. 10. Same place as Figure 9, photographed June 1990.

Fig. 11. Green-rich environment between Park place and the surrounding of the newly constructed SIAM Jusco Rattanathibet Store in Bangkok, two months after the planting of young trees, photographed September 1991.

The use of more than 3 pot-seedlings sometimes brought success. When weeds grow thick after plantation, we have to pull them out and put them on the ground upside down instead of straw. After 3 years, no management is the best management. As the crowns cover the area, the weeds cannot grow thick because not enough sunlight can reach the weeds. We leave the pot-seedlings to natural selection basically do not require management.

In this way, the experiment of recreating the potential natural vegetation has been successful in Japan and Southeast Asia. In Japan, since 1973, international leading companies such as Nippon Steel Corporation, Tokyo Electric Power Co., Kansai Electric Power Co., Honda Motor Co., Tore Company, Mitsui Estate Co., Jusco Co., and many other companies have been successful in planting, the ecology of which we have prescribed. Some examples are presented in Fig. 3–10. We are now proceeding with other places, sponsored by Kanagawa Prefectural Government, Okayama Prefectural Government, Yokohama City, EJR: East Japan Railway, or Japan Construction Ministry (Figs. 1, 3–10; Table 1).

With sponsorship of Mitsubishi Corporation, the restoration in Tropical Rain Forest in Southeast Asia is now making progress under the co-studies with the University of Agriculture in Malaysia. This involves 150000 pot-seedlings of 32 species of trees and more than 100000 germinated seedlings from the forests. We have planted these pot-seedlings on the land after shifting cultivation.

IV. Results

Since 1973 we have successfully planted trees in various places, such as reclaimed land, urban and industry areas, mountains, Hokkaido and

Fig. 12. Newly recreated green environment surrounding the new shopping center Jusco Malacca 6 months after planting of pot-grown young seedlings. Main tree species were selected after a field study of potential natural vegetation: *Shorea* and *Hopea* spp.

North Honshu in summergreen broad leaf forest region to evergreen broad leaf forest (laurel forest) region in Honshu, Shikoku, Kyushu and around the thermal power plant in Ishigaki Island, subtropical Okinawa islands (Miyawaki 1989a). In Japan, we have a rainy season in June, so we plant the pot-seedlings before this season. In June, the roots of the pot-seedlings grow 20–30 cm underground. In August when it is very hot the trees have to overcome these hard conditions. The young trees, after passing the first winter, grow 1 m yr^{-1} and become 10 m tall. After that, the speed of growth is slower but the trunk grows steadily. After 20 years, the trees become 18 m high. The forest ecosystem which recovers is a quasi-natural forest. Though the forest floor needs to develop further, there are shrubs and young trees germinating on the forest floor.

At Bintulu, Sarawak in Southeast Asia, we reached nearly 80% planting success after 6 months. We can find a remarkable growth of 1.5 m Kapur (*Dryobalanops aromatica*) after 1 year (Fig. 20). Around Jusco Shopping Center especially in Malacca, not only shrubs but also Maranti (Shorea spp.), *Dipterocarpus* spp., *Hopea* spp. are growing 1–1.5 m in height (Figs. 11 and 12).

We now monitor the development beyond 10, 20 and 50 years.

V. Discussion

The results of planting the main tree species of potential natural vegetation, from Japan to Malaysia, based on ecological field survey, have been successful. Especially in Japan, where there are typhoons, heavy rains, and a very dry season, densely planted mixed tree species are proof against these dangers and, through natural selection, grow after 10–25 years into multistratal forests with tree layer, low tree layer and shrub

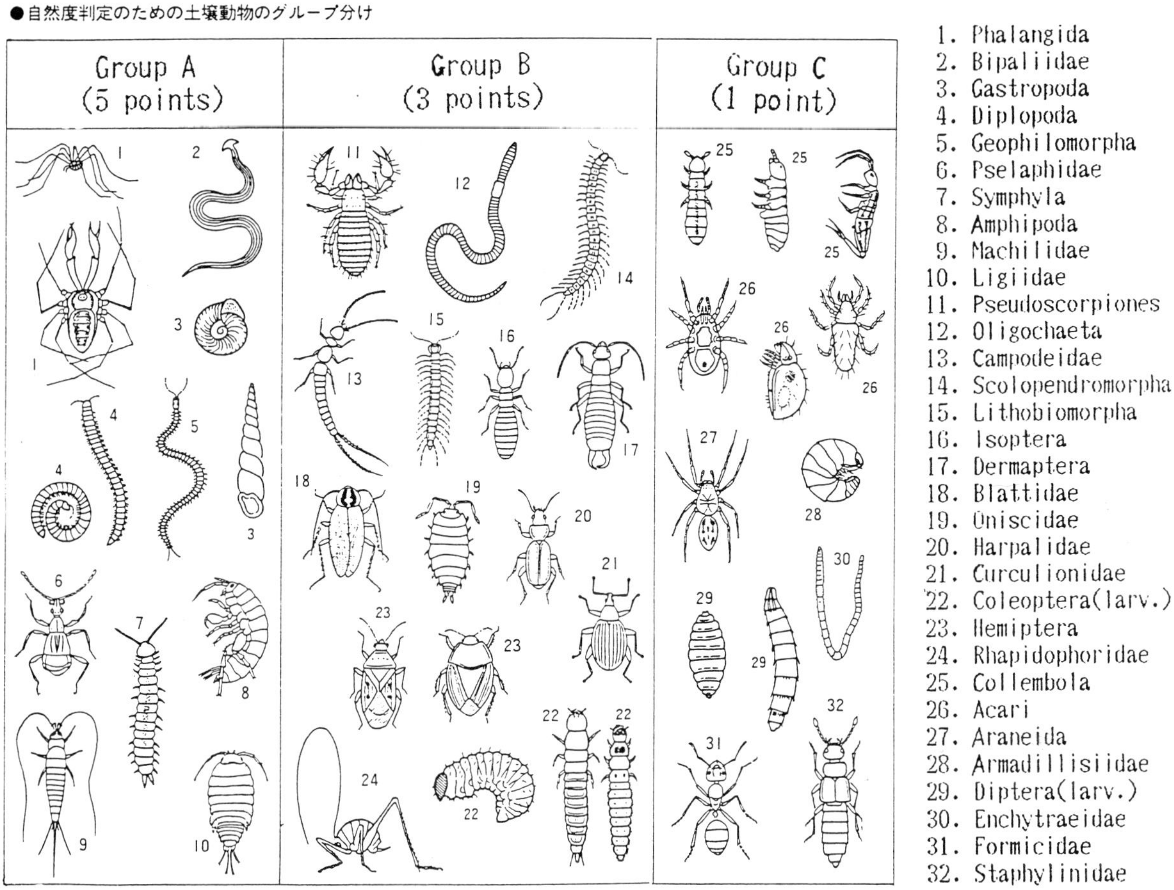

Fig. 13. The grouping of soil animals for calculating nature richness (NR). NR = (5 × nos. of group A) + (3 × nos. of group B) + (1 × nos. of group C) (Aoki 1985).

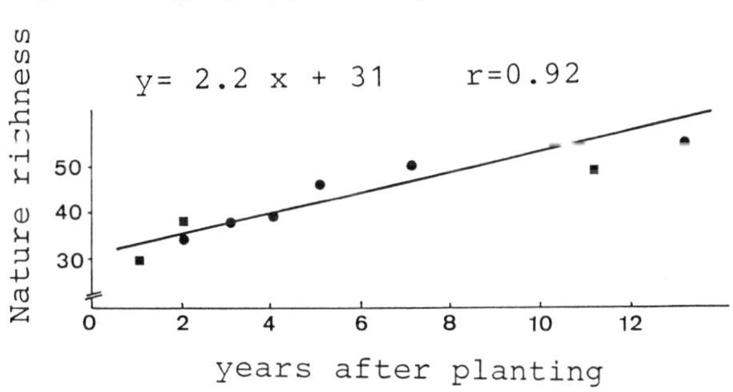

Fig. 14. Nature richness (NR) by soil animals and years after planting in industrial areas around Tokyo (Aoki and Harada 1989, partially modified).

layer. The growth of the forest floor though is still not enough.

According to the regular survey of soil fauna by Prof. Dr. J. Aoki and Dr. H. Harada, underground animal populations as shown in Fig. 13 (see in Aoki and Harada (1985, 1989)), increase in parallel to the growth of the pot-plants. It is very interesting to observe that as soon as we have planted the trees, the number of soil animal populations increases rapidly (Fig. 14).

In Fig. 15 we show that if we leave succession of a tropical rainforest area from its use for shifting cultivation (see Fig. 16) or intensive logging to recover to a full rainforest ecosystem (see Fig. 16) to natural forces, it will take according to Clements (1916) several hundred years. Fig. 17 demonstrates the successional stages which a particular site experiences on its course from a shifting cultivation field to a mixed Dipterocarp forest in S.E. Asia. Several hundred years is too long a wait to restore, rehabilitate and/or recreate forests. The purpose of our study is that we get certain recovery of quasi-natural forest ecosystems sooner by adding some artificial power. We could find the possibility of restoring tropical rain forests in a rather short time if we planted pot-seedlings selected according to a prior ecological survey (Fig. 18). By this method we expect to succeed in the creation or rehabilitation of natural forests as well as we obtained previously modern environmental protection forests in Japan. After 15–20 years the soil fauna

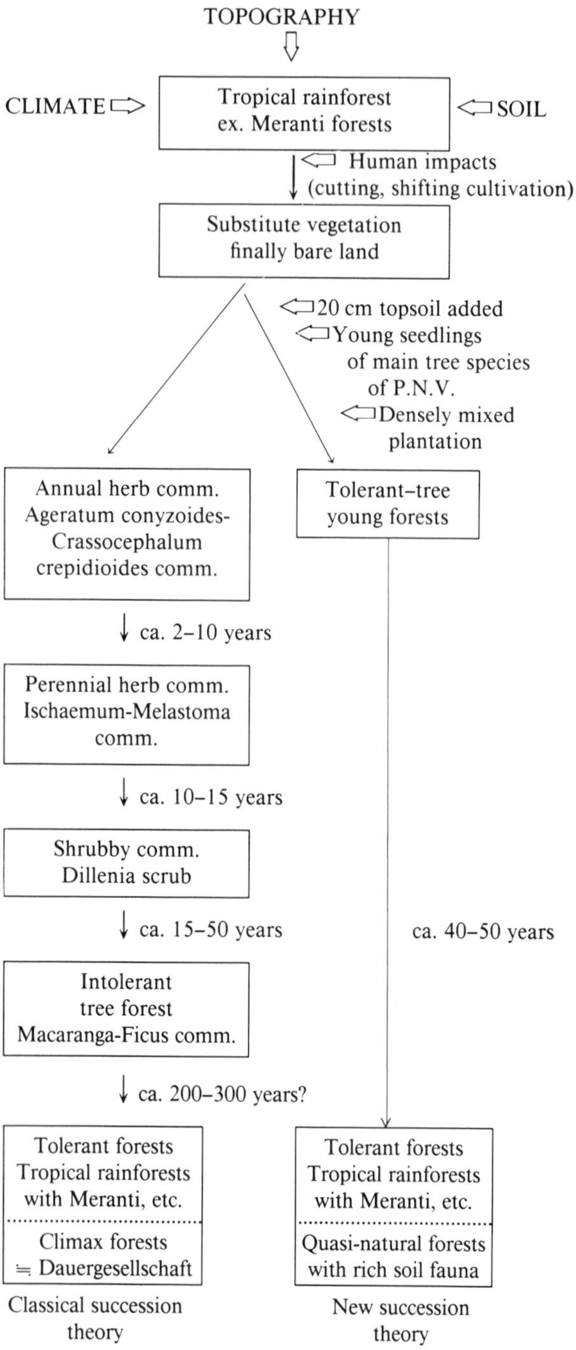

Fig. 15. Comparison between new succession theory and classical theory (Case in Bintulu, Sarawak).

Fig. 16. Preparation for shifting cultivation (Sotek, Kalimantan).

Fig. 17a. A typical rainforest on the mountain slope of Mt. Pemantus (East Kalimantan, Indonesia).

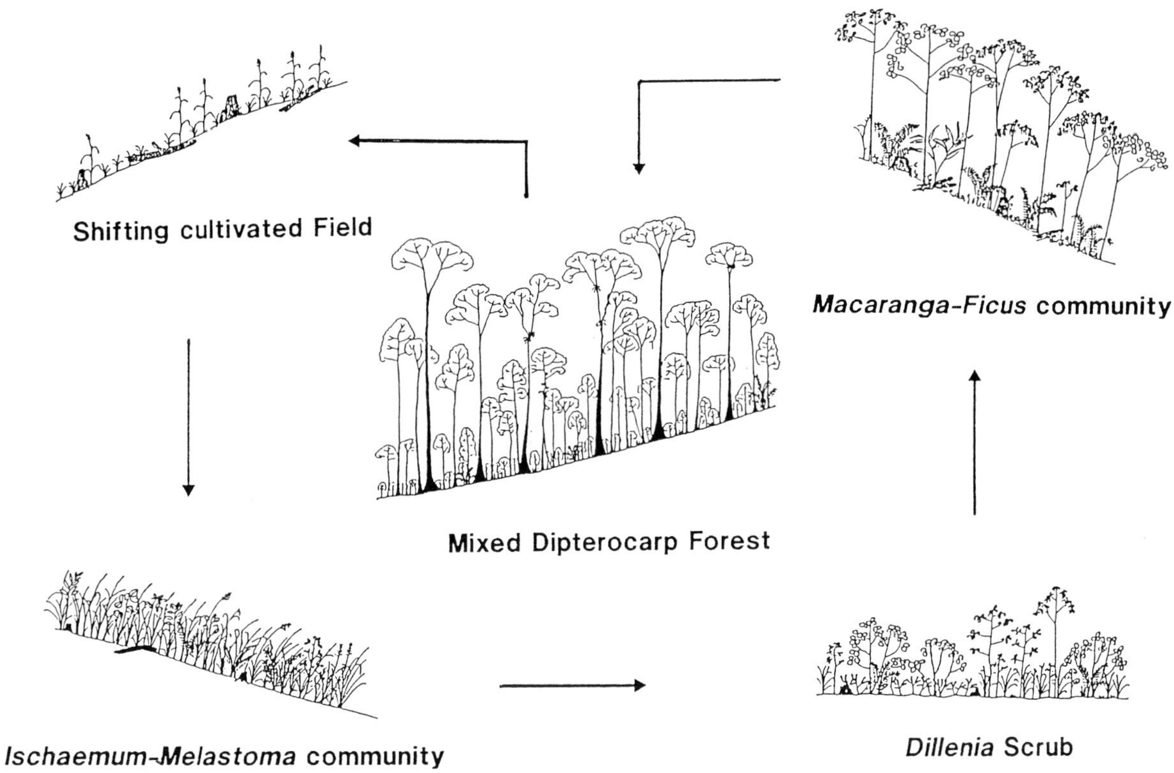

Fig. 17b. A succession of vegetation in the order: natural forest → shifting cultivated field → secondary grassland → scrub → young secondary forest in the vicinity of Bintulu, Sarawak, Malaysia.

Fig. 18. Ecologically guided nursery at the Bintulu Campus of UPM in February 1992. Three hundred thousand seedlings of 32 native tree species from the family *Dipterocarpaceae* are grown to extend our trial plantations near the Campus.

there had almost recovered as shown by Aoki and Harada (1985) (see Figs. 13 and 14), especially the soil mites, which are the main decomposers.

We propose that the method for developing new native or potential natural forests as well as environmental protection forests from bare land be called "man supported succession concept". The method of the author in following this concept is depicted in Fig. 19. We are now studying this concept with colleagues from the University of Agriculture in Malaysia, Centre for Applications (Figs. 20 and 21). We apply this concept there in order to restore the tropical rain forest, which destruction is a most serious problem for modern human activities. Under the sponsorship of Jusco we could restore the rich green urban ecosystem around the shopping centers in

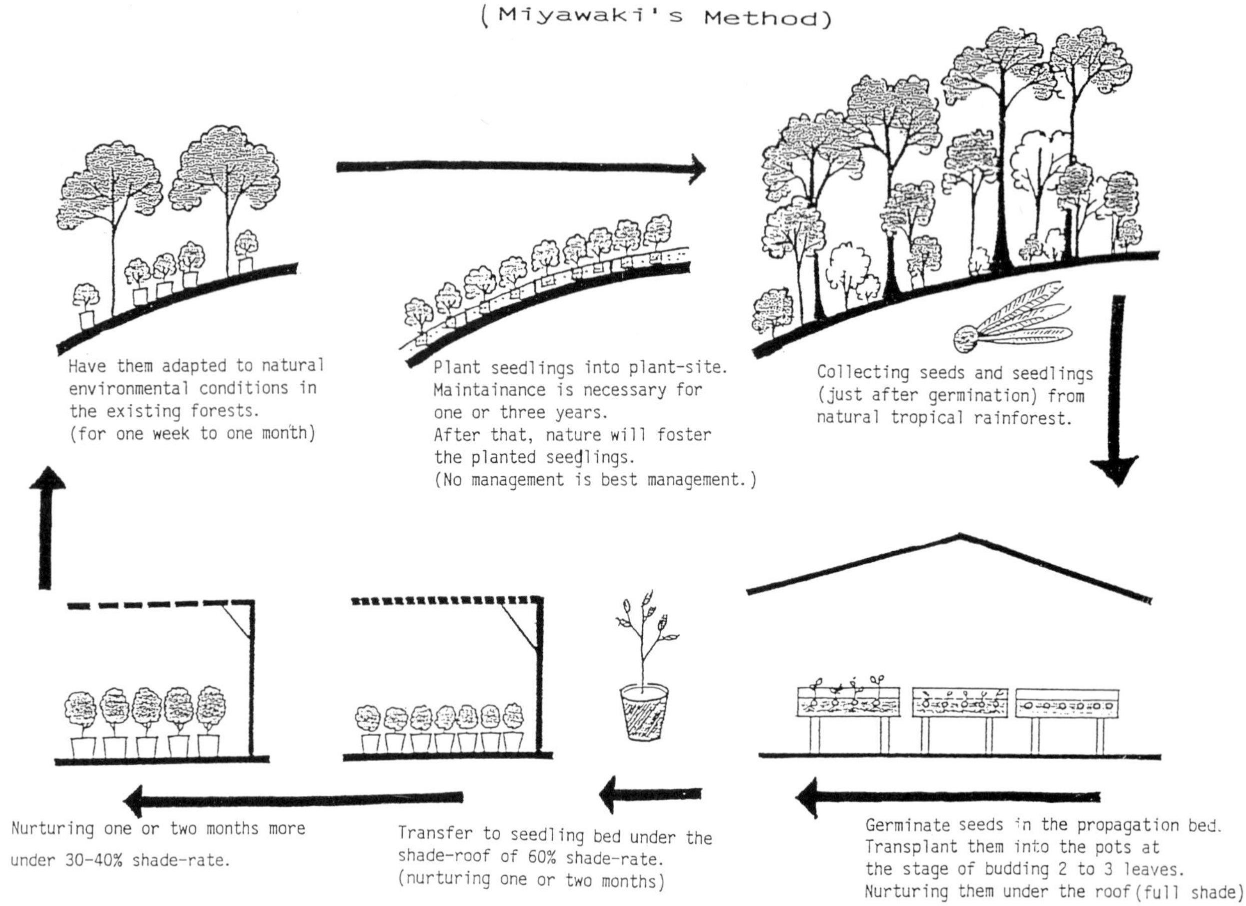

Fig. 19. Working sequence for the rehabilitation of a tropical rainforest ecosystem following the man-supported succession concept explained in Figure 15. The figure demonstrates the steps to be performed from the collection of propagules through their handling in the nursery until the young plants can be put into their later habitat. This procedure is understood when it is described as "Miyawaki's method".

Fig. 20. First trial plantation at the Bintulu site photographed on 15 July 1991 just after 6000 young trees were planted by 1200 persons of the town. Part of our project is the awakening of awareness of the people (especially of young people) for the generation of green environment.

Fig. 21. The same site as shown in Figure 19 on 22 Feb. 1992. The trees, mainly a mixture of Kapur and Maranti, both *Dipterocarpaceae* of the potential natural vegetation of this site, have grown up to 1 m.

Bangkok, Malacca, Kuala Lumpur (see Figs. 11 and 12).

Further we will make an effort to restore, rehabilitate and recreate the forest ecosystem in different tropical regions based on the success of various places, including Japan and Malaysia.

Acknowledgement

The success of our study is founded on all-out support and cooperation by farsighted international companies, prefectural governments and municipal governments in Japan. Mr. Mori and Guest Prof. Ozawa of Mitsubishi Corporation, and Mr. Okada, president of Jusco Corporation, also supported us. M. Abas and his young colleagues at Malaysia Agriculture University (UPM), Dr. Kazue Fujiwara, Dr. K. Ohno at YNU and Mr. F. Maeda and his Mic. co., studied and worked with the sweat on their brows with us. I would like to express my appreciation and respect to all of them.

References

Aoki, J. & Harada, H. 1985. Formation of environmental protection forests and changes in soil fauna, especially *Oribatid* mites. Bulletin of Institute of Environmental Science and Technology, Yokohama National University, 12: 125–135.

Aoki, J. & Harada, H. 1989. A Soil–Zoological Study of Urban Ecosystem around Tokyo. Research Report of the Special Expenditure on Education and Research, 1988. Ecological and Bio-technological Studies for the Conservation, Restoration and Creation of the Environment for Human Life, pp. 23–28. Yokohama National Univ.

Braun-Blanquet, J. 1928. Pflanzensoziologie: Grundzüge der Vegetationskunde. Biologische Studienbücher 7.1, 330 pp, Berlin. 3. Auflage 1951 and 1964, 865 pp, Wien. New York.

Clements, F. E. 1905. Research Methods in Ecology. London.

Clements, F. E. 1916. Plant Succession, an Analysis of the Development of Vegetation. Washington.

Ishi, H. 1988. Report of Global Environments. 258 pp, Iwanami Publ. Tokyo (Japanese).

Lieth, H. 1973. Primary production: terrestrial ecosystems. Human Ecology 1:303–332.

Miyawaki, A. 1975. Entwicklung der Umweltschütz — Pflanzungen und Ansaaten in Japan. Sukzessionsforschung. Bericht über das Internationale Symposium der Internationalen Vereinigung für Vegetationskunde, Vaduz, Cramer, pp. 237–254.

Miyawaki, A. 1976. Die Umwandlung immergrüner in sommergrüne Laubwälder in Japan. Werden und Vergehen von Pflanzengesellschaften. Bericht über das Internationale Symposium der Internationalen Vereinigung für Vegetationskunde, Vaduz, Cramer, pp. 367–381.

Miyawaki, A. 1979. Vegetation und Vegetationskarten auf den Japanischen Inseln. Bulletin of Yokohama Phytosociological Society, Japan (Festschrift für Prof. R. Tüxen), Yokohama, 16: 49–70, mit 2 farb. Vegetationskarte.

Miyawaki, A. 1981. Energy policy and green environment on the basis of ecology. Beyond the energy crisis opportunity and challenge. In: Fazzolage, R. A. & Smith, C. B. (Eds), Oxford & New York, pp. 581–587.

Miyawaki, A. 1982a. Umweltsschutz in Japan auf vegetationsökologischer Grundlage. Bulletin of Institute of Environmental Science and Technology, Yokohama, Yokohama National University, 11: 107–120 (German with English Synopsis).

Miyawaki, A. 1982b. Phytosociological study of the East Kalimantan, Indonesia. Bull. Inst. Environ. Sci. Technol. Yokohama Natn. Univ. 8: 219–232 (Japanese with English Synopsis).

Miyawaki, A. 1984. A vegetation-ecological view of the Japanese archipelago. Bull. Inst. Environ. Sci. Technol., Yokohama Natn. Univ. 11: 85–101.

Miyawaki, A. (Ed.), 1985. Ecological Studies on the Vegetation of Mangrove Forests in Thailand. 152 pp (Japanese with English summary). Yokohama National Univ.

Miyawaki, A. 1988. Green foundation for future cities and civilization — Hokkaido in global perspective. Report of International Symposium on Greenery in Kushiro, Hokkaido, pp. 11–20 (Japanese).

Miyawaki, A. 1985. The montane forests of middle and northern Thailand. Researches related to the UNESCO's Man and the Biosphere Programme in Japan. pp. 3–6. Coordinating Committee on MAP programme. Tokyo.

Miyawaki, A. 1989a. Restoration of evergeen broad-leaved forest ("laurel forest") in Japan. In: Academy, Ch. The World Community in Post-Industrial Society, 5. The human encounter with nature: Destruction and reconstruction. pp. 130–147.

Miyawaki, A. 1989b. Forest types in East Asia and their dynamics under human impact, emphasizing the Japanese Archipelago. Studies in Plant Ecology 18: 180–182. Uppsala.

Miyawaki, A. 1991a. A prescription of green environments. 289 pp. Asahi Shinbun Press. Tokyo (Japanese).

Miyawaki, A. 1991b. New succession theory and its application for creation of environmental protection forests. 34th IAVS Symposium on "Mechanisms in Vegetation Dynamics" 26–30 Aug. 1991. Eger, Hungary. Abstracts. p. 45.

Miyawaki, A. 1992. Ecological perspectives for sustainable development of Southeast Asian forests. Proceedings of International Seminar on Agricultural Change and Development in Southeast Asia (ISACDESA—III), pp. 97–106. Tokyo University of Agriculture.

Miyawaki, A. (Ed.), 1980. Vegetation of Japan. Vol. 1, Yakushima, 376 pp, with 4 color vegetation maps and tables (Japanese), Tokyo Shibundo.

Miyawaki, A. (Ed.), 1981. Vegetation of Japan. Vol. 2, Kyushu, 484 pp, with 4 color vegetation maps and tables (Japanese with German summary), Tokyo Shibundo.

Miyawaki, A. (Ed.), 1982. Vegetation of Japan. Vol. 3, Shikoku, 539 pp, with 4 color vegetation maps and tables (Japanese with German summary), Tokyo Shibundo.

Miyawaki, A. (Ed.), 1983. Vegetation of Japan. Vol. 4, Chugoku, 540 pp, with 4 color vegetation maps and tables (Japanese with German summary), Tokyo Shibundo.

Miyawaki, A. (Ed.), 1984. Vegetation of Japan. Vol. 5, Kinki, 596 pp, with 4 color vegetation maps and tables (Japanese with German summary), Tokyo Shibundo.

Miyawaki, A. (Ed.), 1985. Vegetation of Japan. Vol. 6, Chubu, 604 pp, with 4 color vegetation maps and tables (Japanese with German summary), Tokyo Shibundo.

Miyawaki, A. (Ed.), 1986. Vegetation of Japan. Vol. 7, Kanto, 641 pp, with 4 color vegetation maps and tables (Japanese with German and English summaries), Tokyo Shibundo.

Miyawaki, A. (Ed.), 1987. Vegetation of Japan. Vol. 8, Tohoku, 600 pp, with 4 color vegetation maps and tables (Japanese with German and English summaries), Tokyo Shibundo.

Miyawaki, A. (Ed.), 1988. Vegetation of Japan. Vol. 9, Hokkaido, 563 pp, with 6 color vegetation maps and tables (Japanese with German and English summaries), Tokyo Shibundo.

Miyawaki, A. (Ed.), 1989. Vegetation of Japan. Vol. 10, Okinawa and Ogasawara, 676 pp, with 5 color vegetation maps and tables (Japanese with German and English summaries), Tokyo Shibundo.

Miyawaki, A., Fujiwara, K. & Box, E.O. 1987. Toward harmonious green urban environments in Japan and other countries. Bull. Inst. Environ. Sci. Technol. Yokohama Nation. Univ. 14: 67–82, Yokohama.

Miyawaki, A., Fujiwara, K. Nakamura, Y. & Kiumura, M. 1983. Ökologische und Vegetationskundliche Untersuchungen zur Schaffung von Umweltschutzwäldern in den Industrie-Gebieten Japans. Yokohama Phytosociological Society 22, I. 84 pp, II. 151 pp, with color vegetation maps (Japanese with German and English summaries) Yokohama.

Miyawaki, A., Harada, H. & Ude, H. 1979. Vegetationskundliche Untersuchung zur Schaffung von Umweltschützwäldern um Industrie-Anlagen, erläutert am Beispiel der 11 Fabriken der Tore-Industrie-AG, Bulletin of Yokohama Phytosociology Society 8, 50 pp, with 2 color vegetation maps (Japanese with German summary), Yokohama.

Miyawaki, A., Ohno, K., Suzuki, S. & Nakata, E. 1986. Vegetation of Chatan-cho: For the creation of the new town covered with green, 153 pp, with color vegetation maps (Japanese with English summary), Okinawa, Chatan.

Miyawaki, A., Okuda, S., Suzuki, K., Sasaki, Y. & Mochida, Y. 1982a. Substitutional communities at Sotek Area of East Kalimantan, Indonesia. Bull. Inst. Environ. Sci. Technol. Yokohama Natn. Univ. 8: 265–293 (Japanese with synopsis and German summary).

Miyawaki, A., Sasaki, Y. & Mochida, Y. 1982b. Tropical rain forest in the East Kalimantan, Indonesia. Bull. Inst. Environ. Sci. Technol. Yokohama Natn. Univ. 8: 233–264.

Miyawaki, A., Okuda, S., Fujiwara, K., Sasaki, Y., Suzuki, K. & Harada, H. 1974. Phytosociological Studies on Creation of Environmental Protection Forest around Schools — Based on Field Surveys at 158 Schools throughout Japan. Report of Studies on Creation of Environmental Protection Forest around Schools, 116 pp, with 2 colored vegetation maps (Japanese), Yokohama.

Tüxen, R. 1956. Die heutige potentielle natürliche Vegetation als Gegenstand der Vegetationskartierung, Angewandte Pflanzensoziologie, Stolzenau/Wes. 13: 5–42.

2. Ecosystem rehabilitation and forest regeneration in the humid tropics: Case studies and management insights

MIGUEL CLÜSENER GODT and MALCOLM HADLEY
Division of Ecological Sciences, UNESCO, 7, place de Fontenoy, 75700 Paris, France

Keywords: Forest rehabilitation, humid tropical forest, soil erosion, sustainable management

Abstract. The highest rates of land use change in the world are found in the tropics. For the period 1981–1990, some 17 million hectares of forest have been converted to other uses each year, of which about half are found in the moist tropical zone. By the year 2050, half of the surface covered by tropical forests in 1990 may have been converted. In the years ahead, a greater proportion of tropical lands may be expected to support secondary forest, degraded zones and other human-impacted areas.

Some issues and approaches in ecosystem rehabilitation have been described as part of the process of elaborating one of the four new research orientations within UNESCO's Man and the Biosphere Programme (MAB), in an overview of tropical secondary forests, in several recent synthesis reports on the effects of disturbance on tropical forest ecosystems, and in findings of field projects in such countries as China, Papua New Guinea and Thailand. These various studies suggest that mechanized land clearing and logging using bulldozers that are handled insensitively is a principal course of massive disturbance, soil compaction, increased runoff and damage to residual stands.

Socio-economic perspectives and dimensions are vital to the success of rehabilitation schemes, the ingredients and strategies of which may differ in detail from case to case. However, broad guidelines can be set which suggest that tropical ecosystems can and do recover from intensive and sometimes catastrophic human activities.

1. Introduction

The highest rates of land use change in the world are found in the tropics. The most recent FAO estimates indicate that for the period 1981–1990 some 17 million hectares of forest have been converted to other uses each year, of which about half are found in the moist tropical zone (FAO 1990). In the years ahead, a greater proportion of tropical lands will support secondary forest (those that remain after logging or grew up on abandoned cleared lands), degraded zones and other human-impacted areas. Such systems have received relatively little attention, compared with natural systems, and there is now a growing need for improved scientific understanding on which the effective management of impacted systems (including rehabilitation of degraded areas) could be based. These systems may well hold the key for long-term solutions to human environment in the tropics.

This contribution has three parts. First, we discuss some issues in ecosystem rehabilitation and forest regeneration, drawing upon case studies mainly from the Asia-Pacific region. Second, we present some strategies and principles that have been proposed by researchers with a view to shaping rehabilitation schemes and forest regeneration programmes in the humid tropical regions. Third, we refer to criteria for measuring the success of a rehabilitation scheme.

In so doing, we draw heavily on such materials as (a) those prepared as part of the process of elaborating one of four new research orientations within UNESCO's Man and the Biosphere Programme (MAB) (concerned with the management and restoration of human impacted resources), (b) recent overviews of tropical secondary forests and the effects of disturbance on tropical forest ecosystems, and (c) findings from field observations and research projects in such countries as China, Nigeria, Indonesia, Papua New Guinea and Thailand. Finally, future projects in Colombia, Cuba, Mexico and Venezuela, in which UNESCO may be involved are presented.

H. Lieth and M. Lohmann (eds.), Restoration of Tropical Forest Ecosystems, 25–36.

2. Case studies

2.1. *Improving the biological productivity of tropical wastelands in Thailand*

Shifting cultivation in both upland and lowland areas, over-cutting of forests and tin mining are among the principal causes of land degradation in Thailand. With a view to providing a better technical basis for reclaiming such wasteland areas, researchers from Thailand and Japan have been undertaking a cooperative project under the auspices of the MAB National Committees of the two countries. Experimental trials have been carried out since 1985 in four study areas, using fast-growing trees (*Eucalyptus, Pinus,* etc.), slow-growing native commercial timber trees (e.g. *Pterocarpus*), fertilizing trees (Leguminosae), various crops (cassava, maize, pineapple, peanuts, sorghum, etc.), medicinal plants and ground cover plants. Treatments have included ploughing, weeding, soil dressing, mulching (e.g. with *Imperata* grass), exclusion of fire and grazing, and application of organic and inorganic fertilizers. Some of the findings and tentative conclusions from this work have been reported by co-principal investigators Choob Khemnark of Kasetsart University in Bangkok and Kyoji Yoda of Osaka City University (Khemnark and Sahunalu 1989; Yoda 1990).

Rehabilitation following shifting cultivation in lowland areas has focussed on the Somdet Reforestation Station in Karasin Province in northeastern Thailand, where the most recent land use had seen repeated cultivation of cassava more than fifteen years after slash-and-burn of a mixed deciduous forest on a sandy area.

Studies at the Huey Tung Jew in Chaing Mai Province in northern Thailand have aimed at recovery of **upland areas following shifting cultivation**, specifically *Imperata cylindrica* grassland after repeated cultivation of upland rice, maize and poppy for more than a decade in an area where the original vegetation was hill evergreen forest. The main limiting factors for tree establishment and growth are strong winds and ultra-violet radiation during dry periods, combined with the effects of ground fires. Promising results have been obtained with *Calliandra,* a fast-growing tall shrub tolerant of strong wind, high radiation and low temperature, and seemingly apt to use as a shade tree for tea and coffee cultivation.

At Ratchaburi in central Thailand, the challenge is that of rehabilitating thorny scrub and bamboo bush areas resulting from **repeated cutting of dry dipterocarp forest on heavy lateritic soils** with thick hardpan. Protection against grazing and fire and crushing the shallow hardpan using bulldozers are among the ingredients for satisfactory establishment of most trees from dry dipterocarp and mixed deciduous forest. Mycorrhizal symbiosis has a marked effect on *Eucalyptus* growth. The mycorrhiza populations found at Ratchaburi are not found elsewhere in Thailand, suggesting that the Royal Forest Department might concentrate at Ratchaburi the production of most or all of its country-wide needs for *Eucalyptus* saplings. Coppice management may shorten the cultivation cycle of *Eucalyptus* after the second generation. In terms of agriculture, control of grazing and fire and hardpan-crushing enables many crops to be grown in the rainy season, with high furrow cultivation and application of organic matter countering the tendency to waterlogging and soil compaction.

Reclamation of **abandoned wastelands of tin-mine tailings** (which cover some 32 000 ha in Thailand) has entailed studies at the Takuapa Experimental Area and Student Training Centre of Kasetsart University in the Phang-Nga Province. Experimental treatments (e.g. mulching, soil dressing, ploughing, fertilization) on a completely exposed silica sand area about seventeen years after tin mining were accompanied by measurements on changes in soil fertility and texture, micro-climatic conditions, floristic composition, forest structure, biomass, productivity, decomposition, etc. Treatments of slime soil dressing and mulching with *Imperata* leaves produced the best results in terms of growth of *Eucalyptus* saplings. Pineapple and cassava yields were generally poor, even with considerable inputs. A conclusion is that revegetation of these areas is costly and cannot be justified in terms of economic return on investment.

2.2. *Using exotic species in tropical ecosystem rehabilitation in China*

The use of exotic species in ecosystem rehabilitation is illustrated by projects in three areas of subtropical China, each undertaken under the auspices of the Chinese-MAB National Committee and involving collaborative links between Chinese and foreign scientists.

2.2.1. *Rehabilitating degraded coastal areas at Xiaoliang*

Until relatively recently, Xiaoliang in the coastal area of Dianbai County in western Guangdong Province was a typical example of severely degraded land: barren-eroded, over-used yet unproductive, the remaining agricultural area constantly threatened by flood, storm and lack of irrigation water. In 1959, a cooperative programme for wasteland restoration was begun by the South China Institute of Botany and the Chinese Academy of Sciences and the Xiaoliang Water and Soil Conservation Station. The objective was to study the recovery of vegetation by natural and artificial means on the badly eroded soils of severely degraded coastal terrace lands.

A 433 ha area was replanted with eucalyptus and pine (1959–1965), to be replaced from 1973 onwards by multiple species, multi-storied plantations of broad-leaf mixed forest, which in turn were replaced by economic crops and fruit trees. The successional sequence was accompanied by changes in such parameters as soil organic matter (from 0.6% in barren land to 1.7% in mixed forest), pH (from 4.5 to 5.7), soil moisture (9.8% to 13.2%, at 10 cm depth), annual loss of soil in runoff (from 19.9 t ha^{-1} to 3 kg ha^{-1}) (Yu Zuo-Yue and Wang Zhu-Hao 1989).

Work on the man-made forest at Xiaoliang has continued recently with an ecosystem research perspective, as part of the Cooperative Ecological Research Programme (CERP), launched in 1987 as a three-year cooperative venture between China, the Federal Republic of Germany and UNESCO-MAB, aimed at strengthening ecological research in China. The recent focus has been on the quantification of the ecological, environmental and socio-economic effects of the various crop types along the rehabilitation gradient. Research results are being used to draw up guidelines on sustainable management for the villagers. The results should also make it possible to determine the levels of diversity which must be achieved in order to secure ecological self-regulation and viability.

2.2.2. *Restoration of intensively used pine forests at Dinghu Biosphere Reserve*

Large areas of badly degraded forest lands in the subtropical region of China have been recently reforested with a native pine, *Pinus massoniana,* to prevent further degradation of the landscape. Cutting of the trees is now prohibited, but harvesting of the understorey and litter is allowed to satisfy human needs. This removal of litter and understorey has the potential to retard the return of the landscape to productive forest lands because it interferes with the restoration of soil fertility. The challenge in this subtropical, humid region of China is to develop strategies for rehabilitating forest lands and, at the same time, satisfy human needs for fuels.

Providing a research response to this challenge is the underlying aim of a three-year collaborative project sponsored by the MAB Committees of China and the USA. The project started in late 1989, with two co-principal investigators: Kung Guo Hin of the South China Institute of Botany at Guangzhou and Sandra Brown of the University of Illinois at Urbana. The overall goal of the study is to determine the effect of organic matter and nutrient removal by harvesting understorey and litter on soil organic matter and fertility, nutrient cycling and productivity in pine forest of southern China.

The basic experimental design of the project is to quantify the storages and flows of organic matter and nutrients for control (without harvest), and treatment (with harvest) plots. These storages and flows include the timing, frequency and amount of organic matter and nutrients removed by litter and understorey harvesting; patterns and amount of organic matter and nutrients returned to the forest floor via litter; the rates of decomposition and nutrient release of litter; fine root biomass and dynamics; organic matter production and nutrient accumulation by the understorey and trees; and the amount of organic matter, nutrients, and nitrogen availability in the soil. The effects of treatment on plant diversity are also being determined. The results from these studies will be integrated to determine the effects of litter and understorey removal on the rehabilitation processes of the pine forests and to suggest sustainable management options that minimize negative effects on soil fertility, forest production and plant biological diversity.

The project focuses on the Dinghu Shan Biosphere Reserve in Guangdong Province, with information being compared with that obtained at Luquillo Biosphere Reserve in Puerto Rico. Ecological measurements are using the methods elaborated within the IUBS/MAB initiative on Tropical Soil Biology and Fertility (TSBF); these methods have previously been tested by one of the co-principal investigators (Sandra Brown) in a TSBF project, comparing pine plantation with a paired secondary forest in Puerto Rico.

2.2.3. Comparing water and mineral cycling in pine and broad-leaved forest in Jiangxi

Pinus massoniana plots also feature in a joint China-Japan project in Xingguo prefecture of Jiangxi Province, which seeks to compare water and mineral cycling in severely eroded pine forest and in evergreen broad-leaved forest (Iwatsubo 1990). Estimates of above-ground biomass in *Castanopsis* and *Schima* forest were 396 and 308 t ha^{-1}. Stem increment estimates of 10.8 and 9.9 t ha^{-1} yr^{-1} suggest that growth rates in the 75–90 year old broad-leaved forest were still high. Nitrogen contents were high in comparison with evergreen broad-leaved forests in Japan, though rates of carbon and nitrogen accumulation in mineral soil were similar to forests in southern Japan. Measurements in experimental plots of pine have confirmed the debilitating effect of litter and undergrowth removal on the rate of recovery of soil and on pine growth.

2.3. Regeneration after clear-felling at Gogol, Papua New Guinea

In the late 1960s, a change in paper making technology occurred that was to lead to a radical new approach to logging in several countries where the forests contain few commercially saleable tree species. Paper chemists devised a way of using a mixture of tropical hardwood species woodchips to make paper pulp. This change in technology opened the way to overcome the problems of species diversity and commercial acceptability. Instead of selective logging, whole areas of forest could be clear-felled giving a much higher wood yield. The cleared land could be replanted with fast-growing tree plantations as well as be used for various agricultural projects.

In Papua New Guinea (PNG), the suitability of a variety of timber species for pulpwood was successfully tested. Pulpwood logging, or woodchipping as it became known, seemed, in the word of one observer, a 'golden opportunity' for the industry to break through the silvicultural and utilization limitations experienced up until that time. It would allow the achievement of economies of scale and profitability, two attributes mostly absent from PNG's timber industry. As well, markets in Japan seemed assured with the prospect of other markets elsewhere in Asia. At last it seemed the forests of PNG might be used to help fund the rapidly increasing process of development then underway in the country. Negotiations commenced with a Japanese consortium in 1968. These culminated in 1971 when a woodchipping industry was established based on the forest resources of the Gogol Valley near Madang.

The industry was to be based on two main logging areas, together covering 66 000 ha, with a clear-felling of lowland rain forest for producing mixed-species wood pulp. From the outset of the project, the intention was that part of the logged area should be used to establish plantations of fast-growing tree species suitable for wood pulp. The remainder of the land was to be used for various agricultural projects, or to be allowed to regenerate naturally back to forest.

The project has been controversial from many points of view, not least in the shortfall between what the different groups involved hoped would result from the project, and what actually happened. These different perceptions are examined in an overview of the Gogol project, published in 1990 as Volume 3 in the *Man and the Biosphere Series* (Lamb 1990). A broad review of the general problems of rain forest management includes current problems arising from forest resource management in different parts of the world and difficulties faced by silviculturalists. Chapters on PNG, on the forest of the country and on the Gogol Valley lead on to an account of the logging project and to the main problems encountered and the lessons learned, examined in turn from the viewpoint of the government, the local people and the logging companies.

In terms of the effects of exploitation on forest regeneration, David Lamb offers the following conclusions:

— It seems unlikely that the long-term productive capacity of the lands within the Gogol Timber Project have been seriously damaged by logging.
— Many of the short-term impacts are apparently drastic but the disturbance is such that most components of the ecosystem should recover in time. The time required is probably 50–200 years, depending on how recovery is defined.
— Such a second growth forest would not be identical to the original forest, but it could contain many of the original species and develop a similar structure. The situation with animal species is less clear since many are dependent on particular types of forest.
— In fact, however, large areas of forest in the Gogol Timber Project will be subject to further disturbance before this recovery is complete. The extent to which this will happen is impossible to predict because of the complex pattern of land ownership.
— The likely long-term consequence of logging is, therefore, a disappearance of much of the forest cover in the well-drained valley areas and on gentle slopes. Disturbed forests in other topographic positions will probably recover.

As such, the observations from the fertile lands of the Gogol would seem to support the conclusions of Ariel Lugo and others, that tropical ecosystems can and do recover from intensive and sometimes catastrophic human activities, that the costs to be paid are usually associated with time, and that a cornerstone strategy for tropical forest rehabilitation is the use of natural processes and natural subsidies to the greatest extent possible.

In a more general social context, David Lamb offers some thoughts on lessons that can be learned from the Gogol Timber Project:

— The first and most obvious lesson is that in any similar large timber project, whether involving clear-felling or selective logging, some way must be found to resolve or diminish the differences in perception and attitude between the various parties. It goes without saying that this is likely to be extremely difficult to do.
— A second lesson follows from the first and concerns post-logging land use. If the success of a project depends on reforestation following the initial logging, then it must be clear from the day the first tree is felled just how reforestation is to be carried out, who will do it and on what land it shall be done? Again, the problem is more easily stated than solved, especially in areas with complex patterns of land ownership like the Gogol Valley.
— A third lesson is that the spread of economic benefit cannot be left to chance. In the case of the Gogol Timber Project, it was assumed that the economic gains would follow more or less automatically. But they didn't. While the Gogol Timber Project was not exactly an economic enclave, the desired multiplier effect was much smaller than originally hoped for.
— The final lessons from the Gogol concerns the need to have environmental safeguards form an integral part of the overall management programme and not simply added on once a project starts. In one sense, it is almost irrelevant whether or not a forest can recover after some long time period because the short-term damage can be severe, and it is the short-term that is of most concern to villagers. Hence, protection forests and other reserves should be identified before logging commences; if clear-felling is to be carried out, coupe limits and a dispersal mosaic should be defined, environmental controls such as stream-side buffer strips should be unambiguous and logging and roading plans should be established well enough in advance such that a monitoring programme can be established to detect if environmental stan-

dards are being maintained or not. Given the uncertainties of fulfilling all post logging land use plans, logging operations involving clear felling should be carried out in such a way as to help facilitate the recovery of natural ecological processes if these land use plans do not eventuate.

Are these "lessons" confined to Papua New Guinea, or might they also apply elsewhere? Some might argue that PNG is too much of a special case and that its land laws in particular prevent any lessons being applied more widely. But, even in regions where land tenure arrangements are very different to those in PNG, some of David Lamb's conclusions would appear to warrant close examination: for example, that people actually living on the land of a forestry project should be much more involved in the planning and decision-making process than was the case in the Gogol Timber Project, and that new forms of management need to be developed which accommodate both long-term silvicultural prescriptions and the aspirations of the villagers who live in the area.

2.4. *Effects of land clearing and logging technology in Indonesia and Nigeria*

Traditional ways of cutting trees or clearing land in shifting cultivation — by hand, using long knifes or (more recently) chain saws — cause relatively little damage to the top of the soil structure (even though the fall of any single tree may cause considerable damage to its neighbours). This contrasts with the impact of mechanical land clearing using bulldozers, which may induce massive disturbance and soil compaction. The contrast in the impact of different methods and technologies used in logging and forest clearance is illustrated by data from Nigeria (Table 1) and Indonesia (Table 2). As the data from Indonesia suggest, insensitive use of modern technology (e.g. in transmigration schemes) may lead to very high rates of soil erosion with the soil sun-baked and hardened before settlers arrive (Sinukaban 1989). Clearly, effects will vary from one place to another, for example the higher erosion rates from land clearing in Indonesia compared to Nigeria (indicated in Tables 1 and 2) can be attributed to higher soil erodibility, higher rainfall intensity and steeper land slopes in Indonesia (Sinukaban 1989). Also noteworthy is a wide ranging review of logging technology in tropical forests (Jonsson and Lingren 1990), which concluded that "present logging technology is, in most cases, acceptable and appropriate if properly applied and used". The study also noted that "More efficient and less damaging equipment exists, but it is often more expensive and complicated, and therefore rarely used".

Table 1. Effects of methods of deforestation and tillage techniques on soil erosion in an Alfisol in Nigeria. [*Source:* Lal 1987]

Method of vegetation removal	Sediment concentration ($g\ l^{-1}$)	Runoff ($mm\ yr^{-1}$)	Soil erosion ($ton\ ha^{-1}\ yr^{-1}$)
Traditional farming, incomplete clearing, no till	0	3	0.01
Manual clearing, no till	3.4	16	0.4
Manual clearing conventional tillage	8.6	54	4.6
Shear blade, no till	5.7	86	3.8
Tree pusher, root rake, no till	5.6	153	15.4
Tree pusher, root rake, conventional tillage	13.0	250	19.6

3. Principles and guidelines

Experience from field observations and experiments in different regions is leading to an emerging set of guiding principles on ecosystem rehabilitation in the tropics, proposed or distilled by such authors as Ewel (1987), Lamb (1988), Lugo (1987), Schreckenberg et al. (1990). A sampling of some of these guiding principles and tenets is given in Tables 3 and 4 and in the following paragraphs:

Table 2. Effects of method of deforestation on soil compaction and erosion in Indonesia. [*Source:* Suwardjo and Barus, 1985, cited from Sinukaban 1989]

Method of vegetation removal	Shear strength (kpa)*	Soil resistance (KF cm^{-2})	Aeration (% volume)	Erosion (ton ha^{-1} yr^{-1})
Manual, complete clearing	57	7	34	4.1
Manual, conventional	57	7	34	4.1
Manual, K.6-blade rake-harrow	57	9	9	20.8
Manual, K.6-blade-rake	80	17	9	61.1
Manual, straight blade-rake	70	13	5	47.8
Manual, straight blade-rake-harrow	70	11	6	143.8
Manual, rake-harrow	68	17	7	59.6
Manual, rake-rake	78	77	9	44.6

* The values of shear strength, resistancy and soil aeration were measured at an average depth of 5–35 cm.

3.1. *Diversity of situations and approaches*

TENET 1:
Each ecological region offers special challenges and opportunities for rehabilitation efforts

The range of ecological and socio-cultural situations that characterizes the tropics suggests that many different approaches to rehabilitation are required. Thus Lugo (1987) describes 72 possible conditions for tropical forest land rehabilitation, and this in a relatively simplified situation in which only three levels of damage, three rainfall zones, four biotemperature regions, and two substrata are recognized. In addition to these differences in the physical and biotic environment, socio-economic conditions may also vary enormously from place to place. Thus the immense diversity of environments and situations that exists in the tropical world requires a similarly diverse range of approaches to rehabilitation which takes into account the favourable and adverse conditions that affect rehabilitation in the humid tropics (Table 4).

In the intertropical belt, growth conditions are optimal in the lowland, humid, wet, and rain

Table 3. Strategies for rehabilitating tropical forests. [*Source:* Hadley and Schreckenberg, 1989, based on Lugo 1987, 1988, and other sources]

- Maintain flexibility in the rehabilitation approach
- Be alert to environmental conditions
- Avoid specificity on the ultimate goals of rehabilitation
- Maximize vegetation cover
- Manipulate existing vegetation before attempting a substitution
- Use fallow to do most of the rehabilitation
- Restore tree cover as rapidly as possible
- Be aware that nutrient cycling strategies may change through a rehabilitation from phosphorus limitation to nitrogen limitation
- Develop species mixtures based on their 'ecological' combining ability
- Keep top soil moist, cool and shaded
- Couple ecosystems to maximize their value and accelerate rehabilitation (e.g. use treated urban sewage as irrigation water to accelerate tree growth)
- Maximize ecosystem complexity to optimize use of site resources, maximize protection from pests, and minimize risks
- Use exotic tree species to foster native species through site rehabilitation
- Use multiple seeding techniques when in doubt as to what to plant and let natural selection processes decide the best species combination for a site
- Create nuclei of biotic activity from which habitat rehabilitation occurs under the influence of biotic agents
- Use stressors to arrest succession at a desirable stage

Table 4. Conditions that influence the rehabilitation of ecosystems in the humid tropics. [*Source:* Lugo 1987]

Favourable conditions	Adverse conditions
High redundancy of function in the biota	Species and their characteristics fully known
Fast biotic processes	Too many species
Many nitrogen fixing species in successional communities	Systems are subject to high rates leaching and loss of soil
Year-round growth is possible	It is difficult to control the succession
High resiliency and elasticity	Too many different environmental conditions

forest climates, but problems emerge due to the leaching potential of the environment. Rapid establishment of a diverse vegetation cover through multiple seeding is a solution to this problem. In the dry tropics the problem is low water availability, which slows down succession by stressing plants and animals. This makes the 'time payment' of land use mistakes more costly. Surface water irrigation, construction of wells or inducing rainfall artificially are not feasible solutions in most cases. These solutions work best in small-scale projects. The best alternatives for the rehabilitation of large areas of dry forest ecosystems are planting, transplantation of genetic material from natural or exotic pools, optimization of land forms, removal of human stressors (i.e. grazing, excessive harvesting, or fire), and conservation of soil organic matter so that only water remains as the limiting factor to recovery. Similar strategies apply to altitudinal gradients in the tropics where low temperatures and low atmospheric saturation deficits slow down biotic responses.

3.2. Ecosystem recovery from human impact

TENET 2:
Most ecosystems can recover from intensive and sometimes catastrophic human impact

Recovery does not imply a return to the original condition, rather that some form of vegetation cover can re-establish itself. However, the fact that ecosystems recover from chronic human impacts does not mean that the cost of such impact is low. Costs of carelessness with the use of the land can be extremely high and usually payable in time (e.g. the period of time lands are not usable and productive). In one Puerto Rican example, it took about 30 years of plant growth for soil organic matter to reach pre-disturbance levels and about 50–60 years before poles of useful native species appeared in the understorey.

3.3. Rehabilitation rates of ecosystem services

TENET 3:
Some ecosystem services may rehabilitate faster than others

The services that take most time are often those that are most difficult to replace by humans. For example, genetic material, soil organic matter and soil fertility are usually the most time-demanding sectors of the ecosystem in terms of rehabilitation. A short supply of any of these components prevents complete rehabilitation. Supplying seed, soil or fertilizers over large areas is extremely expensive and often impractical. Examples of situations that tax the limits of practicality are the rehabilitation of certain mangroves (seed limitations), and bulldozed pastures and landslide areas (soil and fertilizer limitations).

3.4. Rehabilitation and human impact

TENET 4:
The ability to recover any site depends on the intensity and location of the human or natural impact upon the ecosystem

Events that remove only biotic structure (e.g. wood, litter, animals) are less severe and easier to rehabilitate than those that remove substrate, nutrients, or other core factors (e.g. seed sources) of the ecosystem. Therefore, if the fundamental conditions of the ecosystem are sufficiently altered, the rehabilitation strategy must consider the establishment of an ecosystem that is different from the original one. A new ecosystem will be required to cope with a new environment if the rehabilitation is to be economically and ecologically feasible. Exotic species are generally successful in human-created environments because these environments usually include new conditions to which native species are poorly adapted.

3.5. Controlling stressors

TENET 5:
Ecosystem rehabilitation can be considered as a reversal of stressing ecosystems

The simplest rehabilitation activity is to reduce or control a stressor such as fire, overgrazing or cutting. A second activity involves the addition of species (planting or seeding) or materials (water, fertilizers, or soil) to an ecosystem. The third activity is to develop systems by which soil fertility and stability are restored quickly by natural means, such as by the use of temporary leguminous vegetation acting as a nurse for permanent vegetation to follow. Expense and effort is involved in regulating the speed of processes by artificial means (e.g. aeration of soil by reversing soil compaction, pumping air into an anaerobic lake or improving soil fertility solely by the use of fertilizers). The most difficult rehabilitation strategy involves changing or regulating the energy inputs to an ecosystem, whereby the manager controls the conditions for ecosystem growth (e.g. changing the hydrology of a forested wetland or modifying the topography on a hillside).

3.6. Linking intervention and education

TENET 6:
Diverse and imaginative education programmes should be an integral part of rehabilitation schemes

Rehabilitation schemes offer special opportunities for education and public participation. Involving local people intellectually in restoration and management provides a means for putting biology back into their cultural repertoire — back on the same status with music, art and religion (Janzen 1986). Restoration schemes, such as that at Guanacaste National Park in Costa Rica, can give an impetus for developing outdoor, living educational institutions that are 'user-friendly' for every level from the local community to visiting schoolchildren, research students and tourists.

4. Criteria for measuring the success of a rehabilitation scheme

Five criteria have been put forward by Ewel (1987, Table 5) for measuring the ecological success or failure of a rehabilitation scheme. It may be possible that, from a human perspective, some of these criteria can be ignored (at a cost) in order to attain a useful return and yield from a rehabilitation.

Table 5. Criteria for measuring the ecological success of a rehabilitation. [*Source:* Ewel 1987]

Criteria	Query
Can the system be sustained?	Is the new system self perpetuating or must it be subsidized to maintain itself?
Vulnerability to invasions	Does the new community resist invasion by other species?
Productivity	Is the new system as productive as the original?
Nutrient retention	How efficient is the nutrient cycling?
Biotic interactions	Are all key animal and plant species present?

In any event, success can only be measured relative to the objectives of the scheme, so it must be quite clear whether the aim was to restore the original ecosystem as nearly as possible or to promote the development of some other kind of ecosystem. If the objectives are clearly defined, progress towards their achievement needs to be monitored. This would allow for further interventions if change is not occurring at the place or in the direction required. However, it may often be difficult to decide which indicators should be monitored as these will vary from site to site and according to rehabilitation objectives.

Lamb (1988) distinguishes two classes of leading indicators concerning either biophysical or socio-cultural factors (Table 6). In both cases it is necessary to evaluate the stability of the ecosystem that has been achieved, its efficiency, and its flexibility. Biophysical factors would include measuring the rate of erosion, the percentage plant cover (for which remote sensing might be useful), the complexity in terms of species diversity in the new community, the quality of water of rivers and lakes. It would also be important to determine whether rehabilitation required continued inputs in the form of planting, weeding or fertilizers among others.

Socio-cultural factors are particularly concerned with the stability of the ecosystem and its human component. This can be measured in terms of migration rates into and out of the area, the extent to which farmers cultivate cash rather

Table 6. Potential indicators of success or failure for a restoration monitoring programme. [*Source:* Lamb 1988]

Biophysical indicators	Socio-cultural indicators
Stability	
Stability of soil surface	Stable human population
Presence of adequate plant cover and growth	Stable market prices
Adequate crop yield	Adequate food
Appropriate plant species composition and structure	Stable land use pattern
Appropriate animal population	Stable land tenure system
Adequate regeneration or reproduction of preferred species	Appropriate balance between subsistence crops and cash crops
Acceptable water quality (surface and groundwater)	Stable rate of fuelwood consumption
Appropriate albedo	Stable rate of water use (in arid zones)
Efficiency	
Need for imputs of seed or fertilizer	Level of public in involvement or participation in the programme
Need for weed control	
Need for irrigation	
Flexibility	
Extent to which alternative or multiple uses can be made of land	Extent to which economic flexibility possible Accumulation of economic wealth

than food crops, the increase or decrease in livestock numbers and quality. The question of who benefits from the rehabilitation is also important as, while the aim may be to help a certain sector of the population, the actual result may be to promote the interests of quite a different group.

Finally, the rehabilitation scheme must be evaluated in terms of time and cost. In many degraded landscapes, nature alone would eventually rehabilitate the ecosystem but the time required can be reduced greatly by active human intervention. Such intervention costs money and it is necessary to check that rehabilitation efforts remain within proposed budgets and do not use up more resources than they are considered to be worth.

5. Future MAB research projects

5.1. *Research on agroforestry systems in the Colombian Amazon region*

In a project in **Colombia**, participation is foreseen with the Dutch Tropenbos Foundation, in a project dealing with multistrata agroforestry on land of settlers and Indians in the Colombian Amazon region which is located in the colonization zone in the Guaviare region and in the Indian area in Aracuara. The objective of this project in general terms is to help solve problems affecting the development of the Amazon basin, as for example in the deterioration of natural resources, the low productive capacity of the land cultivated by Indian and settler families. This will be accomplished by developing a system of technology appropriate for the Amazon region. The system must be compatible with the region's ecology and the predominant socio-economic characteristics of small farmers operating in this area. It also must guarantee them stability and a better living standard, while considering that the type of land management to be carried out must be sustained in a way that preserves natural resources for future generations. The main purpose of the project is to develop an agroforestry model in marginal areas of the Colombian Amazon basin on land belonging to settlers in the Guaviare region or Indians in the Aracuara region. The model would be managed directly by small Indian farmers or settlers and establish multistrata agroforestry methodology appropriate for use in other regions of the Colombian Amazon.

The specific objectives of this project are:

- The identification of alternative models for sustainable use and management of Amazon ecosystems that can improve living conditions for inhabitants and minimize impact on the natural environment.
- The design of agroforestry models based on an evaluation of traditional native systems applied by colonists and Indians.

- The development of a social and economic evaluation concerning the medium and long-term effects of agroforestry systems on living standards and the socio-family organization of each social group.
- The development of economic and social feasibility studies of agroforestry systems.

Given the socio-economic characteristics of each zone of research, results will differ in both a quantitative and qualitative sense. They are expected as follows:

- Agroforestry models featuring a combination of elements that offer maximum efficiency in economic and ecological terms and suitability with respect to the socioeconomic and cultural conditions of each population group;
- Technological methodologies for implementation of agroforestry systems, based on local and institutional experiences;
- Lines of research on production systems that make it possible to identify the terms under which agroforestry systems constitute a sustainable production alternative for the Amazon basin.

Indians and colonists have to confront how to manage their ecosystems, trade relations and the market economy. Training of these population groups should help to guarantee a solution to problems concerning management and conservation of natural resources, marketing and processing of their products.

5.2. Functioning and integrated management of tropical ecosystems and their replacing systems in Cuba, Mexico and Venezuela

In this regional project there is an attempt to save the conceptual distance between basic and applied research to develop a solid technique on the knowledge of the functioning of the environment of plant and animal populations. Therefore, tropical ecosystems had been chosen and their diverse forms of replacing systems with the objective to understand the ecological dynamics of each part. Two main research areas have been identified:

1. Ecological dynamics and applications for sustainable development.
2. Ethnobotanical studies in tropical regions.

In general terms, this project tries to study the ecological properties of each area to evaluate both the dynamic interactions between each component and the rest of the system and to determine the conditions of stability of each system.

5.3. Asian humid tropic network

In the Asian humid tropics a network of field projects is taking shape following a workshop held in Bogor in August 1988 (Adisoemarto 1989) which among other things adopted an action plan on ecosystem rehabilitation. This network took further shape in late 1991–early 1992 through workshops in Brisbane (November 1991) and Kuala Lumpur (February 1992).

6. Conclusion

As human populations and demands on forested areas increase, the proportion of degraded land and secondary forests in tropical regions is likely to increase. In areas of severe land shortage in particular, management of these human impacted systems has become a matter of significant concern. Though these systems have received less research attention than "intact" systems, management insights and guidelines have been generated by field research and observations. The time is perhaps ripe for elaborating and refining a booklet of guidelines on rehabilitation of degraded land areas in the tropics, somewhat analogous to the *ITTO Guidelines for the Sustainable Management of Natural Tropical Forests* (ITTO 1990) or to the draft guidelines on conservation of biological diversity being prepared for ITTO by IUCN. One underlying theme of such rehabilitation guidelines might be that prevention is better than cure, and that much could be achieved through more sensitive use of present technologies. Another underlying theme could be that proposed by Lugo (1988), that a cornerstone strategy for tropical forest rehabilitation is the use of natural processes and natural subsidies to the greatest extent possible. At the same time as refining and promoting the use of guidelines for the restoration of degraded forest lands, there is a need to promote research targeted at particular ecological and socio-economic settings, given that each region offers special challenges and opportunities for rehabilitation efforts.

References

Adisoemarto, S. (Ed.). 1989. Proceedings of Regional Workshop on Ecodevelopment Process for Degraded Land Resources in Southeast Asia. Indonesian MAB National Committee and UNESCO Regional Office for Science and Technology for Southeast Asia, Jakarta.

Ewel, J. J. 1987. Restoration is the ultimate test of ecological theory. In: Jordan III, W. R., Gilpin, M. E. & Aber J. D. (Eds.), Restoration Ecology: A Synthetic Approach to Ecological Research, pp. 31–33. Cambridge University Press, Cambridge.

FAO. 1990. Interim Report on Forest Resources Assessment 1990 Project. Document COFO-90/8(a), prepared for the tenth session of the FAO Committee on Forestry. Rome, 24–28 September 1990. FAO, Rome.

Hadley, M. & Schreckenberg, K. 1989. Contributing to sustained resource use in the humid and sub-humid tropics: Some research approaches and insights. MAB Digest 3. UNESCO, Paris.

ITTO. 1990. ITTO Guidelines for the Sustainable Management of Natural Tropical Forests. ITTO Technical Series 5. ITTO, Yokohama.

Iwatsubo, G. 1990. Japan–China joint research on water and mineral cycling in mature and deteriorated forest ecosystems in southern China. In: Researches Related to UNESCO's Man and the Biosphere Programme in Japan 1989–90, pp. 29–32. Japanese MAB National Committee, Tokyo.

Janzen, D. H. 1986. Guanacaste National Park: Tropical Ecological and Cultural Restoration. Editorial Universidad Estatal A Distancia, San José.

Jonsson, T. & Lingren, P. 1990. Logging Technology for Tropical Forests. Report from the ITTO Pre-Project 'Improvement of Harvesting Systems for the Sustainable Management of Tropical Rain Forests'. Forest Operations Institute 'Skogarbeter', Kista (Sweden).

Khemnark, C. & Sahunalu, P. 1989. Revegetation trials for tin mine tailings reclamation in Thailand. In: Adisoemarto, S. (Ed.), Proceedings of Regional Workshop on Ecodevelopment Process for Degraded Land Resources in Southeast Asia, 18–22. Indonesian MAB National Committee and UNESCO Regional Office for Science and Technology for Southeast Asia, Jakarta.

Lal, R. 1987. Need for, approaches to, and consequences of land clearing and development in the tropics. In: Tropical Land Clearing for Sustainable Agriculture, Proceedings of an IBSRAM Inaugural Workshop, pp. 15–28. Proceedings No. 3. IBSRAM, Bangkok.

Lamb, D. 1988. IUCN Guidelines for Restoration of Degraded Ecosystems. Draft. IUCN, Gland.

Lamb, D. 1990. Exploiting the Tropical Rain Forest. An Account of Pulpwood Logging in Papua New Guinea. Man and the Biosphere Series 3. UNESCO, Paris and Parthenon Publishing, Carnforth.

Lugo, A. E. 1987. Ecosystem Rehabilitation in the Tropics. Paper presented at Symposium on Ecosystem Redevelopment: Ecological, Economic and Social Aspects. Budapest, April 1987.

Lugo, A. E. 1988. The future of the forest ecosystem: ecosystem rehabilitation in the tropics. Environment 30 (7): 17–45.

Schreckenberg, K., Hadley, M. & Dyer, M. I. (Eds.). 1990. Management and Restoration of Human-Impacted Resources: Approaches to Ecosystem Rehabilitation. MAB Digest 5. UNESCO, Paris.

Sinukaban, N. 1989. Deforestation of tropical rain forest and soil degradation in Transmigration Area, Indonesia. In: Adisoemato, S. (Ed.). Proceedings of Regional Workshop on Ecodevelopment Process for Degraded Land Resources in Southeast Asia, pp. 49–53. Indonesian MAB National Committee and UNESCO Regional Office for Science and Technology for Southeast Asia, Jakarta.

Yoda, K. 1990. Improvement of biological productivity in tropical wastelands. In: Researches Related to UNESCO's Man and the Biosphere Programme in Japan 1989–90, pp. 21–28. Japanese MAB National Committee, Tokyo.

Yu Zuo-Yue & Wang Zhu-Hao. 1989. The method of recovery of forest vegetation in degraded ecosystem in tropical and subtropical waste lowland in Guangdong. In: Adisoemarto, S. (Ed.), Proceedings of Regional Workshop on Ecodevelopment Process for Degraded Land Resources in Southeast Asia, pp. 9–17. Indonesian MAB National Committee and UNESCO Regional Office for Science and Technology for Southeast Asia, Jakarta.

3. Biological characters and plasticity of juvenile tree stages to restore degraded tropical forests

Programme proposition for genetic resource maintenance

G. MAURY-LECHON*
URA 1183 CNRS, Le Chênoy, 89170 Saint Fargeau, France
Tel: 86.74.14.48; fax: 86.74.08.09

Keywords: Biological plasticity, conservation, seed germination, seedling establishment, utilization, valuable trees

Abstract. For long-term ecological and economical reasons, international environmental programmes recognize the need to ensure the regeneration of primary forests and their valuable slow growing timber species. Most attempts to encourage natural regeneration after intensive logging have failed because of the elevated maintenance costs for the remaining standing trees, and the difficulties in obtaining high rates of survival for the young plants. In consequence, short rotation intensive monoculture plantations using exotic species have been favored over residual forest or secondary forest management.

A basic research programme is necessary to determine the physiological responses of trees during their initial growth stages, the optimal timing for enrichment plantings and the complementarity of species in mixtures.

The proposed international programme would consist of the following:

a) researches on factors controlling seed germination and young seedling development in order to secure the replacement of the valuable species selectively logged to increase production of the species in high demand through mixed planting of multipurpose species, and to ensure higher growth and survival of planted and naturally regenerating seedlings;

b) the creation of a network of research teams and laboratories specializing in particular research fields: genetic resource centers (collections, conservation areas, arboreta, protected forests, laboratories and nurseries), amelioration of development and survival constraints, definition of the basic species development characteristics (specific and individual plasticity, interspecific complement between indigenous species and also together with exotics), publication of the results and training activities.

1. Rationale

As the practice of the utilization of natural forests has been largely incompatible with the maintenance of tropical timber trees, or with the survival of the forest formation itself, there is an urgent need to find, throughout the tropics, efficient means to facilitate natural regeneration. The root causes of the disequilibrium between utilization and maintenance of forest resources seem to be linked to the fact that man removes from the forest what he needs, without enough respect for the biological requirements for regeneration and survival of these living systems. Often the reason for this lies in the simple lack of knowledge of natural laws, which emphasizes the need for research and training. Tropical Asian rain forests, for example, can be renewed when management practices take into account their functioning cycles (Appanah et al. 1990; Appanah and Salleh 1991; Bruenig and Poker 1989; Borhan and Abdull Rahman Kassim 1989; Kartawinata et al. 1989).

Degradation is not an inevitable product of utilization. Convincing examples exist in correctly managed modern forests or plantations such as certain *Quercus* forests in France,[1] or traditional planted formations of *Cassia siamea* that communities of Thai minority plant and maintain for their firewood production in Yunnan, China.[2]

However, the level of forest degradation and the socio-economic environment appear to be deeply connected. When human conditions become extreme (overpopulation and/or wars with consequent misuses, poverty and resource limitations) the ancient forest zones are transformed into wastelands. Examples include Xiaoliang re-

*c/o Prof. Dorffling, Allgemeine Botanik und Botanischer Garten, Ohnhorststr. 18, D-2000 Hamburg 52, Germany. Tel: (040) 82.22.335; Fax: (040) 82.22.254.

H. Lieth and M. Lohmann (eds.), Restoration of Tropical Forest Ecosystems, 37–46.

gion in South China (Wang et al. 1987), Ba Hao region in South Vietnam (Trung 1987), many parts in India (Maithani et al. 1989). The numerous experiments undertaken in the different continents clearly show that the elevated cost of human interventions represent the major impediment both to the conservation of natural forests and to the restoration of degraded zones.

With the continuous growth of human populations and material expectations, there is no reason to expect that the utilization of trees and forest formations will decrease in the future. Thus we must learn how to better produce (and maintain) trees. This means that it is necessary to find cheaper and more efficient methods to produce larger numbers of trees with the favourable survival and growth characteristics. The juvenile stages (seeds and seedlings) often reflect future growth potential, provided one knows the key characteristics. It should also be possible to utilize ecological knowledge to reduce the costs of manpower for forest establishment or renewal. Examples include the possibility to use wild or introduced fauna (wild or domesticated) for "enrichment sowings" of the disappearing species, according to the results of ecological studies (Maury-Lechon and Poncy 1986a) with the rodent Acouchi "nurserymen" for Carapa sp. Meliaceae, in French Guyana and to experimental trials (Janzen and Vasques-Yanes 1991) with horses and cattle in a National Park of Central America. The biological relations existing within fungal flora may equally be utilized to protect seeds and seedlings of valuable trees from their fungal parasites by the introduction of other fungi predators to the seed parasites (Elouard 1991). Mycorrhizas and tree relations are other examples (Lee 1988; Louis 1988; Gunatilleke and Maheshwaran 1988). Above all, we should intervene more efficiently by respecting and following the "lines of lesser resistance" of the natural evolution of the exploited species.

It is not possible to preserve the remaining forests when local human population have no other source of domestic energy, nor is it possible to create artificial formations according to a fixed model without taking into consideration the local ecological and human needs. Monospecific plantations can respond to immediate human needs for fuel wood and allow the progressive re-establishment of other tree species, native or exotics, to restore the environment and thus the conditions of human life.

The global equilibrium and its requirements can only be respected if we know them. Therefore, we first have to analyse them, and then utilize their major elements to manage their maintenance (Ng 1988; Whitmore 1989; Maury-Lechon et al. 1984, Maury-Lechon 1991).

To try to preserve both the maintenance and diversity of forests and the continuity of production, foresters have focused on two critical levels: the "adult", and the "seminal and juvenile" material. In the former they have tried to improve the growth of larger trees within forest stands. In the latter, three aspects were considered: 1) storage technics to improve seed viability for periods long enough to be of practical value for forest planting, 2) improvement of germination and early survival, 3) increases in young plant height and stem diameter growth.

Only seminal and juvenile material will be taken into consideration here.

2. What we know

2.1. Biological types and species plasticity

The behaviour of juvenile stages is related to their structural characteristics, which define the possible functioning modes, as well as the environment by which the species can germinate, establish and survive (Maury 1978; Galy 1984; Rousteau 1986; Bras and Maury-Lechon 1986; Kohyama et al. 1989; Kohyama and Hotta 1990). From the seed characteristics we predict the behavioural types and consequently the environment. This knowledge facilitates provenance selection for trials of new and less known species.

The moisture content of fresh seed represents the most fundamental and synthetic character. It determines the seed eco-physiological type: "recalcitrant" and "orthodox" seeds (Roberts and King 1980; Chin and Roberts 1980; Fenner 1985; Farrant et al. 1988), correspond in fact to "very hydrated" and "slightly hydrated" seeds (Maury-Lechon et al. 1980, 1991).

Scientific analyses have proven both the existence of large biological (Maury-Lechon and

Poncy 1986a; Rousteau 1986; Gould and Mahedi Andau 1989; Gauthier-Hion 1990; Charles-Dominique et al. 1981) and eco-physiological types of seeds and seedlings, and the specificity of the reactions within these large types (Maury-Lechon et al. 1981, 1991; Maury-Lechon and Poncy 1986a; Awang and Mohammad 1987; Khare et al. 1987; Perl 1988; Bazzaz 1991; Hladick and Miquel 1991; Sasaki and Mori 1978; Wang 1988). Henceforth experimentation is necessary to define the methods of intervention at the specific level.

Experimental studies of ripe seeds and initial seedling stages have allowed us to quantify the extreme and optimal environmental conditions (temperatures, light, moisture) for seed survival and germination, and for seedling development, growth and survival. These limiting environmental values determine the range of biological plasticity of each species, and consequently the possibilities for manipulating seeds and seedlings for utilization within forest management programs (Bras and Maury-Lechon 1986). The potential plasticity includes the expressed plasticity (exhibited under natural conditions) plus the inductible plasticity only exhibited under extreme experimental or natural conditions.

Under the climatic constraints of temperate or cold zones, seeds have acquired the capacity to remain dormant during the unfavourable periods. In humid tropics and particularly in primary forests (Maury 1978, 1979; Rousteau 1986; Maury-Lechon 1982a-b; Maury-Lechon and Poncy 1986a; Ng 1980, 1988; Ashton 1989; Turner 1990; Turner and Newton 1990) the constancy of temperature and moisture conditions permits a continuous uninterrupted process of flowering, seed development and dissemination, germination, early seedling growth, and transition from seminal heterotrophy to juvenile autotrophy. At this stage, the young plant from the primary rain forest can confront the first actual limiting factor: i.e. the lack of light. This explains why at that stage the seedling has gained the possibility to put itself into a "sleep", waiting for a canopy opening (natural or anthropic), for several weeks, months or years according to the species. This also demonstrates the importance of understanding natural evolutionary trends; new biotechnologies fail when acting against them.

In the tropical rain forest ecosystem the above behavior characterizes the species from stable primary forest (Maury-Lechon and Poncy 1986a). Secondary species which characterize gaps, and still more the species which colonize the large openings (pioneers), differ from this model and function in a comparable way with those from temperate regions: their seeds, better protected, may remain dormant until external conditions become favorable for their germination. For these species, once germination has started the seedling imperatively requires light, otherwise it dies (Maury 1978, 1979; Maury-Lechon et al. 1986b; Maury-Lechon 1991).

The diversity of these biological characteristics may be applied to specific regions and situations, to select more appropriate species for forest restoration. For example, light-demanding species from subtropical zones (adapted to open/dry habitats and poor soil conditions) could be introduced into degraded equatorial regions that were previously closed wet forests. In this way a more humid environment should develop in which the more fragile shade-demanding species could be later on replanted (and/or naturally reintroduced by local fauna), provided the introduced species are not likely to persist at the expense of wet zone flora. Practically-oriented publications giving the lists of valuable species (already used, plus their possible local and exotic substitutes), together with their biological useful characters, would stimulate the development of new, efficient solutions to restoration challenges.

2.2. *Valuable species and utilizable biological characters*

Species value and biology

The most highly valuable timbers are most often derived from slow-growing hardwood species of humid, dense and stable primary forest. In this type of environment tree species do not flower yearly (intervals are usually irregular), and their seeds, fleshy and hydrated, lose their viability when dehydrated, so that they cannot be conserved ("recalcitrant" seeds; Bras and Maury-Lechon 1986; Corbineau and Côme 1988;

Tompsett 1990; Maury-Lechon et al. 1991). Furthermore cuttings from adult branches (Hallè and Kamil Hanif 1981; Srivastava and Penguang Mangil 1981; Scott et al. 1988) often develop into plagiotropic axes (laying down), and juvenile tissues must be used to obtain standing axes. In most of these cases trees do not produce shoots when cut or damaged.

Soft woods, generally fast-growing species, come from well-lit sites. These species flower yearly and regularly and their protected seeds tolerate a certain degree of dehydration, and often can be conserved ("orthodox" seeds); their cuttings develop standing axes and when cut or damaged their stumps tend to form shoots.

Between these two extreme species types, numerous intermediary types exist (Maury-Lechon 1979, 1982a-b; Maury-Lechon and Poncy 1986a; Maury-Lechon et al 1986b), which are adapted to the great variety of possible environmental conditions that exist within the rain forest canopy and under wood. This large genetic pool of species and biological diversity contains the new (or forgotten?) useful multipurpose species which could replace the ones in danger of extinction.

Species behavioral types and species condition
It can be considered that forest tree species are broadly put into three behavioural categories (Alexandre 1982, 1989): (1) the forest-forest type, (2) the forest-gap type, (3) the gap-gap type.

The species of the first category (forest-forest type) develop their whole life cycle inside the close stable forest. This group contains the species of very high commercial value, which up to now have been removed according to the "creaming off" method, without real replanting, because of the difficulties mentioned above. As soon as the forest is intensely exploited, their natural regeneration is blocked by the resulting opening of the canopy, and as the cost of the potential regeneration is too high they tend to disappear.

The species of the forest-gap type, probably produce the larger quantity of commercialized timbers in reason of their high production (due in part to their fast growth under well-lighted conditions) and of their commercial value (due to their coloured woods of intermediate hardness between hardwoods and softwoods). These species are naturally much more "plastic", thus "malleable" and "domesticable", than those of the two other groups. After a selective logging they could constitute the essential elements in restoring (or replanting) these degraded forest zones. The species that succeeded in large-scale plantations, in African and South American zones, mostly belong to that group, and these include *Cedrela odorata*, *Khaya ivorensis*, *Swietenia sp.* (three Meliaceae), *Terminalia superba* (Combretaceae), *Aucoumea klaineana* (Burseraceae), *Pterocarpus sauyaxii* (Fabaceae). The very young stages of these species can be relatively sensible under too high light/dryness conditions; however in natural forests, they are often gregarious and this facilitates their exploitation.

The species able to resist the very large openings and the more disturbed soil conditions resulting from clear cuts, correspond to the gap-gap type. They require full openings and degenerate when "over-crowned"; this is the case for *Goupia glabra* (Celastraceae) and also for *Laetia procera* (Flacourtiaceae), both from French Guyana. In spite of the low commercial value of their timbers, this type of tree (as well as *Cecropia sp.*, Cecropiaceae, which has no commercial value at all), may be very useful when the degree of degradation is most severe: i.e. lacking all forestcover (Maury-Lechon 1982a–b; Maury-Lechon and Poncy 1986a; Maury-Lechon et al. 1986b). Thus it is possible to reforest these severely degraded zones by using multipurpose trees. This is the case for the following species: *Didymopanax morototoni* from French Guyana provides wood for matches and may also produce insecticide substances; Australian *Eucalyptus* introduced all over the tropics give firewood, pulp-paper, carpentry wood, etc.; exotic pines play a comparable role; Australian and African *Acacias* give fodder, firewood, pulp-paper, etc. As one moves further into the sub-tropical zone, away from the equator, the gap-gap species became more preponderant in natural environments and their plantations become more frequent.

2.3. *Mixtures of complementary species*

"Optimal" mixtures of species can be variously defined to respond to the characteristics of the local situations, utilizing a balanced combination of species diversity and natural or induced plasticity. The choice of species should give priority to the local species corresponding to the requested ecological types and the socio-economical profiles. However it does not exclude "a priori" the exotics with strong positive socio-economic impact, adapted to local climate without disturbance for the local ecological equilibrium. In this way, for example, a plantation of Eucalyptus, pines, or acacias, may be recommended in case of urgency (for example on bare soils in South China), to allow for a subsequent regeneration of native vegetation in the long run.

Some traditional and experimental examples can be cited:

The Indonesian traditional "damar gardens" (Michon and Bompard 1987; Michon and Jafarsiduk 1989) illustrate the success of the traditional agro-forestry methods based on these same principles. They rely on an experimental approach and the verification of the effects.

In China precise studies for 25 years (Yu manuscript 1987) have demonstrated how Eucalyptus acts on ground water layers.[3] These trees quickly produce firewood, pulp-paper and timber, and the resulting income allows for financial support of other rotations. It should thus be interesting to try, without elevated cost, to intermix these ultra-fast growing exotic species with some trees per hectare of other much more slowly growing species. They could benefit from the shade of the former species (eventual alterations of their production should be analyzed), and when they need more light, they could give shade to other agricultural or forest cultures. Eucalyptus should then be suppressed.

Similarly, it can be thought that by planting valuable timber trees, of slow-growing type, between the rubber-trees, in the Malaysian industrial plantations (which are at present reaching the limit of their productivity, about 40 years), a second rotation would perhaps be rapidly possible while the newly planted rubber-trees would develop until their productive stage. Vietnamese experiments (Trung 1987) on Dipterocarps intermixed with various food plants, shrubs and trees encouraged new trials within the International Working Group on Dipterocarps (Kostermans 1987). Some were undertaken a few years ago with fast-growing species of Dipterocarps and Acacia (Soerianegara et al. 1991).

3. What we need

3.1. *Fundamental research and possible applications*

The above analysis of tree biological characters and their utility to restore degraded forests, underlines the present needs and defines some biological and experimental principles on which a research programme could be conducted.

Fundamental research is lacking for the local valuable species, mainly in the fields of biology, ecology, physiology and biochemistry of seed germination and seedling initial development and survival. As these stages represent the first essential step in forest reconstitution, research works are needed to determine factors which can control them.

3.2. *Biological and experimental principles*

— Biological diversity may be used:
- according to the natural characteristics of the different species,
- in function of the fixed goal (socio-economic goal),
- and of the future evolution of the forest formation (ecological aim).

— Natural processes can be modified according to biological plasticity.
— Natural characters must first be analyzed and then used.
— Effects of treatments need to be controlled on the ulterior growth.
— Booklets of synthesis easy to handle are needed.

3.3. *Research programme proposition*

Questions

Common research methods are essential, but the results are not always extrapolable to other situations. For a given country or region it is not yet known:

Table 1. Conservation and utilization of genetic resources of valuable tropical trees, research of factors controlling seed germination and seedling establishment (Proposed programme of cooperative research)

Base principles
1 To utilize biological diversity
2 To qualify and quantify natural plasticity
3 To define methods for action

Double approach

1 *Scientific study:*
(fundamental research)
factors susceptible to inhibit/stimulate
• *Seed germination,*
• *Seedling development/Survival*

2 *Application* (applicable results from 1):
creation of centers for the genetic resources of valuable tropical trees with:
• *Seed/Seedlings banks* in + ex situ
• *Selected mother trees* in + ex situ.

Scientific Study
Biological plasticity in young stages of valuable tropical trees.
Utilization to restore tropical forests and preserve genetic resources of valuable species.

Successive phases:
(1) Eco-physiological characters of natural and experimental ontogenesis at tree and root-symbiont level:
— tree:
• critical stages for germination,
• evolutive study before, during and after germination,
• ecophysiological and biochemical analyses of natural processes.
•• seed moisture content, presence of abscissic acid and germinative ability, during seed maturation and after harvest,
•• optimal and limit conditions for seed germination.

— root symbiont characterization:
• species descriptions,
• infection dynamism,
• trials to define the degree of dependence of tree-root symbioses.

(2) Factors controlling development, survival and growth of juvenile regenerative material:
• seed conservation: biological, physical, chemical inhibition/stimulation, research of efficient ways to conserve seminal and juvenile material ex and in situ to facilitate silviculture practices,
• production of performant seedlings (experimental inoculation of symbionts).

(3) Potential complement of valuable species to create plurispecific and pluristratified formations:
• definition of species to be mixed in these populations,
• tests on the "optimizing" effect of species mixtures on plant development and contribution to maintain local equilibrium.
(4) Comparative dynamics of juvenile populations into natural and man-utilized zones.
(5) Modelization of growth (individuals and populations) and dynamism of populations (juvenile competition and symbioses).

Goals
(1) Conservation: seminal + regenerative material
(2) Production of performant plants
(3) Definition of complementary species
(4) Modeling of individuals and populations

Global aims
(1) Define methods to restore degraded tropical forests,
(2) Create genetic resource pools for valuable species.

— which local species could efficiently replace the introduced ones,
— which are the biological characteristics, mainly the plasticity, of locally important species,
— which could be the benefits when these local important species are used into mixed populations with other local species or with exotics (each species responding to specific local needs),
— which are the key species for the correct functioning of local ecosystems, according to local diverse constraints,
— how species selected for a given situation, and used in pure or mixed populations, do fit the local soil/climatic conditions,
— how does the presence of exotics influence the behavior of local valuable species and the local ecology,

— how to efficiently manage the regenerative material (seeds and seedlings) for local species of ecological or socio-economical value.

Answers: programme proposition
Elements to answer the above questions could be found through the present proposal for a research programme based on three main points and a double approach (Table 1).

4. Conclusion

From the presently available selected scientific data, biological and experimental principles have been here underlined, and consequently a research programme has been structured to investigate how much biological characters and plasticity of juvenile forest tree stages may be controlled and utilized to restore degraded tropical forest zones.

This project is not only a simple scientific exercise. It corresponds to a real need, clearly expressed by the countries concerned (Malaysia, Indonesia, India, the Philippines, Thailand, Viet-Nam, Japan, Germany, Great Britain, America and France) during the 4th Round Table Conference on Dipterocarps in BIOTROP of Bogor (Indonesia) in December 1989.

The same demand exists in the other tropical zones, which would justify the creation of a network of centers for experimentation and studies. Diverse research institutes and universities are already in contact, forming a nucleus from which such a programme could develop in cooperation with local researchers from countries in demand, and securing at the same time a training component.

In this way the programme could be applied in different tropical zones to demonstrate its validity and applicability in forest reconstitution. The programme proposal could be treated either entirely or by isolated sections according to the specialization of the scientific teams and to the local priorities. An international group of scientists from Malaysia, France and the United Kingdom, is already developing a cooperative programme entitled "Domestication of Dipterocarps in Malaysia: factors controlling establishment of valuable trees", funded by the Commission of European Communities (TS2∗0298 M (EDB): 1990–1994), and corresponding to the proposal presented on Table 1 (phase 3 excepted for financial reasons).

Acknowledgements

Acknowledgements are addressed to Professor K. Dörffling for the present 6 months' stay in his laboratory (Institute für Allgemeine Botanik, Universität Hamburg), to French C.N.R.S. and to D.F.G. for support and agreement of this scientific exchange (German–French Convention), to the Commission of the European Communities (CEC) for the ongoing programme TS2∗-0298 M(EDB) which allowed the author to take part in this symposium, and to Malaysian Airline System who facilitated the France-Malaysia flights of European scientists participating in the CEC project.

Notes

1. Successfully managed French forests exist as a result of a silviculture of high quality oak: sessile oak in the Central-East region of Bourbonnais and Bourgogne, the Bentranges forest (Nièvre department) and the communal forests of Corcelles-les-Citeaux and Saulon-la-Chapelle (Côte d'Or Department) (Office National des Forêts 1991). In Fontainebleau another type of example is given: as the "natural" forest was degraded after 8 centuries of use, a preservation management started during the King Louis XIV period; important plantations were later incorporated during the XVIIIth and the first half of the XIXth centuries so that it has now become an appreciated "cultivated" forest (Tendron 1983).
2. Author's visit in Xishuangbana forest region to the protected park of Chun-Huan (Manjinglan village) which includes a traditional Thai village with a harmonious ecosystem formed by 4 main parts: 1) the houses close to the river, shaded by planted fruit trees and surrounded by 3 other parts. Buffalos and other domesticated animals share that space with humans; 2) the cultures: rice and other foods in lower humid parts; 3) the *Cassia siamea* plantation on more undulated and higher part; branches of trees are cut in every 2 years up to the highest height accessible by men from the ground, a rotation among the trees ensure a regular supply of firewood; the exudations of black sap strangely dye the boles; tree shape and "forest" result entirely from man activity; 4) the temple on the highest, more beautiful and more aerated part, surrounded by fragrant flowering trees, bushes and flowers. Because most land is being converted into Hevea

plantations, and because other ethnic groups such as Haini leave their native mountains bringing shifting cultivation into Thai lowlands, the region equilibria are destabilized.

3. During author's consultancy to CERP seminar (MAB Cooperative Research Program between China and Germany) in 1987, Professor Yu Zuo-Yue (South China Institute of Botany, Academia Sinica, Guangzhou) presented and distributed his results entitled "The methods and effects on the recovery of vegetation in tropical eroded coastal area in China". In 1959 experiments on water and soil conservation started at Xiaoliang Station, Dian Bai county, Guangdong province. The aim was to provide "scientific bases for improvement and utilization of tropical and subtropical hilly wastelands". In 1980 permanent plots of forest climate, forest soil and forest hydrology were built respectively in barren slope, pure eucalyptus forest and broad-leaf mixed forest where the conditions of geomorphy, geology and pedology are similar. The biological and ecological effects of plants, animals, insects, soil animals and soil microbes were studied simultaneously. The results were as follows: "... (4) The water capacity of surface soil increased gradually from barren wasteland to mixed forest. For example in January 1982 the water capacity in 10 cm soil layer were 9.8% and 13.2% in wasteland and mixed forest respectively. But on the other hand, in subsoil of 100 cm or more, the water capacity was highest in barren wasteland, and then in mixed forest, and was lower in eucalyptus forest. The reason for the difference was the covering action of the forest in the former, and was the absorption of water by the root system of trees and the transevaporation of the canopy in the latter. (5) The ground water was highest in mixed forest with the vertical variation from 1 to 4 m and lowest in eucalyptus with the variation from 9 to 11 m, and middle in wasteland with vertical variation from 3 to 5 m. The reason for the difference was the low water loss in wasteland because there was no canopy transevaporation, although the water supply was low for most rainfall lost from run off, so the ground water in wasteland was rather stable. In mixed forest, the waste of water was high and the ground water table was high and stable because this forest had high ability of water preservation and the water supply was larger than loss. The ground water table was lowest in eucalyptus forest because the forest floor was bare with poor ability of water preservation and the water supply was lower than the evaporation and transpiration."

References

Alexandre, D. Y. 1982. Aspects de la régénération naturelle en forêt dense de Côte d'Ivoire. Candollea 37(2): 579–588.

Alexandre, D. Y. 1989. Dynamique de la régénération naturelle en forêt dense de Côte d'Ivoire. Stratégies écologiques des arbres de la voûte et potentiels floristiques. O.R.S.T.O.M. (Ed.), Col. Etudes et Thèses, 102 pp.

Appanah, S., Weinland, G., Bossel, H. & Krieger, H. 1990. Are tropical rainforests non-renewable? An enquiry through modelling. Jour. Trop. For. Sc. 2(4): 331–348.

Appanah, S. & Salleh, M. N. 1991. Natural regeneration and its implications for forest management in the dipterocarp forests of Peninsular Malaysia, pp. 361–369. In: Gomez-Pompa, A., Whitmore, T. C. & Hadley, M. (Eds.), Rain Forest Management. UNESCO, Paris.

Ashton, P. S. 1989. Dipterocarp reproductive biology, pp. 219–240. In: Lieth, H. & Werger, M. J. A. (Eds.), Tropical Rain Forest Ecosystems. Ecosystems of the World 14B. Elsevier Science Publishers B.V., Amsterdam.

Awang, K. & Mohammad, A. 1987. Status of ecophysiological studies on forest species in Malaysia, pp. 19–26. In: Doley, D., Osmond, C. B., Wongkaew, W., Suselo, T. B., Torquebiau, E. & Tjitrosomo, S. S. (Eds.), Proc. Int. Conf. Tropical Plant Ecophysiology. BIOTROP special Pub. 31, SEAMEO-BIOTROP, Bogor.

Bazzaz, F. A. 1991. Regeneration of tropical forests: physiological responses of fast growing tres, in: Gomez-Pompa, A., Whitmore, T. C. & Hadley, M. (Eds.), Rain Forest Regeneration and Management. UNESCO, Paris.

Bras, P. & Maury-Lechon, G. 1986. Graines forestières tropicales de type fortement hydraté: la conservation et ses effets, exemple du Symphonia globulifera L.f. de Guyane française. B.F.T. 212, 35–46.

Borhan, M. & Kassim, Abdull Rahman, 1989. The growth of a planted stand of Dipterocarpus baudii in Forest Research Institute of Malaysia (FRIM), Peninsular Malaysia, pp. 214–217. In: Wan Razali Wan Mohd, Chan, H. T. & Appanah, S. (Eds.), Proc. Seminar on Growth and Yield in Tropical Mixed/Moist Forests. FRIM, Malaysia.

Bruenig, E. F. & Poker, J. 1989. Management of Tropical Rainforests: Utopia or chance of survival? In Proc. Inter. Symp. Food and Agriculture Development Centre, German Foundation for International Development (DSE), Feldafing, FRG. 112 pp. Nomos Verlagsgesellschaft (Eds.). Baden-Baden.

Charles-Dominique, P., Atramentowicz, M., Charles-Dominique, M., Gerard, H., Hladik, A., Hladik, C. M. & Prevost, M. F. 1981. Les mammifères frugivores nocturnes d'une forêt guyanaise: inter-relations plantes-animaux. Rev. Ecol. (Terre et Vie) 35: 341–435.

Corbineau, F. & Côme, D. 1988. Storage of recalcitrant seeds of four tropical species. Seed Science and Technology 16: 97–103.

Chin, H. F. & Roberts, E. H. 1980. Recalcitrant Crop Seeds, 160 pp. Tropical Press Sdn. Bhd., Kuala-Lumpur, Malaysia.

Elouard, C. 1991. Contribution à l'étude du parasitisme fongique des Dipterocarpaceae indonésiennes. Thèse de Doctorat (Ecologie), 287 pp., 7 annexes (15 tab.), Université Paul Sabatier, Toulouse.

Farrant, J. M., Pammenter, N. W. & Berjak, P. 1988. Recalcitrance: a current assessment. Seed Science & Technology 16(1): 155–166.

Fenner, M. 1985. Seed ecology. Chapman and Hall, 151 pp.

Galy, M. 1984. Ontogénie des stades juvéniles des arbres tropicaux: contribution à la connaissance des premiers stades de régénération de deux Meliaceae africaines et d'une parcelle en forêt dense humide de Guyane. 46 pp. D.E.A. Univ. Pierre et Marie Curie, Paris.

Gauthier-Hion, A. 1990. Fruit characters, fruiting phenology and fruit consumption by vertebrates in an African rain forest. In: Bawa, K. S. & Hadley, M. (Eds.), Reproductive ecology of tropical plants. UNESCO-MAB, Paris & Parthenon Publ.

Gould, K. & Mahedi, Andau 1989. Selection and rejection of five species of dipterocarp fruits by wild and captive primates, squirrels and hornbills. Malayan Nature J. 42(4): 245–249.

Gunatilleke, I. A. U. N. & Maheshwaran, J. 1988. Nutrient changes of litter decomposition in a lowland rain forest and deforested area in Sri Lanka, pp. 291–305. In: Ng, F. S. P. Trees and mycorrhiza. FRIM, Malaysia.

Hladick, A. & Miquel, S. 1991. Seedling types and plant establishment in the African rain forest, pp. 269–290. In: Bawa, K. S. & Hadley, M. (Eds.), Reproductive ecology of tropical forest plants. UNESCO-MAB, Paris & Parthenon Publ.

Halle, F. & Hanif, Kamil 1981. Vegetative propagation of Dipterocarps by stem cuttings and air layering. Mal. For. 44(2–3): 314–318.

Janzen, D. H. & Vazques-Yanes, C. 1991. Aspects of seed ecology of relevance to management of tropical forested wildlands, pp. 137–157. In: Gomez-Pompa, A., Whitmore, T. C. & Hadley, M. (Eds.), Rain Forest Regeneration and Management. UNESCO-MAB series, vol. 6, Paris and Parthenon Publishing.

Kartawinata, K., Jessup, T. C. & Vayda, A. P. 1989. Exploitation in Southeast Asia, pp. 591–610. In: Lieth, H. & Werger, M. J. A. (Eds.), Tropical Rain Forest Ecosystems. Ecosystems of the World (14B, Series Ed. Goodall, D. W.) Elsevier, Amsterdam.

Khare, P. K., Yadav, V. K. & Mishra, G. P. 1987. Collection, germination and storage of *Shorea robusta* Gaertn. f. seeds, pp. 154–157. In: Proc. Inter. Symp. on Forest Seed Problems in Africa. Department of Forest Genetics, Swedish Agricultural Univ., Umea, Sweden.

Kohyama, T. & Hotta, M. 1990. Significance of allometry in tropical saplings. Functional Ecol. 4(4): 515–521.

Kohyama, T., Hotta, M., Ogina, K., Syahbuddin, F. & Mukhtar, E. 1989. Structure and dynamics of forest stands in Gunnung Gadut, West Sumatra, pp. 33–47. In: Hc ta, M. (Ed.), Diversity and plant–animal interaction in equatorial rain forests, Occasional Papers No. 16, Kagoshima Univ. Res. Cen. for South Pacific.

Kostermans, A. J. G. H. 1987. (Ed.) Proceedings of the Third Round Table Conference on Dipterocarps. UNESCO-MAB: 657 pp. (Samarinda, 1985).

Lee, S. S. 1988. The ectomycorrhizas of *Shorea leprosula* (Dipterocarpaceae), pp. 189–206. In: Ng, F. S. P. (Ed.), Trees and mycorrhiza. FRIM, Malaysia.

Louis, I. 1988. Ecto- and endomycorrhizae in tropical dipterocarp, *Shorea parvifolia*. Mycol. 80(6): 845–849.

Maithani, G. P., Sharma, D. C. & Bahuguna, V. K. 1989. Problems of Sal forests, an analysis. Ind. For. 115(8): 513–525.

Maury, G. 1978. Dipterocarpacées: du fruit à la plantule. Thèse Dr. d'Etat, Univ. Toulouse, IA-B: 432 pp., II: 344 pp.

Maury, G. 1979. Plantules et régénération forestière en Guyane française: premières constatations sur une coupe à blanc de 25 ha. Bull. Soc. bot. Fr. 126, Actualités bot. 3: 165–171.

Maury-Lechon, G. 1979. Dipterocarpacées: Taxonomie, Phylogénie, Ecologie. Entretiens du Muséum, Mem. Mus. nat. Hist. nat. B XXVI, Paris, 162 pp.

Maury-Lechon, G. 1982a. Régénération forestière sur 25 ha de coupe papetière en forêt dense humide de Guyane française. C. R. Acad. Sc. Paris 294(III): 975–978.

Maury-Lechon, G. 1982b. Régénération forestière en Guyane française II: recrû sur 25 ha de coupe papetière en forêt dense humide (ARBOCEL). B.F.T. 197: 3–21.

Maury-Lechon, G. 1991. Comparative dynamics of tropical rain forest regeneration in French Guiana. In: Gomez-Pompa, A., Whitmore, T. C. & Hadley, M. (Eds.), Rain Forest Regeneration and Management UNESCO, Paris.

Maury-Lechon, G., Corbineau, F. & Côme, D. 1980. Données préliminaires sur la germination des graines et la conservation des graines et des plantules de *Symphonia globulifera* L.F. (Guttifère). B.F.T. 193: 35–41.

Maury-Lechon, G., Hassan, A. M. & Bravo, D. R. 1981. A preliminary study on seed storage of *Shorea parvifolia* Dyer and *Dipterocarpus humeratus* van Slooten. Malay. For. 44(2–3): 267–280.

Maury-Lechon, G., Hadley, M. & Younes, T. 1984. The significance of species diversity in tropical forest ecosystems. Inter. Biol. special issue 6, 74 pp. IUBS, Paris.

Maury-Lechon, G. & Poncy, O. 1986a. Dynamique forestière sur 6 ha de forêt dense protégée à partir de quelques exemples d'espèces de forêt primaire et d'espèces de cicatrisation. Mem. Mus. nat. Hist. nat. A 132: 211–242.

Maury-Lechon, G., Betsch, J. M. & Betsch-Pinot, M. Ch. 1986b. Dynamiques comparées de la végétation et de la pédofaune dans un recrû en zone forestière tropicale. In: Vertébrés et Forêts Tropicales d'Afrique et d'Amérique. Mém. Mus. nat. Hist. nat. Entretiens du Muséum, A 132: 243–255.

Maury-Lechon, G., Lenoir, C., Bras, P., Soyza, N. D., Prasad, S. N., Tomboc, C., Duc, P. H., Sudiyani, Y., Chaisurisri, K. & Yap, S. K. 1991. Tropical forest seeds of the very hydrated type: the conservation and its effects. In: Sorianegara, I. et al. (Eds.), Proceedings of the Fourth Round Table Conference on Dipterocarps, BIOTROP Spec. Publ. 41: 479–498.

Michon, G. M. & Bompard, J. M. 1987. The damar gardens (*Shorea javanica*) in Sumatra, pp. 3–17. In: Kostermans, A. J. G. H. (Ed.), Proc. 4th Round Table Conference on Dipterocarps. UNESCO-MAP (Paris, Jakarta).

Michon, G. & Jafarsiduk, D. 1989. *Shorea javanica* cultivation in Sumatra, an original example of forest management strategy. In: Bruenig, E. F. & Poker, J. (Eds.), Management of Tropical Rainforests. Nomos Verlagsgesellschaft.

Ng, F. S. P. 1980. Germination ecology of Malaysian woody plants. Malay. For. 43(4): 406–419.

Ng, F. S. P. 1988. Forest tree biology, pp. 102–125. In: Malaysia (Ed. Earl of Cranbrook), Trehene, J. E. (Ed), Key Environments. Pergamon Press, Oxford.

Office National des Forêts. 1991. Une richesse pour le futur: les rouvres de Bellême et de Réno-Valdieu, 32 pp. Congrés Forestier Mondial; Paris. ONF.

Office National des Forêts. 1991. Forêts, un héritage pour le futur. 10ème Congrés Forestier Mondial; Paris. ONF.

Perl, M. 1988. Recent developments in seed physiology and biochemistry. Seed Science & Technology 16(1): 135–138.

Roberts, E. H. & King, M. W. 1980. Storage of recalcitrant seeds. In: Lindsey, A. & Williams, J. T. (Eds.), Crop genetic resources. The conservation of difficult material. Proc. Inter. Workshop, Univ. Reading, U.K., 8–11 Sept. 1980, I.U.B.S. Serie B42. pp. 39–45.

Rousteau, A. 1986. Les plantules d'arbres forestiers de Guadeloupe: adaptations structurales et dimentionnelles. Mém. Mus. Nat. Hist. Nat. 132: 185-191.

Sasaki, S. & Mori, T. 1978. Process of protein synthesis during storage and dormancy in tree seeds, pp. 54–63. In: Akihama, T. & Nakajima, K. (Eds.), Long term preservation of favourable germ plasm in arboreal crops. Fruit tree research station, Ministry Agri., For. and Fish.

Scott, E. S., Rao, A. N. & Loh, C. S. 1988. Production of plantlets of Shorea roxburghii G. Don from embryonic axes culture in vitro. Ann. Bot. 61(2): 233–236.

Soerianegara, I., Tjitrosomo, S., Umaly, R. C. & Umboh, I. 1991. Proceedings fourth round-table conference on Dipterocarps, BIOTROP Spec. Publ. 41.

Srivastava, P. B. L. & Penguang, Mangil 1981. Vegetative propagation of some dipterocarps by cuttings. Mal. For. 44(2–3): 301–313.

Tendron, G. 1983. La forêt de Fontainebleau: de l'écologie à la sylviculture. 96 pp. Office National des Forêts. Fontainebleau.

Tompsett, P. B. 1990. Predicting the storage life of orthodox tropical forest tree seeds, pp. 93–98. In: Proc. UIFRO Symp. Seed problems of multipurpose trees and other tropical and sub-tropical species, ACIAR, Canberra.

Trung, T. V. 1987. Perspectives de réhabilitation de l'écosystème forestier de Diptérocarpacées détruit par la guerre chimique dans les provinces de Dongnai (Ma Da) et Tay Ninh, Vietnam, pp. 355–366. In: Kostermans, A. J. G. H. (Ed.), Proc. 4th Round Table Conference on Dipterocarps. UNESCO-MAB.

Turner, I. M. 1990. The seedling survivorship and growth of three Shorea sp. in a Malaysian tropical rain forest. Jour. Trop. Ecol. 6(4): 469–478.

Turner, I. M. & Newton, A. C. 1990. The initial responses of some tropical rain forest tree seedlings to a large gap environment. J. Appl. Ecol. 27: 605–608.

Wang, B. S. P. 1988. Review of new developments in tree seeds. Seed Science & Technology 16(1): 215–226.

Wang, Z. H., Yu, Z. Y. & Bruenig, E. F. 1987. Conservation, utilization and development of natural resources in South China, pp. 42–49. In: Ecologic–Socioeconomic system analysis and simulation: a guide for application of system analysis to the conservation, utilization and development of tropical and subtropical land resources in China. Federal Research Centre for Forestry and Forest Products, Hamburg. ISBN 3.926301-01-7.

Whitmore, T. C. 1989. Southeast Asian tropical forests, pp. 195–217. In: Lieth, H. & Werger, M. J. A. Tropical Rain Forest Ecosystems. Ecosystems of the World 14B. Elsevier Science Publ., Amsterdam.

Yu, Z. Y. 1988 (typed document). The methods and effects of the recovery of vegetation in tropical eroded coastal area in China (South China Institute of Botany, Academia Sinica, Guangzhou).

4. Soil factors influencing the reforestation on mining sites in Amazonia

JOÃO B. S. FERRAZ
INPA – Instituto Nacional de Pesquisas da Amazonia, C.P. 478, 69011 Manaus – AM, Brazil

Keywords: Amazon, degraded areas, mining, reforestation, soil rehabilitation

Abstract. Reforestation with predominantly native species to redevelop areas degraded by opencast bauxite mining showed marked spacial variation in tree growth. In areas without uniform plant cover, erosion and washing away of surface organic matter had already begun. In order to determine the factors responsible for differences in tree growth and degree of vegetation cover in the plantations, physical and chemical soil parameters were analyzed and tree growth measured in selected areas. Low levels of soil organic matter, and N, P, K, Ca, Mg and Mn were found to be the principal factors for reduced growth of planted trees.

Introduction

Opencast mining results in the formation of surfaces lacking in any vegetation cover. Surfaces are covered with rock and/or soil, but these surface layers many times represent deeper horizons brought up or exposed by the mining activities, and thus do not correspond to original soil morphology.

As a general rule, such drastic anthropic alterations of the vegetation and soil cover, and hence of animal life, make the natural recuperation of the environment very difficult. The resulting process is generally a continuation of the degradation, via erosion, ending up with an area of low biological activity, and in extreme cases, deserts. A reversal of this process, on a short- or medium-term basis, therefore, can only be effected by human intervention.

The redevelopment of a degraded area, via reforestation, implies first its rehabilitation (reclamation), i.e., the actions necessary to restore the relief and soil, which often involve mechanical processes.

The processes of rehabilitation seek to renovate soil site factors, with stocks of macro- and micronutrients, water retention capacity, organic matter, physical structure, and also a topography adequate for plant growth.

The processes of recuperation, in turn, imply all the actions necessary to restore biological activity in rehabilitated areas.

The opencast mining of bauxite in Porto Trombetas, Para, results in areas without any vegetation and with soils having the aforementioned characteristics of degraded sites. Since the beginning of the mining activities, Mineração Rio do Norte S.A. has developed a program of recuperation of degraded areas, by means of plantations of predominantly native species (Knowles 1988).

Such plantations, carried out in the area of the Saraca Mine, present great variability in growth and degree of vegetation cover, even in areas of apparently homogeneous soils, which were rehabilitated with basically the same techniques.

This situation was such that certain areas were insufficiently covered with vegetation, with empty spaces between the trees, which for their part, had very reduced growth. The litter produced in such areas is carried away by water or wind, thus preventing the establishment of a cycling of nutrients which would sustain plant growth. This situation aggravates even further the nutrient supply to the trees, and consequently, their growth.

The areas not covered by new vegetation also present clear signs of superficial erosion by rain

H. Lieth and M. Lohmann (eds.), Restoration of Tropical Forest Ecosystems, 47–52.

water, and along some edges, some gullying. Such a situation, undesirable in itself, also presents the danger of degrading surrounding areas which were reforested successfully.

In view of this situation, the following questions were raised:

- what were the physical and chemical characteristics of the soils after mining and before reforestation?
- which pedological factors are influencing the growth of trees in plantations?
- why is it that in some areas tree growth is very slow, while in nearby areas, with supposedly identical soils, tree growth is much greater?
- which measures can be taken to invigorate the growth of trees in areas where vegetation cover and growth is unsatisfactory?

Objectives

Short-term

- establish the relationship between the morphological and physical-chemical characteristics of the mined soils and the growth of different tree species used in reforestation at the Saraca Mine;
- determine the site factors responsible for the differences in tree growth and degree of vegetation cover in different areas (= ages) of plantations;
- determine the most appropriate measures to revigorate tree growth and increase vegetation cover in areas where it is reduced.

Long-term

- define and organize the stages necessary for the recuperation of a degraded soil through reforestation;
- provide a basis for a re-evaluation of the physical and chemical parameters of the soil in plantations in order to estimate, on a long-term basis, the influence of reforestation on rehabilitated and recuperated soils;
- define techniques to evaluate disturbances detected in reforestations of degraded areas, as well as develop a "Plan for Ecological Monitoring" for reforestation in such areas; and
- demonstrate that mining (evaluate tree growth, plant succession and soil fauna) activities in the tropics do not necessarily imply environmental degradation.

Description of the study area

Location

The city of Porto Trombetas is located on the right bank of the Trombetas River and is part of the municipality of Oriximina, Para. The region is located between the coordinates 56°–58° W and 0°–2° S, belonging to, by international geographic division, Sheet SA.21 – Santarem (Projeto Radambrasil 1976). The Saraca Mine is situated between the edge and center of a plateau (tabular interfluve), 25 km NE of Porto Trombetas.

Climate

According to Koeppen's classification, the regional climate is of the AW type (A.C.A.B. 1984). According to data from the MRN Meteorological Station, in Porto Trombetas, the mean annual temperature between 1972 and 1985 varied between 23.8 and 27.5 °C (mean for the period: 26.5 °C).

During this same period, mean annual precipitation varied between 1427.4 mm and 2602.9 mm (mean for the period: 2125.8 mm). The months of greatest rainfall were March (monthly mean for the period: 347.9 mm) and April (288.7 mm), and the lowest were August (63.5 mm) and September (60.1 mm). These data indicate a seasonal climate, with a well-defined dry season.

Geology

The region is characterized by sediments of the Barreiras Formation, of Cenozoic age (since 65 million years; Projeto Radambrasil 1976). These sediments, deposited in the Cretaceous period to the Pliocene, are distributed in plateaus and lowered surfaces. On the plateaus, more clayey sediments derived from poorly consolidated, variegated schist clays predominate. In the lower surfaces, sandy sediments predominate, and they are derived from friable, kaolinitic or consolidated, fine- to medium-textured sandstones, with cross stratification and oxide impregnation.

It is supposed that the bauxites originated from purely chemical alteration processes, under tropical and subtropical climates, and generalized peneplain conditions (Projeto Radambrasil 1976). Bauxite shields are encountered in residual relief of more elevated planation surfaces (Plio-Pleistocene pediplanes) of the Rio Negro–Rio Trombetas dissected plain. These tabular landforms are cut into sediments of the Barreiras Formation, with ravined slopes.

Soils

The region presents an association of latosols and podsols. Through field, photointerpretation, and laboratory studies, Ranzani et al. (1982) produced a soils map (1:50,000 scale) of the right margin of the lower Trombetas River, between the coordinates 1°25′ and 1°47′ 50″ S and 56°25′ and 56°55′ W.

The following pedological units were found to predominate:

1. yellow latosol, with clayey texture;
2. red-yellow podsol, with medium to clayey texture;
3. concretionary red-yellow podsol;
4. quartz sands; and
5. alluvial and hydromorphic soils.

The soils found on the plateau studied are latosols, coinciding with observations made on other plateaus of the region (Monte Branco, Rebolado, and Cruz Alta) by Ranzani et al. (1982). They present a latosolic B horizon, clayey texture, and are acid, deep and well-drained, and are plastic and sticky. They are soils with low fertility, low cation exchange capacity, and low levels of base saturation.

In certain projections of the Saraca Plateau (such as Block C-East) patches of "Indian black earths" are found, which from the presence of potsherds and stone artefacts, indicates past occupation by Amerindians. Research done in these areas (Lopes 1981; Hilger 1988) raised the hypothesis that these patches were ancient agaricultural sites, of up to 500 years ago. It is interesting to note that the revegetation of such sites is such that at present, no sign of human influence can be seen in the vegetation (Knowles, pers. comm. 1987).

Vegetation

The region has been visited by several botanical expeditions (P. Cavalcante 1957; Cid Ferreira 1980; J. Revilla 1981 and 1982 and P.I. Soares 1981; apud Revilla and Silva, in Alcoa 1982). The vegetation can be divided into four types: dense "terra firme" (unflooded) forest, "campinarana" (low forest on sandy soils), "campina" (low, open forest on sands), and "igapo" (flooded) forest, with terra firme forest predominating. Between the campina and igapo forest, "capoeira" (secondary forest) is frequently encountered.

Revilla and Silva (in Alcoa 1982) describe the tera firme forest as being "uniform, with mean height of 30 m, dense physiognomy, moderately sparse trees with straight trunks, high and broad branching with globose or umbrella-shaped crowns, few epiphytes and lianas, with a uniform canopy accompanying changes in topography". They observed a litter layer of 10–20 cm depth, and a sparse understory, dominated by stemless palms.

The families which present the greatest number of species are the Leguminosae (Caesalpinioideae 12 species, Mimosoideae 5, and Papilionoideae 4); Sapotaceae (11 species); Olacaceae, Lecythidaceae, Moraceae, and Chrysobalanaceae (6 species each) and Apocynaceae, Annonaceae, and Euphorbiaceae (5 species each).

A forest inventory was carried out in the same area by Higuchi et al. (1982). They found a total of 291 species with DBH greater than or equal to 20 cm. According to these authors, the heterogeneity and floristic composition of the forest is comparable to that found near Manaus, among other locations.

The original vegetation (terra firme forest) of the plateau is very similar to that described for the region (Knowles, pers. comm. 1989). However, the litter layer on the Saraca Plateau is much thinner than that of the dense forest in lower areas.

The present vegetation consists of the reforestation done with predominantly native species,

Table 1. Height (mean, minimum and maximum, in meters) of trees planted in 1981, 1982, and 1984, in reforestation at Saraca Mine. Site = Soil profile number. Trees species are described in Ferraz, 1990

Year	Area	Parcel	Site	Mean	Minimum	Maximum
1981	C-Leste	C	9	0.62	0.22	2.10
1981	C	A	1	5.81	2.70	9.40
1982	C	C	7	0.74	0.30	2.08
1982	C	A	6	3.98	0.80	7.30
1984	A-3	–	21	1.44	0.25	3.50
1984	A-3	–	20	6.32	2.50	9.30

after rehabilitation of the mined area, and of colonizing species. A complete list of the species employed can be found in Knowles (1988).

Methods

Study sites were chosen in areas reforested in 1981, 1982, 1983, 1984, 1985, 1986, and 1987. The list of the species planted in the study sites, details of plantation establishment and soil amendment, are given in Ferraz (1990). For purposes of comparison, a site in primary forest was chosen near a mined area. The data discussed in this paper refer to the oldest plantations (1981, 1982, and 1984), which presented the greatest differences in growth.

In each area chosen, locations were selected for placing soil profiles, for morphological description and sampling, with a total of 26 profiles. In these sites, the type of tree species planted and their growth was recorded. All soil samples, as well as tree measurements were taken between December 1987 and January 1988.

The following parameters were analyzed in the soil samples: pH (H_2O and $CaCl_2$); organic matter; N (Kjedahl); P (resin); K and Na (flame photometry); Ca, Mg, Al, Fe, Mn, Cu, Zn (atomic absorption spectophotometry); S and B (photocolorimetry); texture; real and apparent density and color (Munsell).

Results and discussion

The greatest variations observed in the mean growth of trees planted in recuperated areas are presented in Table 1. Comparing the total height for trees planted in February 1981, it can be seen that the site where growth was the least (Area C-East, mean height = 0.62 m) differs by a factor of 9 from trees planted in the same year in Area C (mean height = 5.81 m). The differences between the Muiracatiaras (*Astronium lecointei*) planted on both sites differs by up to a factor of 10 (0.30 m and 2.95 m, respectively).

The trees in thc arcas reforested in 1982 and 1984 also present significant differences in mean height. In the first case the variation was of 0.74 m (Area C, Parcel C) to 3.98 m (Area C, Parcel A); in the second case, 1.44 m (Area A-3) to 6.32 m (Area A-3; see Table 1).

The analyses of the surface horizons of the reforested soils are presented in Table 2. Comparing the two plantations made each year, it can be seen that those which had good height growth and poor height growth, grew in soils with similar levels of acidity. In general, the acidity of these soils can be classified as high or very high. Only the soils in the 1984 plantation (site 20) reach a medium level of acidity, which may be explained by the higher levels of calcium observed (Table 2).

The plantations which presented the greatest growth were those whose soils had the greatest levels of organic matter (O.M.). Some of these levels (i.e. site 20) were greater than the levels observed in the surface soil (4.9 and 2.5% O.M.) of the adjacent primary forest (Table 3). This may be explained by the great quantity of organic matter deposited by chance in some locations during the rehabilitation of degraded soils. It is important to note that the beneficial effect of the accumulation of organic matter is not only the result of its presence on the surface

Table 2. Chemical analysis of surface horizons of areas reforested in 1981, 1982, and 1984, at Saraca Mine

Year	Site	Depth (cm)	pH ($CaCl_2$)	O.M. (%)	N (%)	P (µg g^{-1})	K	Ca	Mg	CEC	S	Fe	Mn	Cu	Zn	B
							(meq 100 ml^{-1})				(µg g^{-1})					
1981	9	0.0–2.0	4.9	0.1	0.01	2.3	0.01	0.3	0.1	1.7	44.0	52.0	1.9	0.1	4.5	0.1
		2.0–25.0	4.9	0.1	0.01	4.7	0.01	0.2	0.1	1.6	35.2	55.0	1.0	0.1	4.5	0.1
1981	1	0.0–3.0	4.7	4.2	0.21	1.0	0.10	0.5	0.2	4.2	5.6	125.0	16.3	0.1	4.0	0.2
		3.0–12.0	4.4	1.0	0.05	1.0	0.06	0.5	0.2	3.6	19.6	158.0	9.3	0.1	3.2	0.1
1982	7	0.0–2.0	4.3	0.1	0.01	1.0	0.01	0.3	0.1	2.4	70.0	148.0	4.6	0.1	3.1	0.1
		2.0–5.0	4.2	0.1	0.01	3.5	0.01	0.3	0.1	2.2	68.0	108.0	3.5	0.1	2.4	0.1
1982	6	0.0–2.0	4.4	2.5	0.13	3.5	0.10	1.7	0.6	4.6	34.2	174.0	16.4	0.3	22.2	0.1
		2.0–5.0	4.3	1.3	0.07	1.0	0.06	0.6	0.4	3.6	27.0	180.0	8.4	0.1	3.3	0.1
1984	21	0.0–2.0	4.3	2.6	0.13	8.3	0.03	0.5	0.2	4.9	21.0	198.0	2.0	0.2	5.7	0.6
		2.0–18.0	4.3	2.5	0.13	10.8	0.05	0.2	0.2	4.6	30.0	286.0	3.0	1.1	1.3	0.3
1984	20	0.0–2.0	5.2	1.6	0.08	3.5	0.27	2.0	0.5	5.3	8.0	190.0	3.4	0.7	3.0	0.7
		2.0–5.0	4.8	5.8	0.29	7.1	0.24	0.4	0.3	4.7	4.4	173.0	5.3	1.0	1.3	0.6

Table 3. Chemical analysis of surface horizons of yellow latosol from adjacent primary forest at Saraca Mine

Depth (cm)	pH ($CaCl_2$)	O.M. (%)	N (%)	P (ppm)	K	Ca	Mg	CEC	S	Fe	Mn	Cu	Zn	B
					(meq 100 ml^{-1})				(µg g^{-1})					
0.0–2.0	4.0	4.9	0.25	4.5	0.07	0.07	0.3	4.5	22.6	245	4.0	0.1	0.3	0.1
2.0–5.0	3.9	2.5	0.13	4.0	0.04	0.02	0.2	3.5	13.6	296	3.0	0.1	0.9	0.2
5.0–15.0	4.0	2.0	0.10	4.3	0.07	0.02	0.1	3.8	11.0	284	3.8	0.1	0.8	0.1

of the soil, but also includes buried organic matter (site 20; cf. Ferraz 1990).

The levels of nitrogen in the soils accompany the levels of organic matter: the soils richest in organic matter also show the highest levels of nitrogen. In these areas (sites 1, 6 and 20), where tree growth and degree of soil cover is greater, a greater accumulation of litter on the surface was seen, indicating the establishment of nutrient cycling processes. The levels of foliar nitrogen in the litter of these restored sites varies between 0.96 and 1.29% (Ferraz 1990), well below that found in the litter in adjacent primary forest (1.99% N). In other areas of plateau in Central Amazonia, and also on yellow latosols, a level of 1.78% nitrogen was found in the litter (Luizao 1989).

The levels of phosphorus in the restored soils are low, and no relationship was observed between tree growth and levels of this element.

The other macronutrients, potassium, calcium, and magnesium, are present in restored soils in levels which can be considered, for the most part, as deficient. Even so, the highest levels of these nutrients were found on the sites where plantations showed the greatest growth. Cation exchange capacity is very low, and did not reach even the mean values for kaolinite, the most common mineral in those soils. The greatest contribution to CEC, therefore, is from organic matter, added to the soil before reforestation.

The results of the analyses for sulphur and micronutrients are presented in Table 2. The levels of sulphur found in the surface layers of restored soils are markedly higher than those in the original soil (Table 3), which leads to the supposition that there was deposition of this element after reforestation of the area (Ferraz 1990).

The levels of iron are for the most part average (31–200 µg g^{-1}), while levels of copper can be classified, for the most part, as deficient (<0.5 µg g^{-1}). Both elements did not show any positive relationship with tree growth. In the case of manganese, however, higher levels (classified as from low to average) were found on sites which showed better growth of trees. Zinc and boron did not show any relationship to plant growth. Levels of zinc may be classified as low (1–3.9 μg g^{-1}) to average (4–40 μg g^{-1}), while levels of boron were, for the most part, deficient (<0.3 μg g^{-1}).

The fact that the criteria for interpretation of soil nutrient levels were developed for agronomic species, and not for forest species, imposes certain limitations on their application for the evaluation of the nutrient status of forest soils. The plantations analyzed here, which presented greatest growth, did not present any visible symptoms of nutrient deficiency, despite the markedly low levels found in the soils. A more realistic evaluation of this question can only be made through the foliar analysis of the trees used for reforestation.

Conclusions

The reversal of the anthropic processes which form degraded areas on a large scale, with the object of recomposing the forest in the most diverse form possible, necessarily depends on human intervention. This reversal, of course, does not establish the original conditions of biological diversity of the ecosystem. However, even a poorer restored ecosystem is able to avoid a continuing process of degradation and gives rise to natural processes which consequently increases diversity.

The restored areas which received fresh surface soil showed the greatest tree growth, denser plant cover, greater quantity of litter, and a greater diversity of plants.

The surface soil used as an amendment before reforestation, should not be stockpiled in the open. An ideal form for storing such soil has not yet been determined.

In reforested sites which have a compacted subsoil, there may be problems with root penetration, and subsequent toppling of trees due to poor rooting. Some examples of this have been observed in Porto Trombetas (Knowles, pers. comm. 1988).

The frequent occurrence of cutter ants (*Atta sextans*) may slow down the establishment of trees, and causes an increase in costs if replanting becomes necessary.

Acknowledgements

The author acknowledges the support from the Mineração Rio do Norte (Porto Trombetas, Para) and the local staff, and thanks Oliver Henry Knowles for the stimulating discussions.

References

A.C.A.B. (Atlas Climatológico da Amazônia Brasileira), 1984. SUDAM, Belém (PA). Publicação No. 39.

Alcoa Mineração S.A. 1982. Convênio ALCOA-CNPq-INPA, Relatorio. 3 Vols. INPA, Manaus, AM.

Ferraz, J. 1990. Diagnóstico do Comportamento do Reflorestamento na Mina Saraca, Porto Trombetas (PA). Relatório. 5 Vols. Mineração Rio do Norte S.A., Porto Trombetas (PA).

Higuchi, N., Jardim, F. C. S. & Barbosa, A. P. 1982. Inventário Florestal no Rio Trombetas (INPA/SHELL-ALCOA). In: Alcoa Mineração S.A. (1982). Vol. 3, 49 pp. e apêndice. INPA, Manaus, AM.

Hilger, K. 1988. Projeto de Salvamento Arqueológico na Região de Porto Trombetas (PA). III. Relatório de Viagem. 9 pp. Mapas e ilustrações. Mineração Rio do Norte S.A., Porto Trombetas (PA).

Knowles, O. H. 1988. Informações básicas para o exito na reposição florestal em áreas degaradadas. Trabalho apresentado no "Simpósio Internacional Alternativas para o Desmatamento", Belém, PA, 29. Jan. 1988; 20 pp., 7 tabelas anexas. Relatório Mineração Rio do Norte, TSA, Porto Trombetas, PA.

Lopes, D. F. F. 1981. Salvamento Arqueológico em Porto Trombetas (PA). Museu Paraense Emilio Goeldi, Divisão de Arqueologia. Relatório. Março de 1981. 7 pp. Belém, PA.

Luizão, F. J. 1989. Litter production and mineral element input to the forest floor in a central amazonian forest. GeoJournal 19: 407–417.

Projeto Radambrasil. 1976. Levantamento de Recursos Naturais. Vol. 10; Folha SA.21-Santarém. Ministério das Minas e Energia; Departamento Nacional da Produção Mineral, 510 pp. e 7 mapas. Rio de Janeiro, RJ.

Ranzani, G., Scatalini, F. & Podesta, J. A. F., 1982. Inferências para a recuperação de terras em áreas onde a bauxita é explorada, sugeridas pelos solos presentes. In: Alcoa Mineração S.A. (1982). Vol. 1, 34 pp. e 1 mapa. INPA, Manaus, AM.

5. Restoring Caribbean dry forest

A systems framework for site analysis and restoration research

BECKY J. BROWN[1] and GARY J. RAY[2]

[1]*Institute for Environmental Studies and Department of Forestry, 120 Russell Labs, University of Wisconsin-Madison, Madison, WI 53706, USA;*

[2]*Institute for Environmental Studies, 120 Russell Labs, University of Wisconsin-Madison, Madison, WI 53706, USA*

Keywords: Conservation biology, ecosystem science, restoration ecology, tropical, Virgin Islands

Nomenclature is according to Little, E. L., Woodbury, R. O. & Wadsworth, F. H. *Trees of Puerto Rico and the Virgin Islands.* Volume 1 (1964) and Volume 2 (1974)

Abstract. This paper describes a systems approach to the ecological restoration of Caribbean dry forests, in which factors that limit natural regenerative processes are identified, and ecological strategies are developed in order to remove those limitations. The approach is illustrated using information gathered from restoration research in progress on the island of St. John, U. S. Virgin Islands. Information about driving variables, together with data on biotic and abiotic site conditions, define the context and starting point for active management.

Introduction

Subtropical dry forests (*sensu* Holdridge 1967), which were once a major ecological community on Caribbean islands, have been largely destroyed or seriously disturbed by intensive human activity in the region during the past five centuries (Wadsworth 1950). Logging, overgrazing, fires, and agricultural activity have left large tracts of dry forest habitat in a degraded condition, often with low productivity, bare soils, accelerated erosion and low nutrient status (Lugo et al. 1981). It is estimated that half of Puerto Rico consists of land abandoned from farming or grazing (OTA 1984), and the situation on smaller, drier Caribbean islands is even worse (Brown 1982; Watts 1988).

The need for rehabilitation of these impoverished sites is critical and urgent, but successful restoration requires a clear understanding of the ecological processes that control dry forest regeneration. While a limited number of studies have provided a basic understanding of the structure and function of tropical dry forests (Ewel 1977; Reich and Borchert 1982, 1984; Dunevitz 1985; Lugo and Murphy 1986; Lott et al. 1987 and see review by Murphy and Lugo 1986), little is known about either their regenerative ability or techniques for restoring them.

On the island of St. John in the U. S. Virgin Islands, we are developing techniques for restoring the diversity and productivity of severely disturbed Caribbean dry forest. The objectives of this paper are: (1) to describe a systems framework for restoration research that we have found helpful and (2) using data from a site analysis, to illustrate how this framework has guided our restoration efforts in Caribbean dry forest.

Address correspondence: Becky J. Brown, Dept. of Forestry, 120 Russell Labs, University of Wisconsin-Madison, Madison, WI 53706, USA. Telephone 608-262-0249.

Systems framework

Ecological restoration involves an attempt to control the direction and speed of succession on a degraded site. Without active intervention, a damaged ecosystem may or may not return in time to its original, pre-disturbance state. Some disturbed ecosystems may never return to their

H. Lieth and M. Lohmann (eds.), Restoration of Tropical Forest Ecosystems, 53–61.

pre-disturbance states without active management, and in fact may deviate even further over time from their original conditions. An example would be the replacement of tropical wet forest vegetation by shrub thicket or savanna following disturbance (Uhl 1982; Buschbacher 1986). In cases where the vegetation changes after disturbance are undesirable, restoration efforts must actively redirect the course of succession.

In other situations, disturbed ecosystems that are not actively managed may recover (i.e., return to pre-disturbance state) by means of naturally-occurring processes. In some cases an ecosystem may appear *not* to be recovering simply because succession is occurring at a slow rate relative to the time scale of human interest. This situation is likely to be a problem when the disturbance has altered important processes in the ecosystem, such as nutrient cycling or seed dispersal. When the recovery process occurs at an unacceptably slow rate, ecological restoration becomes essentially a tool for accelerating the rate of vegetation change in a desired direction. For example, enrichment plantings in species-poor forests with limited local seed sources would hasten plant community recovery.

To accelerate succession in an ecological system, one must first identify the factor(s) that limit succession, and then develop ecological strategies to remove the limitations. Three general categories of factors may limit succession: (1) driving variables, (2) abiotic site variables, and (3) biotic site variables.

Driving variables (also called forcing functions) provide the environmental context within which an ecological restoration occurs. Driving variables are those variables not altered by disturbance, but which are important in determining the nature of the recovery process. Generally, driving variables are factors over which the restorationist has little or no control, such as climatic variables or topography. Driving variables often constrain restoration efforts, and recognizing these absolute limits is important in planning a restoration project.

Abiotic site variables are physical and chemical factors that are altered by disturbance and that are potentially important in the recovery process (e.g., soil structure and soil nutrient content). In planning a restoration project, it is important to evaluate the extent to which abiotic site variables have been altered, and whether they have been altered in ways that limit succession. Ameliorating damage to abiotic site variables may be relatively easy (e.g., adding fertilizer to improve soil quality) or extremely difficult (e.g., replacing topsoil lost through erosion or removing toxic soil contaminants).

Biotic site variables refer to the biological organisms present on the degraded site. Together with abiotic site variables, they define the starting point for a restoration effort. It is useful in planning a restoration to understand in detail how the disturbance has altered the major plant and animal populations on the site. It is particularly important to evaluate whether or not sufficient genetic material is available for the recovery of the biotic community.

A detailed site anlysis has been recommended as one of the first steps in planning an ecological restoration (Nuzzo and Howell 1990). A careful evaluation of driving variables, biotic variables and abiotic variables defines the context and starting point for active management, and gives direction to a restoration effort. Below we describe a site analysis conducted on St. John, U. S. Virgin Islands, and discuss its implications for the restoration of dry forest vegetation on this island.

Methods

Study site

The site for this study is the Virgin Islands National Park and Biosphere Reserve on the island of St. John in the U. S. Virgin Islands. Established in 1956 on land donated by Laurence Rockefeller, the Virgin Islands National Park currently occupies 2950 ha of St. John, encompassing approximately 56 percent of the island. The center of the 19 square mile island is located at 18° 20′ N and 64° 44′ W. The topography is steep and rugged, with numerous rock outcroppings. More than four-fifths of the island landscape is composed of slopes greater than 30% (CH2M Hill 1979).

Rainfall on the island ranged from 580 to 1880 mm at one sea-level station over the past 100 years. Mean annual rainfall over this period is 1120 mm. Prevailing winds are from the east and the east-northeast, blowing consistently at two to twelve knots (Stehle 1944). Major hurricanes are estimated to occur once in every 33 years, according to National Park Service records.

St. John soils are derived primarily from basic volcanic rocks; Cramer is the predominant soil series (USDA 1970). These are red or reddish-brown soils occurring on steep slopes. They are shallow, dry, moderately permeable, well-drained clayey soils of moderate fertility and basic reaction with a gravelly or stony surface layer (Teytaud 1986).

The impact of aboriginal inhabitants on St. John's preexisting plant communities is unknown. The earliest botanical investigation of the Virgin Islands was conducted in the late eighteenth century by West (1793). Plantation agriculture, with accompanying forest clearing and broad-scale cultivation, had largely run its course on St. John by the time original floras were being assembled.

Five forest stands on St. John were included in initial site analyses: Mary Point (MP), Lameshur Bay (LM), Hawksnest (HN), Caneel Hill (CH), and Cinnamon Bay (CB). The youngest stand, at Mary Point, is former pastureland marginally protected from grazing ungulates since park establishment 35 years ago. The forest on the south-facing slope of Mary Point is regenerating slowly following more than two centuries of intensive livestock grazing (Edwards 1991 unpubl.). This species-poor site (six tree species) at 40 m elevation is dominated by a dense thicket of *Leucaena leucocephala,* a rapidly-growing exotic tree, and by tangled masses of thorny vines. Mary Point remains a favorite habitat for roving bands of feral asses, which are likely retarding natural recovery processes with their grazing activities and their erosion-inducing network of trails. A 70-year-old forest stand above Lameshur Bay, a cattle ranch abandoned in the 1920s, contains 18 native tree species (including *L. leucocephala*). The forest inventory plots at 80 m elevation face southeast. At Hawksnest, a half-hectare plot established in 1986 by The New York Botanical Garden contains a dry semievergreen forest community that has not been cleared for cultivation for at least 100 years and has received only moderate to low intensity grazing disturbance during its recovery (Tyson 1987). Species richness at Hawksnest is high (55 tree species on 0.5 ha), in spite of the site being dominated by the exotic tree, *Melicoccus bijugatus* (Reilly et al. 1990). The fourth stand, at Caneel Hill, is a dry semi-evergreen forest. We estimate stand age at Caneel Hill to be greater than 100 years, based on sketchy land use documents, but the historic loss of topsoil and site accessibility to cattle following abandonment are likely to have retarded recovery processes. We recorded 18 species in two 0.05-ha plots. The stand at Cinnamon Bay is a dry semi-evergreen forest transitional to upland moist forest. We estimate a century or more of recovery since the site was last disturbed by humans. Two 0.05-ha plots contain 23 tree species.

Abiotic site variables

Abiotic site variables were assessed beginning in January 1988 at each of four sites: Mary Point, Lameshur Bay, Hawksnest and Cinnamon Bay. Surface rockiness, forest litter accumulation, and canopy cover were estimated.

We estimated the density of exposed rock on the soil surface by a point sampling method. We used a grid of 100 sample points spaced at 10 cm intervals within a one meter square quadrat. The number of rock or gravel interceptions per 100 sampling points was used as an estimate of surface rockiness. At each site, 10 quadrats were sampled, located at 10 m intervals along transects running lengthwise through a 10 × 50 m (0.05 ha) forest plot.

Litter on the forest floor was estimated by collecting accumulated litter in three, one meter square quadrats at each of the four sites. The material was separated into leaf and woody litter portions, dried at 70 °C, and weighed.

Canopy cover was estimated from photographs taken from ground level looking vertically upward through the tree canopy. Ten photo points were located at equally spaced intervals along a transect in each of the four stands. The photographs were digitally scanned and converted to computer images, from which canopy coverage values were calculated.

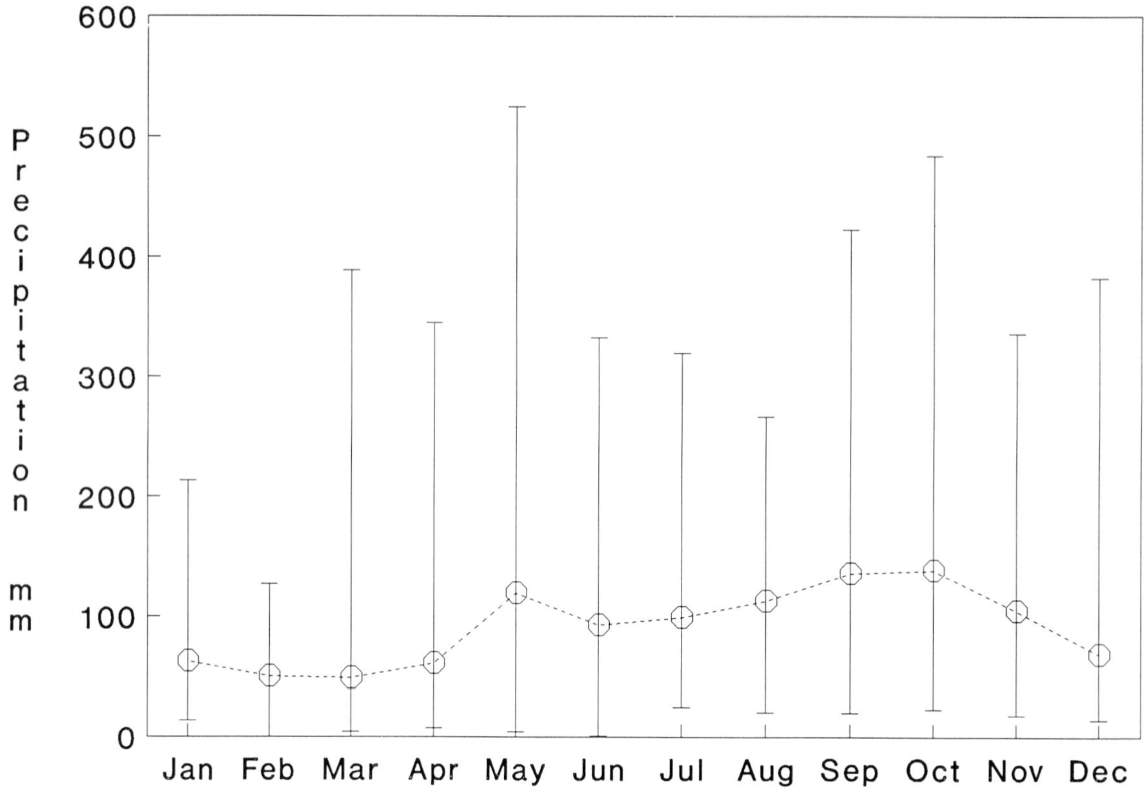

Fig. 1. Precipitation at Cruz Bay, St. John, U.S. Virgin Islands, 1871–1989. Circles are mean values. Vertical bars show ranges for the 119-year period.

Soil organic matter and percent total nitrogen of the topsoil were estimated from soil samples collected during July 1988. Ten samples to a depth of 5 cm were taken at randomly selected locations at each of the four study sites. Soil analyses were done at the Institute of Tropical Forestry in Puerto Rico. Organic matter was determined using the modified Walkley and Black method (Page 1982). Total nitrogen was determined using the modified Kjeldhal to include nitrate and nitrite (Page 1982). Soil pH was determined in a 1:1 solution using an Orion Ionalyzer Model 901 with a combination pH electrode.

Vegetation inventories

Tree seedlings, saplings and mature trees were inventoried in two 10 × 50 m plots at each of four study sites, beginning in June 1988. All individuals with stems measuring less than 1 cm diameter at a height of 10 cm were defined as seedlings. Seedlings were inventoried in ten, one square meter plots, spaced at equal intervals along two transects in each 10 × 50 m plot (i.e., two percent of the plot area was sampled for seedlings). Seedlings were identified to species and tallied. We classified stems greater than 1 cm diameter (at a height of 10 cm) but less than 4.0 cm diameter at breast height (dbh) as saplings. Saplings occurring in a two meter wide zone centered on transects were tagged and identified to species. Sapling diameters were measured at 10 cm above the ground, and heights were recorded. Forty percent of the plot area was sampled for saplings. Stems greater than 4.0 cm dbh were designated as mature trees. All trees were tagged and identified, and the stem diameters were measured.

Table 1. Characteristics of four dry forest sites on St. John, U.S. Virgin Islands. Sites are Mary Point (MP), Lameshur Bay (LM), Hawksnest (HN), and Cinnamon Bay (CB). Values are mean ± s.d. unless otherwise noted. Means not followed by the same letter are significantly different at the 0.05 level

	Site			
	MP	LM	CB	HN
Stand age (yr)	35	70	100+	100+
Surface rockiness ($n = 10$)				
% rock	2.0 ± 1.5[B]	9.0 ± 2.1[A]	9.3 ± 1.2[A]	14.3 ± 0.9[A]
% gravel	0.0 ± 0.0[B]	6.7 ± 1.7[A]	2.0 ± 1.5[AB]	2.7 ± 1.7[AB]
Total rockiness (mean, rock+gravel)	2.0	15.7	11.3	17.0
Soil organic matter (%) ($n = 10$)	10.1 ± 2.2[A]	7.2 ± 0.7[B]	10.6 ± 2.3[A]	9.2 ± 2.1[AB]
Total nitrogen (%) ($n = 10$)	0.48 ± 0.12[A]	0.46 ± 0.05[A]	0.44 ± 0.12[A]	0.56 ± 0.14[A]
Soil pH (median)	6.4	6.4	6.5	6.6
Litter (g/m^2; $n = 3$)				
Leaf litter	58 ± 13[B]	207 ± 18[A]	211 ± 29[A]	266 ± 77[A]
Woody litter	426 ± 98[A]	346 ± 139[A]	318 ± 88[A]	406 ± 104[A]
Total (leaf + woody litter)	484 ± 101[A]	553 ± 154[A]	529 ± 103[A]	672 ± 179[A]
Percent canopy cover ($n = 15$)	87.9 ± 2.0[A]	89.5 ± 1.2[A]	84.8 ± 1.0[A]	89.7 ± 0.8[A]

Results

Driving variables

(1) Rainfall

To evaluate rainfall distribution over time, we summarized precipitation data from Cruz Bay, St. John, 1871–1989 (Figure 1). Precipitation ranged from approximately 600 to 1900 mm/year and was highly sporadic, without well-defined wet or dry seasons. These long-term data showed that total monthly rainfall on St. John is unpredictable and extremely variable from year to year. Rainfall during each month varied more than an order of magnitude during this 119-year period. Daily rainfall data (not shown) confirmed our field observations that high rainfall for a month does not necessarily mean that the entire month was rainy; it was not uncommon for most of the precipitation during a one-month period to occur during one or two rainfall events.

(2) Hurricanes

Hurricanes have dramatic short-term effects on vegetation. The long-term impacts of hurricanes are often more subtle and difficult to evaluate. In September 1989 Hurricane Hugo passed over the U. S. Virgin Islands, giving us the opportunity to observe hurricane impacts on Caribbean dry forest. The major visible effects of the hurricane on the vegetation of St. John were: (1) complete defoliation of most trees and shrubs; (2) extensive breakage of trunks and branches, especially on hillsides that received the full force of the storm; and (3) some uprooting of large trees. Most trees resprouted rapidly, and there was relatively little tree mortality. The major impact of the hurricane appeared to be a temporary alteration of the canopy, but not a long-term impairment of the plant community's ability to process energy. Obviously, there may have been important long term impacts as well, such as soil erosion and long distance seed dispersal, that we were not able to measure.

Abiotic site variables

(1) Canopy cover

Mean percent canopy cover in the young, severely degraded forest at Mary Point was similar to that in older, more well-developed dry forest stands on the island (Table 1). Mean canopy cover at Mary Point was 88%. Thus, on average, vegetative cover on this degraded site appeared to have reestablished well without active management. However, careful analysis of the data revealed that the average size of canopy openings, through which light could penetrate, differed significantly between the four stands ($p < 0.01$), and the Mary Point stand had the largest mean size of canopy openings. Moreover, the standard deviation of hole sizes (used here as a measure of canopy patchiness or heterogeneity) differed among sites ($p < 0.05$). Mary Point had the most heterogeneous canopy using this indicator variable.

(2) Surface rockiness

The proportion of the ground surface covered by rocks and gravel was lower at the degraded Mary Point site than at other locations on the island (Table 1). On a rugged island characterized by rocky slopes, only two percent of the ground surface at Mary Point was exposed rock or gravel. Exposed rock may reflect a number of different factors, such as geological features, position, and history of site manipulation. The low estimate of exposed rock at Mary Point may reflect past land use. For instance, erosion from Mary Point's upper slopes, augmented by intensive grazing of steep upper slopes, deposited soil that covered rocks on lower slopes (where soil rockiness was measured). Field observations on St. John indicate that upper slopes tend to contain shallower soils and more exposed rock than do lower slopes.

(3) Litter accumulation

We estimated the amount of woody and leaf litter on the ground as an indicator of how well the soil surface was protected. Ground cover reduces splash erosion and helps to buffer the soil against temperature fluctuations. We found that the degraded Mary Point site had significantly less leaf litter than did other sites ($p < 0.05$, Table 1). Amounts of woody litter did not differ between sites ($p < 0.05$).

(4) Soil quality

We measured soil organic matter and nitrogen content as indicators of soil quality. Brown et al. 1984 predicted that approximately 50 years are required after a disturbance for soil organic carbon to approach the levels in undisturbed tropical forests. We anticipated that the 30-year-old Mary Point site would have poorer soil quality than the older sites, but this was not the case. Percent soil organic matter at Mary Point was similar to soil organic matter in the more well-developed stands and was significantly higher than in an older forest on a drier site (Table 1). Total percent soil nitrogen did not differ among sites. Our estimates of soil organic matter were higher than values from a tropical dry forest in Puerto Rico (Lugo and Murphy 1986), and our total nitrogen estimates were similar to those in the Puerto Rican forest. Soil deposited from erosion on upper slopes may explain the relatively good status of Mary Point topsoil. Our results may also be explained in part by high productivity and nitrogen fixation by *Leucaena leucocephala,* a weedy, non-native tree that dominates the Mary Point site.

Biotic site variables

Results from our forest inventories on several St. John sites showed that the woody vegetation at the degraded Mary Point site was very different from that at three other sites on the island.

Table 2 summarizes the distributions of tree species at the four sites. The Mary Point site (MP) supported a species-poor plant community, with only six tree species present as mature individuals. Moreover, only three of the six species at Mary Point were present in the older stands (Table 2). These data, together with information about seedlings and saplings (not shown), suggest that the Mary Point stand is changing very slowly, if at all, into a more species-rich plant community. Nearly half of the tree species (15 of 33 species present as mature individuals, saplings or seedlings) in the >100-year-old Cinnamon Bay forest were reproducing in this older stand (i.e., 15 species were present on that site in both mature and immature stages). In comparison, only four weedy species were reproducing at Mary Point. Only five spe-

Table 2. Species present in four dry forest stands on St. John, U.S. Virgin Islands. Sites are Mary Point (MP), Lameshur Bay (LM), Caneel Hill (CH), and Cinnamon Bay (CB). Data for each stand include all individuals >4 cm dbh in two 0.05-ha plots

Species	STAND (Age in years)			
	MP (35)	LM (70)	CH (100+)	CB (100+)
Acacia macracantha	*			
Adelia ricinella	*			
Croton astroites	*			
Exostema caribaeum	*	*		
Leucaena leucocephala	*	*		
Citharexylum fruticosum	*	*	*	*
Acacia muricata		*		
Amyris elemifera		*		
Capparis flexuosa		*		
Capparis indica		*		
Cassine xylocarpa		*		
Cephalocereus royenii		*		
Krugiodendron ferreum		*		
Pithecellobium unguis-cati		*		
Schaefferia frutescens		*		
Piscidia carthagenensis		*	*	
Bourreria succulenta		*	*	*
Bursera simaruba		*	*	*
Erythroxylum brevipes		*	*	*
Eugenia biflora		*	*	*
Tabebuia heterophylla		*	*	*
Coccoloba microstachya			*	
Eugenia cordata			*	
Guettarda parviflora			*	
Randia aculeata			*	
Sabinea florida			*	
Eugenia monticola			*	*
Guapira fragrans			*	*
Guettarda scabra			*	*
Maytenus elliptica			*	*
Nectandra coriacea			*	*
Tetrazygia elaeagnoides			*	*
Andira inermis				*
Ardisia obovata				*
Bumelia salicifolia				*
Casearia decandra				*
Chrysophyllum pauciflorum				*
Coccothrinax alta				*
Cordia alliodora				*
Daphnopsis americana				*
Faramea occidentalis				*
Myrciaria floribunda				*
Zanthoxylum martinicense				*

cies from the older site had colonized the degraded site (as indicated by their presence at Mary Point in immature stages only). Although seedling densities at Mary Point were relatively high (nearly 50 seedlings per square meter), more than 80% of the seedlings were of a single species: *L. leucocephala,* a weedy invader of disturbed sites.

Discussion

The site analysis has guided our strategies for forest restoration on St. John in several ways. Our analysis indicated that the degraded Mary Point site is both spatially and temporally variable. Unpredictable rainfall patterns, combined with a patchy canopy, rocky surface on upper slopes, and a sparse litter layer, create a precarious environment for plant regeneration. Because of heterogeneous site conditions, there is likely a shortage of safe sites (i.e. favorable microenvironments) for seed germination and establishment. Based on this analysis, we are considering a long-term restoration strategy using an approach suggested by Lugo (1988). This approach focuses restoration efforts on small areas in the landscape with favorable site conditions, where the probability of successful reintroduction of native plants is greatest. As these small areas develop, plant regeneration can then spread gradually across the landscape. Our analysis has led us to recognize that the restoration work (including the identification of safe sites) will be tedious, and the recovery of dry forest across large land areas will be a slow process.

Water shortage is a major problem on this site. Irrigation would help to accelerate vegetation recovery and may be necessary in early stages, but irrigation is impractical and costly and would not be feasible on a large scale. Because our analysis indicated that the soil surface at Mary Point lacks a well-developed litter layer, we are investigating two ways to better protect the soil surface from excessive drying and erosion. One approach is to take advantage of the vegetation presently on the site. We propose to use the non-native tree that dominates the site, *L. leucocephala,* as a nurse crop under which native trees will be planted. Our analysis suggests that this fast-growing tree has probably already enhanced soil nitrogen content, and the litter produced by this species will help to protect the soil surface. Research is needed to determine *Leucaena*'s productivity and nitrogen fixing status on this site. Additional mulch may be necessary for adequate surface protection. To address this need, we are currently investigating the possibility of developing a composting facility in collaboration with a private enterprise on the island that will provide composted organic material for our degraded site and simultaneously reduce the serious solid waste disposal problem on St. John. The idea of coupling systems for mutual benefit (Lugo 1988) is one that has yet to be adequately tested in ecological restoration projects in tropical regions.

The vegetation inventory indicated that a shortage of propagules of native species may be limiting vegetation recovery at Mary Point. The paucity of seedlings and saplings of native dry forest trees on the degraded site is the basis for our ongoing field experiments. It is critical in planning a restoration strategy to know whether the regeneration of native trees on a degraded site is limited (1) by a paucity of seeds, or (2) by the inability of seedlings to become established. If the lack of seeds is a primary factor limiting forest recovery, then importing seeds to the site may hasten the regenerative process. We are testing this hypothesis in a seed rain experiment, in which we broadcast seeds of target species into experimental plots and monitor germination and seedling survival. We are also transplanting shadehouse-grown plants into the field to test the hypothesis that forest recovery in a dry climate can be accelerated by eliminating the uncertainties of seed germination under field conditions.

Conclusions

A carefully planned site analysis can help to guide an ecological restoration effort by focusing attention on critical limiting factors. The approach described in this paper provides a framework for collecting data on important site variables and ecological processes for use during the planning phase of a tropical forest restoration project. A site analysis need not be a time-consuming, long-term study, but will be most useful if it examines the site from a systems perspective, including an analysis of outside factors that constrain ecosystem redevelopment as well as a careful evaluation of biotic and abiotic site conditions.

Acknowledgements

This work was supported by the U. S. Man and the Biosphere Program, The University of Wisconsin Graduate School, The Nave Fund, The Latin American and Iberian Studies Program at the University of Wisconsin, and a Hewlett Foundation grant to the U.W.-Madison Institute for Environmental Studies. We thank the National Park Service and Virgin Islands National Park for logistical and in-kind support of this project, the Institute of Tropical Forestry for soil analyses, and the Institute for Environmental Studies Environmental Monitoring and Remote Sensing Laboratory for image processing. We especially wish to thank C. Rogers, Research Biologist at the Virgin Islands National Park, for help in all aspects of this project. We are grateful to E. Gibney for assistance with species identifications. We thank K. Kearns for field assistance, and D. Siewert and M. Whiting for help with computer analyses.

References

Brown, S. 1982. An overview of Caribbean forests. In: Lugo, A. E. & Brown, S. (Eds.), Forestry in the Caribbean pp. 8–21, U.S. MAB Report No. 7. Proceedings of the First Workshop of Caribbean Foresters, Castries, St. Lucia.

Brown, S., Glubczynski, A. & Lugo, A. E. 1984. Effects of land use and climate on the organic carbon content of tropical forest soils in Puerto Rico. In: New forests for a Changing World pp. 204–209, Proc. Convention of Society of American Foresters. SAF, Washington, D.C.

Buschbacher, R. J. 1986. Tropical deforestation and pasture development. BioScience 36: 22–28.

CH2M Hill, Inc. 1979. A sediment reduction program. Report to Dept. Conservation & Cultural Affairs, Government of Virgin Islands, St. Thomas, VI. In: U.S. Department of the Interior, National Park Service. Marine and Terrestrial Ecosystems of the Virgin Islands National Park and Biosphere Reserve. Biosphere Reserve Report No. 29, VIRMC, St. Thomas, VI. 112 pp.

Dunevitz, V. L. 1985. Regrowth of clearcut subtropical dry forest: mechanisms of recovery and quantification of resilience. M.S. Thesis, Michigan State University. 110 pp.

Ewel, J. 1977. Differences between wet and dry successional tropical ecosystems. Geo-Eco-Trop 1: 103–117.

Holdridge, L. R. 1967. Life zone ecology. Tropical Science Center, San Jose, Costa Rica. 206 pp.

Lott, E. J., Bullock, S. H. & Solis-Magallanes, J. A. 1987. Floristic diversity and structure of upland and arroyo forests of coastal Jalisco. Biotropica 19: 228–235.

Lugo, A. E. 1988. The future of the forest: ecosystem rehabilitation in the tropics. Environment 30: 17–20, 41–45.

Lugo, A. E. & Murphy, P. G. 1986. Nutrient dynamics of a Puerto Rican subtropical dry forest. J. Tropical Ecology 2: 55–72.

Lugo, A. E., Schmidt, R. & Brown, S. 1981. Tropical forests in the Caribbean. Ambio 10: 318–324.

Murphy, P. G. & Lugo, A. E. 1986. Ecology of tropical dry forest. Ann. Rev. Ecol. Syst. 17: 67–88.

Nuzzo, V. A. & Howell, E. A. 1990. Natural areas restoration planning. Natural Areas J. 10: 201–209.

Office of Technology Assessment of U. S. Congress (OTA). 1984. Technologies to sustain tropical forest resources. OTA-F-214. Washington, D.C. 344 pp.

Page, A. E. 1982. Methods of soil analysis, Part 2. American Society of Agronomy, Madison, WI.

Reich, P. B. & Borchert, R. 1982. Phenology and ecophysiology of the tropical tree, *Tabebuia neochrysantha* (Bignoniaceae). Ecology 63: 294–299.

Reich, P. B. & Borchert, R. 1984. Water stress and tree phenology in a tropical dry forest in the lowlands of Costa Rica. J. Ecology 72: 61–74.

Reilly, A. E., Earhart, J. E. & Prance, G. T. 1990. Three subtropical secondary forests in the U.S. Virgin Islands: a comparative quantitative ecological inventory. Advances in Economic Botany 8: 189–198.

Stehle, H. 1944. Location map and forest-type zones of the Caribbean Islands. Caribbean Forester Vol. 6–7, Appendix (Map Insert). USDA Forest Service, Tropical Forest Research Service. Rio Piedras, Puerto Rico.

Teytaud, R. 1986. Environmental factors and potential natural vegetation in St. John, U.S. Virgin Islands. Report to the National Park Service. 55 pp.

Tyson, G. F. 1987. Historic land use in the Reef Bay, Fish Bay, and Hawksnest Bay watersheds of St. John, U.S. Virgin Islands, 1718–1950. Biosphere Reserve Research Report No. 19, VIRMC. National Park Service, MAB. 54 pp.

Uhl, C. 1982. Recovery following disturbances of different intensities in the Amazon rain forest of Venezuela. Interciencia 7: 19–24.

United States Department of Agriculture (USDA). 1970. Soil survey of the Virgin Islands of the United States. USDA Soil Conservation Service. Washington, D.C. 78 pp.

Wadsworth, F. H. 1950. Notes on the climax forests of Puerto Rico and their destruction and conservation prior to 1900. Caribbean Forester 11: 38–47.

Watts, D. 1988. Development and renewable resource depletion in the Caribbean. J. Biogeography 15: 119–126.

West, H. 1793. Bidrag til beskrivelse over Ste. Croix, med en kort udsigt over St. Thomas, St. Jean, Tortola, Spanishtown og Crabeneiland. Copenhagen. 363 pp.

6. Secondary forest regeneration on degraded tropical lands
The role of plantations as "foster ecosystems"

JOHN A. PARROTTA
Institute of Tropical Forestry, USDA Forest Service, Southern Forest Experiment Station, Call Box 25,000, Rio Piedras, PR 00928-2500, USA

Keywords: *Albizia lebbek*, biodiversity, canopy-understory interactions, seed dispersal, seedling dynamics, tropical coastal dunes

Nomenclature: follows Liogier (1982)

Abstract. Forest plantations established on degraded sites can accelerate natural succession through their effects on vegetation structure, microclimate, and soils. Spatial and temporal patterns of secondary forest species regeneration were studied in permanent quadrats in *Albizia lebbek* plantation plots and control areas at a degraded coastal pasture in Puerto Rico. Approximately 6.7 years after plantation establishment, a total of 22 tree and shrub species were found in the plantation plots, compared with only one species (*Albizia lebbek*) found in the control plots. The majority of tree species in the plantation have seeds that are dispersed by either birds or bats, suggesting that the plantation canopy plays a key role in the regeneration process by providing roosting habitat for seed-dispersing animals. Spatial variations in plantation understory seedling populations were found to be associated with both distance from probable parent trees and understory light intensity. These results indicate profound differences between the plantation and adjacent control plots with respect to their provision of regeneration habitat for secondary forest species, and suggest several factors that should be considered in the design of "foster ecosystems" for the rehabilitation of severely degraded tropical forests.

Introduction

The rehabilitation of damaged tropical forests is a major challenge facing ecologists, foresters, and those who depend on productive and diverse forest ecosystems for their sustenance and livelihood. The scale and severity of forest damage, as well as the difficulties associated with their rehabilitation, vary greatly from site to site, and are a function of initial forest structure and species composition, soils, climate, complexity, and resilience of underlying ecosystem processes such as energy flows and biogeochemical cycling, history of disturbance, and the persistence and intensity of ecosystem stresses which inhibit or slow natural successional processes (Lugo 1988).

Severely degraded forest ecosystems, such as those occurring in mining areas throughout the tropics, *Imperata* grasslands of Southeast Asia, and degraded or abandoned pastures in Eastern Amazonia, sub-Saharan Africa, and many parts of India, require special attention if they are to quickly return to productivity. Severely degraded lands are typically characterized by eroded or nutrient-deficient soils, hydrologic instability, reduced primary productivity and low biological diversity. Persistent physical, chemical, and biological barriers often prevent natural succession processes from operating on a time scale compatible with short- and medium-term human needs (Uhl et al. 1982, 1988; Uhl 1987; Nepstad et al. 1991). These barriers to natural forest regeneration may include one or more of the following: low propagule availability (seeds, root stocks), seed predation, non-availability of suitable microhabitats for plant establishment, low soil nutrient availability, absence of obligate or facultative fungal or bacterial root symbionts, seedling predation, seasonal drought, root competition with old field vegetation (particularly grasses), and fire.

An appropriate goal of forest ecosystem rehabilitation is to facilitate, accelerate, and direct natural successional processes so as to increase biological productivity, reduce rates of soil erosion, increase soil fertility (including soil organic matter), and increase biotic control over biogeochemical fluxes

H. Lieth and M. Lohmann (eds.), Restoration of Tropical Forest Ecosystems, 63–73.

within the recovering ecosystem. Forest plantations, using carefully selected tree and shrub species, can play an important role in tropical forest rehabilitation (Lovejoy 1985; Lugo 1988). Numerous studies have demonstrated that plantations of native and exotic species adapted to the harsh conditions of degraded sites can reverse degradation processes by stabilizing soils, increasing soil organic matter through enhancement of above- and below-ground litter production and turnover (Lowry et al. 1988; Montagnini and Sancho 1990), moderation of soil pH, and improvement of soil nutrient status (Sanchez et al. 1985; Chakraborty and Chakraborty 1989; Parrotta 1992).

Through their localized effects on microclimatic and biotic conditions, plantations can facilitate recruitment, survival, and growth of native forest species in the understory, thus acting as "foster ecosystems" (Lugo 1988) for species-rich secondary forests. *Albizia lebbek*, a nitrogen-fixing, leguminous tree species chosen for this study, exhibits a number of characteristics that make it a good candidate for restoration of degraded tropical lands. Native to South Asia and naturalized in many parts of the Caribbean, including Puerto Rico, the tree grows well in semiarid and moist tropical and subtropical regions under a wide range of edaphic conditions including saline, alkaline, and very rocky or sandy soils. Currently used in a variety of agroforestry systems, it is an excellent source of small timber, fuelwood, livestock fodder, tannins, gums, and traditional medicinal products (Parrotta 1987).

This paper presents the results of a study of secondary forest regeneration within *Albizia lebbek* plantation plots and control areas at a degraded coastal site in Puerto Rico. Patterns of recruitment, survival, and growth of woody seedlings are evaluated over a two-year period between 4.5 and 6.7 years after plantation establishment. The factors affecting spatial distribution of secondary forest recruits, and their implications for the design and management of foster ecosystems, are discussed.

Site description and history

The investigation was conducted at the University of Puerto Rico's experimental farm in Toa Baja, located on the northern coast of the island (66° 10′W, 18° 27′N). Rainfall at the site averages 1600 mm per year, and is moderately seasonal in distribution, with two to five months per year in which rainfall is less than 100 mm. Average daily temperatures range from 23.8 °C in January to 29.4 °C in August. Winds at this site are predominantly northeasterly to southeasterly throughout the year.

The soils are well-drained, moderately alkaline (pH 7.9–8.4), calcareous sands of marine origin (isohypothermic Typic Troposamments). Surface soil organic matter is highly variable within the experimental site (0.5–2.5%) and soil nitrogen (<0.1%) is low (Parrotta 1992). Once part of a coastal dune system, soils and vegetation at the site have been subject to frequent, often severe disturbances in recent decades, including cultivation and large-scale soil perturbations such as sand extraction and levelling. For approximately 2 to 3 years prior to plantation establishment, the site had been subjected to moderate grazing and occasional small-scale fires. At the time of establishment of experimental plantations in 1984, the site was dominated by grasses, principally *Panicum maximum* and *Tricholaena repens*, and a wide variety of forbs, primarily Compositae, Leguminosae, and Euphorbiaceae; woody plants were absent. A wooded pasture with a much less severe disturbance history, comprised of approximately 20 secondary forest tree species, borders the experimental area to the south and east (Fig. 1). Urban areas border both the northern and western site boundaries.

Methods

Plantation establishment

Two months prior to the start of the experiment, the experimental area was cultivated using a disk harrow. In December 1984, plantation and control plots were established using a randomized block experimental design, each of three blocks containing one replicate plot of three tree spacing treatments and a control. In the plantation plots, 324 four-month-old *Albizia lebbek* seedlings were planted at spacings of 2 × 2-m, 1 × 1-m, and 0.5 × 0.5-m. Plot size varied depending

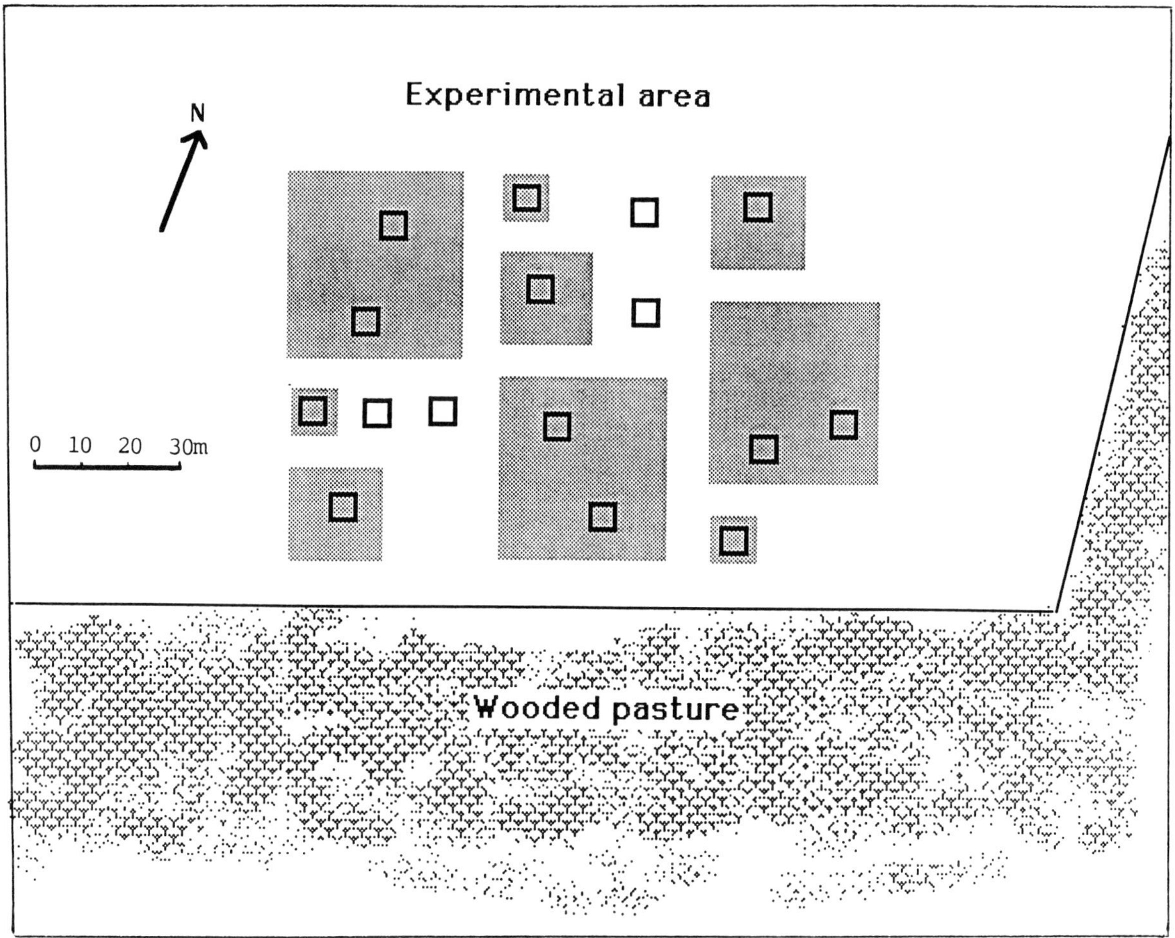

Fig. 1. Experimental site showing locations of permanent quadrats within *Albizia lebbek* plantation plots (shaded squares) and control plots (unshaded).

on tree spacing: 1296 m^2, 324 m^2, and 81 m^2, for the low, intermediate and high density treatments, respectively, and 200 m^2 for the control plots. After planting, the site was protected from grazing and fire for the duration of the study. Approximately every two months during the first year of the experiment, plantation plots were manually weeded. Plantation plots were neither irrigated nor fertilized. Within one year of plantation establishment, vegetation within the control areas was structurally and floristically similar to that which occurred on the site prior to site preparation, and was dominated by grasses and forbs.

In May 1989, biomass, vegetation and soil macronutrient stores, and understory floristics were evaluated in both the plantation and control plots (Parrotta 1992). At this time, plantation stand parameters in the low, intermediate, and high density treatments, respectively, were as follows — mean tree height: 5.0, 4.2, and 2.6 m; total basal area: 17, 25, and 44 m^2 ha^{-1}; total above-ground tree biomass: 32, 42, and 61 Mg ha^{-1}. Total understory biomass, dominated by grasses, averaged 185, 92, and 203 g m^{-2} in the low, intermediate, and high density plantation treatments, and 420 g m^{-2} in the control plots. Seedlings of secondary forest tree species, absent in the control plots, occurred in the plantation plots at an average density of 1.14 m^{-2}. No consistent differences among plantation density treatments were noted with respect to understory floristics, understory biomass, or seedling density.

Study design

In May 1990, a total of twelve 16-m^2 permanent quadrats were established in the plantation plots: two in each of the largest three plots, and one in each of the remaining six plots. Quadrats were located randomly within the central zone of each plot (excluding the 4 outer tree rows). An additional four quadrats were located in the control plots (Fig. 1). Within each quadrat, all woody seedlings taller than 5 cm were identified, mapped, and tagged. Basal diameters and total heights were recorded. At four-month intervals thereafter, until August 1991, all seedlings present earlier were measured, mortality noted, and new recruits identified, measured, mapped and tagged.

Over several days in July and August 1990, understory light intensity was measured between 8:00 and 13:00 hours using a portable light meter at numerous (10–20) points within each quadrat at a height of 1.0 m in both the plantation and control plots. As the vegetation height within the control plots averaged 60 to 90 cm, these values represented open light conditions. Light intensity readings were averaged over all measurement dates to yield a mean value for each quadrat.

Seed dispersal mechanisms and minimum dispersal distances were evaluated for all woody species represented in the permanent quadrats. Likely dispersal mechanisms (e.g. wind, birds, bats) were based on an evaluation of seed characters and a review of the literature. Minimal dispersal distances, estimated for all species in each quadrat, were based on a survey of reproductively mature individuals found in the adjacent wooded pasture or elsewhere within a radius of 200 m from the study site.

Results and discussion

Species diversity, recruitment, and population dynamics

Throughout the study period, dramatic differences were observed between the *Albizia* plantation and control plots with respect to both population densities and diversity of secondary forest species. In the control quadrats, the only species represented in the seedling flora was *Albizia lebbek*, which occurred at an average density of 0.03 to 0.11 individuals per m^2 across all measurement dates, with no individuals larger than 80 cm in height. In contrast, the plantation understory quadrats included seedlings of a total of 21 to 23 tree and shrub species during the period from May 1990 to August 1991 (corresponding to plantation ages of 5.4 and 6.7 years). During this period, seedling densities increased from 1.93 to 2.82 individuals per m^2, of which 17 to 25% were *Albizia lebbek* and 35 to 41% were the successional forest species *Citharexylum fruticosum* (Fig. 2a). Also during this period, the mean number of secondary forest species represented in the seedling flora of the twelve permanent quadrats increased from (all values are mean $\pm$ SE) 6.5 $\pm$ 0.9 to 8.4 $\pm$ 0.6 per 16-m^2 quadrat. These data suggest that while the total number of tree and shrub species in the plantation remained more or less constant, the spatial distributions of these species were becoming more even over time.

Differences among plantation density treatments were noted with respect to understory seedling density. At 6.7 years, mean seedling densities for secondary forest species were 1.72 $\pm$ 0.41, 2.08 $\pm$ 0.96 and 2.77 $\pm$ 0.90 individuals per m^2 in the low, intermediate, and high density treatments, respectively. An opposite trend was noted for *Albizia lebbek* seedlings which occurred at densities of 0.81 $\pm$ 0.22, 0.67 $\pm$ 0.23, and 0.50 $\pm$ 0.38 individuals per m^2 within the three plantation density treatments. However, as these differences were not statistically significant, data from all plantation understory quadrats were pooled for subsequent analysis. Density, mean seedling basal diameters and heights, and importance values (Curtis 1959) for each of the species found in the plantation and control quadrats at the end of the study period are given in Table 1.

At 6.7 years of age, the plantation understory seedling basal area was 5.45 cm^2 per m^2. The corresponding value for the control plots was 0.026 cm^2 per m^2. As indicated in Fig. 2b, the dominant tree species in the plantation understory, *Citharexylum fruticosum*, accounted for 75 to 77% of the total seedling basal area. While

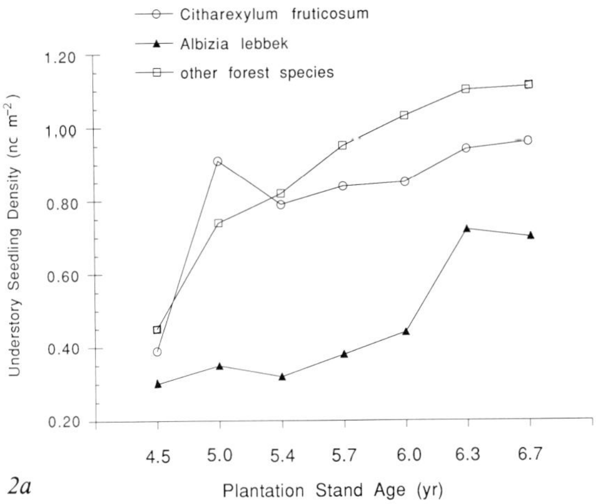

2a

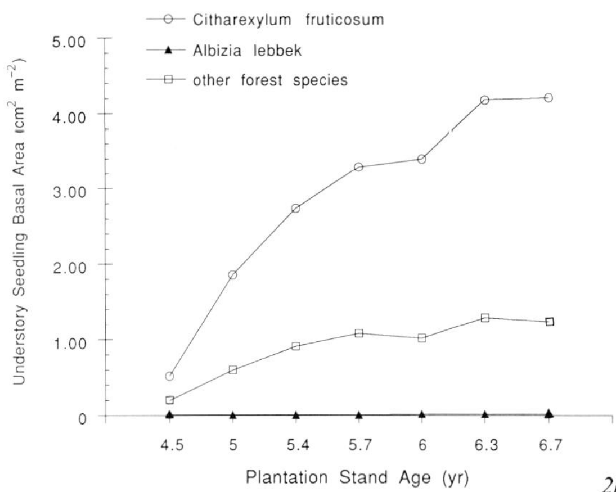

2b

Fig. 2. Tree and shrub seedling/sapling population trends in plantation understory 4.5 to 6.7 years after plantation establishment. Data at 4.5 and 5 years derived from surveys involving 46 m^2 and 54 m^2 total quadrat area; 5.4 to 6.7 year data derived from permanent quadrat surveys (192 m^2). a. Understory seedling density. b. Understory seedling basal area.

Table 1. Understory seedling population characteristics and importance values (Curtis 1959), for tree species occurring in plantation and control quadrats in August 1991, 6.7 years after plantation establishment. Values following means indicate standard errors

Species	Seedling density (No. m^{-2})	Mean basal diameter (cm)	Mean seedling height (m)	Importance Value (I.V.)
Plantation				
Citharexylum fruticosum	0.974	1.76 ± 0.12	1.72 ± 0.12	124.0
Albizia lebbek	0.724	0.17 ± 0.01	0.17 ± 0.01	36.5
Cordia polycephala	0.146	1.11 ± 0.17	1.48 ± 0.25	20.5
Rauvolfia nitida	0.161	1.26 ± 0.15	1.23 ± 0.15	19.1
Lonchocarpus latifolius	0.234	0.79 ± 0.08	0.80 ± 0.10	15.4
Spathodea campanulata	0.089	1.19 ± 0.20	0.88 ± 0.27	14.6
Bourreria succulenta	0.115	0.92 ± 0.20	1.04 ± 0.24	12.9
Eugenia monticola	0.099	0.36 ± 0.04	0.33 ± 0.04	11.7
Schinus terebinthifolius	0.053	0.63 ± 0.19	0.87 ± 0.34	8.6
Roystonea borinquena	0.057	0.52 ± 0.12	0.37 ± 0.07	8.2
Calophyllum brasiliense	0.047	0.59 ± 0.15	0.49 ± 0.14	7.8
Bursera simaruba	0.036	0.81 ± 0.14	1.06 ± 0.28	7.1
Andira inermis	0.026	0.50 ± 0.05	0.34 ± 0.02	3.2
Psidium guajava	0.010	0.70 ± 0.10	0.85 ± 0.10	2.5
Trichilia hirta	0.005	0.50	0.16	1.4
Terminalia catappa	0.005	1.50	0.80	1.4
Albizia procera	0.005	1.10	1.25	1.3
Unidentified species (5)	0.036	0.79 ± 0.35	0.59 ± 0.26	3.8
Pasture control				
Albizia lebbek	0.109	0.43 ± 0.14	0.31 ± 0.10	

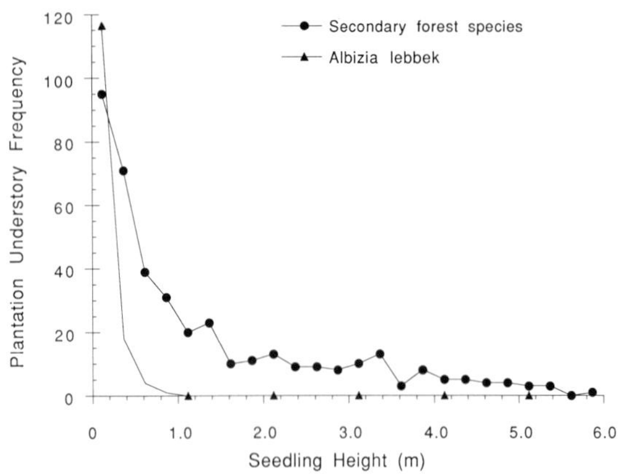

Fig. 3. Height distribution for secondary forest species and *Albizia lebbek* in plantation understory quadrats 6.7 yrs after plantation establishment. Total quadrat area: 192 m^2.

seedlings of *Albizia lebbek* were common within the plantation, they comprised less than 2% of the total seedling basal area throughout the study period. In contrast to the secondary forest species, which exhibited a fairly broad seedling height distribution, the population structure of *Albizia lebbek* understory seedlings was sharply skewed toward the smallest height classes (Fig. 3). Despite abundant seed production by the plantation crop since the second year after plantation establishment, the understory *Albizia* seedling population was dominated by small and in most cases very young individuals. Only 17% of the *Albizia lebbek* seedlings in the plantation understory quadrats at 6.7 years were taller than 25 cm. In contrast, 77% of *Citharexylum fruticosum* seedlings and 76% of seedlings of all other secondary forest species exceeded this height.

Seedling mortality patterns also suggest that relative to the secondary forest species, *Albizia lebbek* does not survive well in the plantation understory. Between 5.4 and 6.7 years a total of 264 seedlings were recruited into the permanent quadrats (192 m^2 total area), including 138 *Albizia lebbek* and 44 *Citharexylum fruticosum*. During this period, 96 seedlings died, including 65 *Albizia* and 10 *Citharexylum*. Thus, seedling mortality rates during this 1.3-yr period were 36% overall; 47% for *Albizia*, 23% for *Citharexylum*, and 26% of all other secondary forest species.

Table 2. Dispersal characteristics of secondary forest species occurring in *Albizia lebbek* plantation plots

Species	Importance value rank	Distance* from seed source (m)	Diaspore dispersal mechanism
Citharexylum fruticosum	1	>50	Birds
Albizia lebbek	2	0	Wind
Cordia polycephala	3	>30	Birds
Rauvolfia nitida	4	>90	Birds
Lonchocarpus latifolius	5	>20	Wind
Spathodea campanulata	6	>35	Wind
Bourreria succulenta	7	>50	Birds
Eugenia monticola	8	>200?	Birds
Schinus terebinthifolius	9	>200?	Birds
Roystonea borinquena	10	>20	Birds
Calophyllum brasiliense	11	>150	Bats
Bursera simaruba	12	>50	Birds/ mammals
Andira inermis	13	>200?	Mammals
Psidium guajava	15	>200?	Birds
Trichilia hirta	16	>200?	Birds
Terminalia catappa	17	>65	Bats
Albizia procera	18	>100	Wind
Unidentified species (5)	14,19–22	?	?

* Distance to nearest reproductively mature individual for each species located within a 200-m radius of the plantation site. "?" Indicates that no such individuals were found within this distance of the site.

Seed dispersal and the role of animals as successional catalysts

The absence of secondary forest species in the control plots, in marked contrast to the plantation, strongly suggests that pre-plantation soil seed banks at this site have played only a minor role in the observed patterns of understory seedling regeneration and that the forest succession process has been dominated by post-plantation seed introductions. Of the secondary forest species found in the plantation understory at 6.7 years, most have seeds that are typically dispersed by birds (Table 2). These species, such as *Citharexylum fruticosum*, *Cordia polycephala*, *Rauvolfia nitida*, *Roystonea borinquena*, and *Psidium guajava* (guava) are small-seeded species that produce fleshy fruits consumed by birds. At least two species, *Calophyllum brasiliense* and *Terminalia catappa*, produce large-

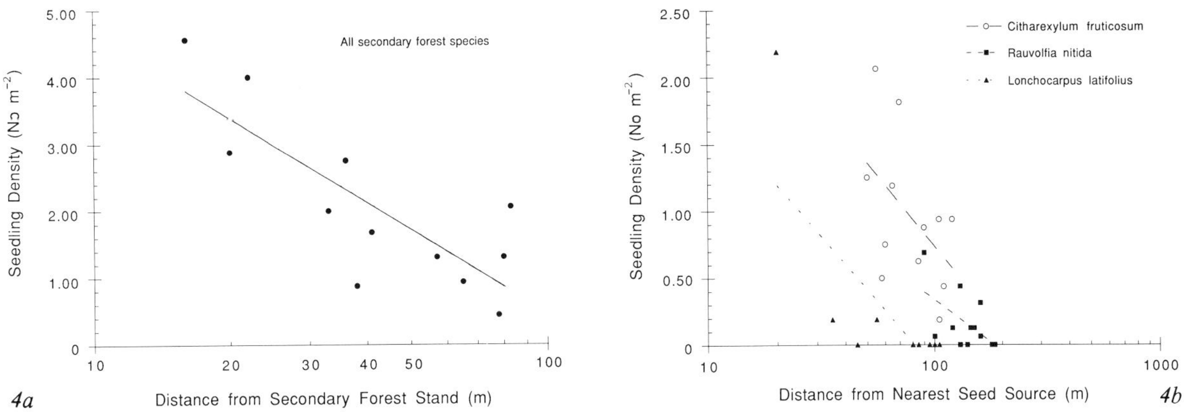

Fig. 4. Relation between seedling density in plantation understory quadrats at 6.7 yrs and distance from nearest potential seed sources. a. All secondary forest tree and shrub species, excluding *A. lebbek* ($r^2 = 0.66$; $P < 0.01$). b. *Citharexylum fruticosum* ($r^2 = 0.25$; $P < 0.10$), *Rauvolfia nitida* ($r^2 = 0.31$; $P < 0.10$), and *Lonchocarpus latifolius* ($r^2 = 0.54$; $P < 0.01$). Correlation coefficients and significance levels are for regressions for the general form: $y = a - b * \log_{10} x$.

seeded fruits that are most commonly dispersed by bats. Only three species, *Lonchocarpus latifolius*, *Spathodea campanulata*, and *Albizia procera*, can be characterized as wind-dispersed species. Thus, animal dispersal appears to be the dominant mechanism of seed input to the plantation system.

Based on a comparison of plantation understory seedling data and a survey of reproductively mature trees found within a 200 m radius of the plantation site, the minimum distances over which the seeds of these species were transported was estimated (Table 2). Seeds of most of the more common species appear to have been transported by animals or wind relatively short distances (<100 m). A few, notably *Eugenia monticola*, *Schinus terebinthifolius*, *Andira inermis*, *Psidium guajava* and *Trichilia hirta*, all bird- or mammal-dispersed species, were carried much further, since these species were not represented in the local flora within 200 m of the plantation site. These data suggest that the plantation is serving an important function in the process of secondary forest regeneration, or succession, by providing roosting and in some cases nesting habitat for frugivorous birds and bats.

Early in the study, it was noted that seedling densities and species distributions showed considerable variation with location within the plantation system. Comparing seedling populations among quadrats located in plantation plots (at 6.7 years), a significant negative correlation ($P < 0.01$) was observed between tree seedling density (for all species) and distance from the secondary forest stand adjacent to the experimental area (Fig. 4a). This relationship between minimum seed dispersal distance and quadrat seedling density appeared to be less pronounced when individual species were evaluated (Fig. 4b). Nonetheless, these trends do suggest that for species whose seeds are dispersed by either wind or animals, proximity to seed sources may be a very important factor controlling the rate of secondary forest species colonization of the plantation plots. It should be noted that within the plantations, the larger individuals of a few species, notably *Citharexylum fruticosum* and *Cordia polycephala*, had begun to produce flowers and mature fruits when the plantations were approximately four years old.

As a larger number of individuals and species within the plantation system become reproductively mature, it could be expected that the relative importance of external seed inputs into the plantation system would diminish over time, although external seed sources would continue to be important for increasing floristic biodiversity.

Understory light intensity, microhabitat structure and competition

A second factor, independent of seed source proximity, which appears to play a role in con-

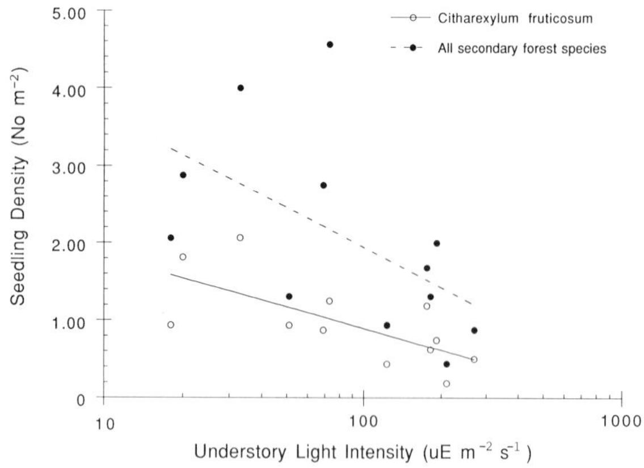

Fig. 5. Relation between seedling density and light intensity in plantation understory quadrats at 6.7 yrs. All secondary forest tree and shrub species, excluding *A. lebbek* ($r^2 = 0.31$; $P < 0.10$) and *Citharexylum fruticosum* ($r^2 = 0.48$; $P < 0.05$). Correlation coefficients and significance levels are for regressions for the general form: $y = a - b * \log_{10} x$.

trolling understory seedling density is light intensity. Understory light intensities, measured in July and August 1990, ranged from 18 to 269 μE m^{-2} s^{-1} (mean $\pm$ SE 118 $\pm$ 24) in the plantation quadrats, or 2 to 29% of the light intensities in the control plots, which averaged 924 μE m^{-2} s^{-1}. The wide range of understory light conditions in the plantation plots, highly correlated with canopy leaf area index (unpublished data), are the result of variable growth rates of the plantation crop and mortality of some larger trees caused by the vascular wilt *Fusarium solani* (Parrotta 1990). Following rapid initial growth of the planted *Albizia lebbek*, culminating in maximum mean annual wood increments between two and five years across all density treatments, canopy structure has become progressively less uniform within the plantation plots, with a general trend toward greater light penetration to the forest floor. A significant inverse relationship between seedling density and understory light intensity was found for all secondary forest species (excluding *Albizia lebbek*) and for *Citharexylum fruticosum* (Fig. 5).

It is unlikely that light intensity is exerting any direct control over seed germination or subsequent seedling survival in the plantation system. Most of the secondary forest species considered here are either pioneer species or otherwise capable of natural regeneration under conditions of full sunlight or light shade. Several inter-related factors could help to explain the apparent negative relationship between understory light intensity and seedling population density. Quadrats characterized by higher light levels tended to occur on sites with coarser sands relatively low in organic matter, and also tended to be more dominated by grasses than were quadrats with lower light levels. Thus the relative paucity of seedlings in the more open plantation quadrats could be the result of some combination of the following factors: (a) increased root competition with other understory floristic components, principally grasses; (b) less favorable soil moisture conditions affected by soil texture and soil organic matter; and (c) lower soil nutrient availability.

These same factors may also help to account for the dramatic differences observed between the plantation and control plots with respect to secondary forest regeneration, or rates of forest succession. Overall differences in light intensity, graminoid dominance, and soil properties between the degraded pasture control plots and the plantation understory, and the effects on seedling populations of microsite differences observed with regard to these factors among plantation quadrats, support this hypothesis. The relative importance of these factors, and the importance of differential seed input rates into plantation and control plots, is unclear and would require further study.

Implications for the design and management of "foster ecosystems"

The results of this study suggest that forest plantations established on degraded tropical lands can serve three important roles — (1) production of wood and other forest commodities, (2) soil improvement (Parrotta 1992), and (3) acceleration of secondary forest succession. The degree to which plantations realize these objectives is influenced by several factors, among the most important being species selection, plantation design, and management practices.

The choice of plantation species can influence both the rate and trajectory of rehabilitation processes. Candidate species may differ greatly in their capacity to resist or survive fire, drought, and other stresses, to stabilize soils, increase organic matter and available soil nutrients (Gill et al. 1987; Sharma and Gupta 1989; Lugo et al. 1990; Wang et al. 1991); and facilitate understory development (Sommariba 1988; Pande et al. 1988). These differences relate to their relative susceptibility to pests and diseases, patterns of above-ground and root biomass accumulation (Jonsson et al. 1988; Parrotta 1989; Lugo et al. 1990), nutrient utilization and allocation (Jorgensen and Wells 1986; Wang et al. 1991), litter production and fine root turnover, rates of litter decomposition, greatly influenced by litter chemistry (O'Connell 1986; Gill et al. 1987), and the presence of secondary compounds that may inhibit the activity of decomposing organisms (Suresh and Vinaya Rai 1988).

Nitrogen-fixing species inoculated with selective strains of *Rhizobium* or *Frankia* can have a dramatic effect on soil fertility through their production of readily decomposible, nutrient-rich litter and turnover of fine roots and nodules (Bernhard-Reversat 1988; Montagnini and Sancho 1990). In the case of planted exotic species, poor natural regeneration, as observed in *Albizia lebbek* in the present study, can be viewed as a positive trait if one of the goals of rehabilitation is the facilitation of secondary forest succession.

Plantation design considerations which may influence rates of forest species colonization include the location of planted forests in relation to primary and secondary forest stands, and plantation density. Proximity to seed-bearing trees and, for native forest species with animal dispersed seeds, the habitat requirements of frugivorous birds, bats, and other seed dispersers, appear to be key factors influencing development of floristic diversity in plantations.

Regardless of initial tree planting density, the recruitment of secondary forest species appears to depend on the maintenance of sufficient canopy cover to ease understory seedling competition with grasses through shading and to provide sufficient habitat diversity for birds and other seed dispersal agents. Plantation harvests should be designed to ensure at least partial canopy closure until the secondary forest species assume a more dominant structural role within the plantation system.

The influence of natural forest proximity on seed recruitment into plantations established on severely degraded sites poses a special problem for the restoration of secondary forests over large areas. On such sites, the establishment of continuous plantation forest cover would not be an economically viable option in most instances, since forest yields would most likely be relatively low given the initial productivity constraints. Thus, more creative strategies that take advantage of natural forest succession processes may be required.

One such approach could be the creation of planted forest "archipelagos" extending into the heart of the area to be rehabilitated from the edges of existing forests. In such a system, it is hypothesized, secondary forest species colonization would be most rapid in plantation "islands" closest to the edges of the natural forest and would proceed gradually toward areas further removed from seed sources. Over time, these forested islands might be expected to enlarge as secondary forest species colonize the edges of planted forest stands, and eventually merge with neighboring islands. Similarly, planted forest strips or corridors, again extending away from the edges of natural forest stands, could be used as a lower-cost alternative to continuous plantations over large areas.

Implicit in such a model for large-scale restoration of forest cover on severely degraded sites is an understanding of the life history requirements of species which one hopes to see established in the planted foster ecosystems. Of particular importance is an understanding of seed ecology, including dispersal characteristics, and microhabitat requirements for germination and seedling development. For complex tropical forests, our knowledge in these areas is fragmentary at best. However, by focusing on the ecology of better known native, early successional species, the initial species composition, size and spatial arrangement of such islands or corridors could, in theory, be designed so as to facilitate their rapid, early colonization, and catalyze forest succession processes at the lowest possible cost. For example, fruit-bearing species likely to attract birds, bats or other animals known to be key dispersal agents for desired secondary forest species, could be included in the plantations.

The inclusion of native forest surface soils in the potting mix of containerized seedlings prior to outplanting could facilitate the development of soil faunal and microfloral diversity. This in turn could improve prospects for successful colonization of secondary forest species and re-establishment of complex community interactions and ecological processes on sites which previously had been biologically impoverished.

While there are not universally applicable techniques for tropical forest rehabilitation and restoration, the use of planted forests as foster ecosystems to facilitate or catalyze natural forest succession, may be one important tool for rehabilitating highly degraded sites. However, the success of this approach would appear to be highly dependent on several factors relating to species selection, plantation density, location with reference to seed sources, availability of dispersal agents, and management approaches. The synthesis of ecological and silvicultural knowledge can be of great value in this process by helping us to identify the key barriers to natural forest succession characteristic of specific degraded sites and to intervene constructively to bypass these constraints.

Acknowledgements

I wish to acknowledge the contributions of D. Lamb and J. Vancley to the development of ideas on plantation design presented here. I thank Alain Henri Liogier of the University of Puerto Rico Botanical Garden for his help in identifying plant specimens. This work was carried out in cooperation with the Center for Energy and Environment Research, University of Puerto Rico, Rio Piedras.

References

Bernhard-Reversat, F. 1988. Soil nitrogen mineralization under a *Eucalyptus* plantation and a natural *Acacia* forest in Senegal. For. Ecol. Manage. 23: 233–244.

Chakraborty, R. N. & Chakraborty, D. 1989. Changes in soil properties under *Acacia auriculiformis* plantations in Tripura. Indian Forester 115: 272–273.

Curtis, J. T. 1959. The vegetation of Wisconsin. An ordination of plant communities. Univ. of Wisconsin Press, Madison WI, U.S.A.

Gill, H. S., Abrol, I. P. & Samra, J. S. 1987. Nutrient recycling through litter production in young plantations of *Acacia nilotica* and *Eucalyptus tereticornis* in a highly alkaline soil. For. Ecol. and Manage. 22: 57–69.

Jonsson, K., Fidjeland, L., Maghembe, J. A. & Högberg, P. 1988. The vertical distribution of fine roots of five tree species and maize in Morogoro, Tanzania. Agrofor. Syst. 6: 63–69.

Jorgensen, J. R. & Wells, C. G. 1986. Tree nutrition and fast-growing plantations in developing countries. Int. Tree Crops J. 3: 225–244.

Liogier, A. H. 1982. Flora of Puerto Rico and adjacent islands: a systematic synopsis. Editorial de la Universidad de Puerto Rico, Rio Piedras, PR, U.S.A.

Lovejoy, T. E. 1985. Rehabilitation of degraded tropical forest lands. Environmentalist 5: 13–20.

Lowry, J. B., Lowry, J. B. C. & Jones, R. 1988. Enhanced grass growth below a canopy of *Albizia lebbek*. Nitrogen Fixing Tree Res. Rep. 6: 45–46.

Lugo, A. E. 1988. The future of the forest — ecosystem rehabilitation in the tropics. Environment 30: 16–20, 41–45.

Lugo, A. E., Wang, D. & Bormann, F. H. 1990. Comparative analysis of biomass production in five tropical tree species. For. Ecol. Manage. 31: 153–166.

Montagnini, F. & Sancho, F. 1990. Impacts of native trees on tropical soils: a study in the Atlantic lowlands of Costa Rica, Central America. Ambio 19: 386–390.

Nepstad, D. C., Uhl, C. & Serrão, E. A. S. 1991. Recuperation of a degraded Amazonian landscape: forest recovery and agricultural restoration. Ambio 20: 248–255.

O'Connell, A. M. 1986. Effect of legume understory on decomposition and nutrient content of eucalypt forest litter. Plant and Soil 92: 235–248.

Pande, P. K., Bisht, A. P. S. & Sharma, S. C. 1988. Comparative vegetation analysis of some plantation ecosystems. Indian Forester 114: 379–389.

Parrotta, J. A. 1987. *Albizia lebbek* (L.) Benth. Silvics of Forest Trees of the American Tropics, SO-ITF-SM-7. U.S. Department of Agriculture, Forest Service, Institute of Tropical Forestry, Southern Forest Experiment Station, New Orleans, Louisiana, U.S.A. 5 pp.

Parrotta, J. A. 1989. The influence of management practices on the biogeochemical stability of tropical fuelwood plantations. Trop. Ecol. 30: 1–12.

Parrotta, J. A. 1990. Occurence of *Fusarium* sp. in plantation monocultures of *Albizia lebbek* in Puerto Rico. Nitrogen Fixing Tree Res. Rep. 8: 118–119.

Parrotta, J. A. 1992. The role of plantation forests in rehabilitating degraded tropical ecosystems. Agric. Ecosystems Environ. 41: 115–133.

Sanchez, P. A., Palm, C. A., Dabey, C. B., Szott, L. T. & Russell, C. E. 1985. Tree crops as soil improvers in the humid tropics? In: M. G. R. Cannell & J. E. Jackson (Eds.), Attributes of Trees as Crop Plants. Institute of Terrestrial Ecology, Natural Environmental Research Council, Abbots Ripton, Huntington, U.K., pp. 327–350.

Sharma, B. D. & Gupta, I. C. 1989. Effect of tree cover on soil fertility in western Rajasthan. Indian Forester 115: 348–354.

Somarriba, E. 1988. Pasture growth and floristic composition under the shade of guava (*Psidium guajavam* L.) trees in Costa Rica. Agrofor. Syst. 6: 153–162.

Suresh, K. K. & Vinaya Rai, R. S. 1988. Allelopathic exclusion of understory by a few multipurpose trees. Intl. Tree Crops J. 5: 143–151.

Uhl, C. 1987. Factors controlling succession following slash-and-burn agriculture in Amazonia. J. Ecology 75: 377–407.

Uhl, C., Jordan, C., Clark, K., Clark, H. & Herrara, R. 1982. Ecosystem recovery in Amazon caatinga forest after cutting, cutting and burning, and bulldozer clearing treatments. Oikos 38: 313–320.

Wang, D., Bormann, F. H., Lugo, A. E. & Bowden, R. D. 1991. Comparison of nutrient-use efficiency and biomass production in five tropical tree taxa. For. Ecol. Manage. 46: 1–2.

7. The ecology of Zambezian dry evergreen forest with recommendations for conservation management

F. P. MALAISSE
U.R. Ecologie, Faculté des Sciences agronomiques, Gembloux, B 5030, Belgium

Keywords: Dry evergreen forest, forest restoration, gap dynamics, Shaba, Zaire, Zambezian region

Abbreviation: (D.E.F.) – Dry evergreen forest

Nomenclature: Flora Zambesiaca, Flore d'Afrique centrale

Abstract. The particularities of the Zambezian regional center of endemism are presented which results in the definition of dry evergreen forests. The state of knowledge of the Zambezian dry evergreen forest in general and the Luiswishi forest in particular is summarized. Regeneration potentialities and observations carried out on gap dynamic are analyzed. The recommended methods of restoration take into account bush fires, logging, clear cutting and enrichment. These strategies are discussed in relation to man's pressure on the Zambezian environment.

Introduction

The basic guidelines for a phytogeographical subdivision of Africa are provided by the A.E.T.F.A.T. map (White 1981) and by the descriptive memoir written by White in 1983. In this work, eleven regional centers of endemism are specified, including that of the Zambezian region, which covers an area of 3 770 000 km^2. It is the richest floristic center of Africa with at least 8500 higher plant species, of which circa 54% are endemic, but with few endemic genera and no endemic families.

White (1983) recognizes 24 vegetation units for the Zambezian region, among them several types of woodland. In the northern part of the region, the wetter Zambezian miombo woodland, which is dominated by the Caesalpiniaceae, can be observed. It is the most widespread type of woodland covering altogether 1 240 000 km^2, some 33% of the Zambezian region and 3.9% of Africa (Fig. 1).

Miombo woodlands are fireclimax. Schmitz (1962) argues that the wetter miombo woodland is derived from the dry evergreen forest, the *climax vegetation*. According to other authors (Duvigneaud 1958; Fanshawe 1960, 1969), dry evergreen forests are confined to edaphically favourable sites, those with the deepest soils, whilst some poor miombo woodlands (Xerobrachystegion) are located on the shallowest soils.

Today, the dry evergreen forests are scattered across the high rainfall belt of the Zambezian region. They are endangered, quickly disappearing, and may only be observed over 1% of the region. Although fires caused by lightning and Stone Age man may have started the destruction of these dry evergreen forests, it is evident that man-made fires of recent centuries as well as clearings for cultivation had by far the greatest effect (Chapman and White 1970). Only isolated patches have survived the destructive agents of man and the ravages of fire and time (Fig. 2). These patches are living communities, able to maintain themselves under existing climatic conditions, and no moribund or relics of a past climatic era (Lawton 1963, 1964). Further evidence to show that dry evergreen forests were at one time more widespread is suggested by the wide distribution of scattered evergreen thickets (Trapnell 1953).

The present paper deals with this particular tropical forest, a "forêt dense sèche tropicale sempervirente à semi-décidue" of the French-speaking authors.

H. Lieth and M. Lohmann (eds.), Restoration of Tropical Forest Ecosystems, 75–90.

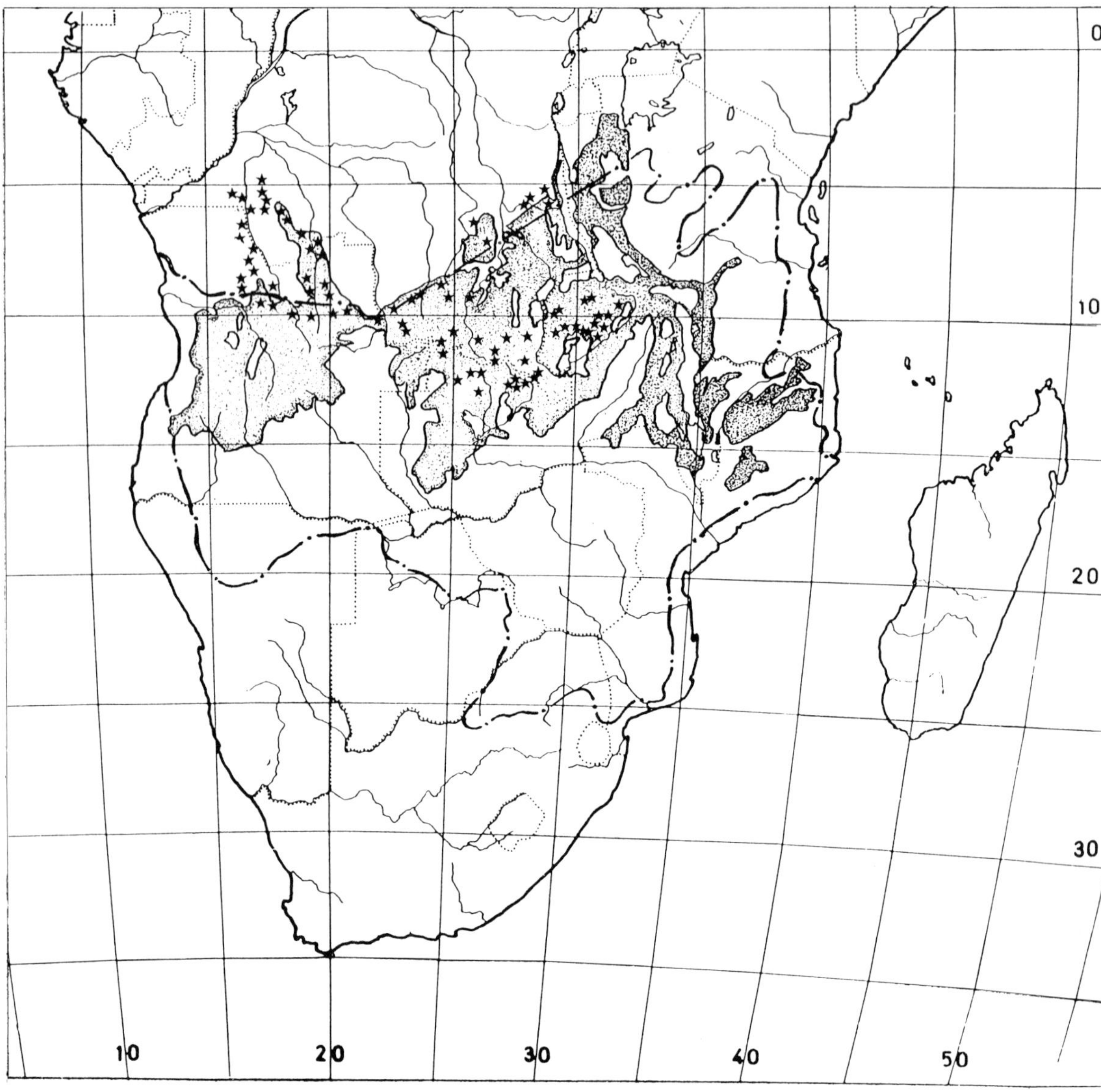

Fig. 1. Location and differentiation of the Zambezian region (---), showing the wetter miombo woodlands and the occurrence of Zambezian dry evergreen forests (*) (after the A.E.T.F.A.T. map 1983, modified).

The Zambezian dry evergreen forest

The Zambezian dry evergreen forest (D.E.F.) "represents part of a transition, both floristic and physiognomic, from Guineo-Congolian rain forest to Zambezian woodland". It is a closed, dense, multilayered forest. It "rarely exceeds 25 m in height, except for a few emergents. . . Zambezian D.E.F. varies greatly in floristic composition from place to place. There are eight dominant and emergent tree species, viz. *Berlinia giorgii, Cryptosepalum pseudotaxus, Daniellia alsteeniana, Marquesia acuminata, M. macroura, Entandrophragma delevoyi, Parinari excelsa* and *Syzygium guineense* subsp. *afromontanum* (White 1983). Below the upper canopy, there is a discontinuous stratum of understorey trees varying from 5 to 12 m in height. The main species in this layer include *Aidia micrantha* var. *msonju, Rothmannia fisheri* and *Tabernaemon-*

Fig. 2. A very small dry evergreen forest patch leant against a termite mound in a savanna environment (photograph F. Malaisse).

tana pachystigma. Immediately below the understorey tree stratum there is a well-developed continuous shrub stratum consisting of evergreen shrubs, scandent shrubs, climbers and saplings (Lawton 1964). Climbers are well-represented. They sometimes form a dense fringe of vegetation and may even climb into the crowns of the upper canopy species (Fig. 3). One such climber is the very abundant *Combretum gossweileri*.

Epiphytic ferns, lichens and mosses are not rare. Leaf litter covers the forest floor; very few herbs grow under the forest canopy, but many seedlings of woody species are present (Lawton 1964; Malaisse 1984). Regeneration of the dominant species is generally excellent, and quite sufficient to perpetuate the forest type (Fanshawe 1960), provided that fire and the destructive activities of man are excluded.

Our knowledge of this ecosystem is relatively scanty. Fanshawe (1960) distinguished three types of forest in Zambia: the *Parinari-Syzygium* forest (vernacular name: mateshi in Zambia, muhulu in Shaba), the *Cryptosepalum pseudotaxus* forest (vernacular name: mavunda in Zambia, mufuka in Angola) restricted to the Kalahari sands, and the *Marquesia acuminata* forest occurring on very deep, reddish sandy loams derived from granite. Malaisse (1984) and Cottrell and Loveridge (1966) provide transects of the first two types respectively (Fig. 4), while bird's communities of the last two types are analyzed by Cottrell and Loveridge (l.c.), Oatley (1969) and Benson and Stuart Irwin (1965), respectively.

The ecology of the *Marquesia acuminata* D.E.F. was documentated by Lawton (1964). Schmitz (1962) provides a phytosociological study, recognizing one alliance, the *Diospyro-Entandrophragmion delevoyi*, two associations and four sub-associations. The same author (Schmitz 1971) set up a synthesis of the connections between the different phytosociological units — including the D.E.F. climax — recog-

Fig. 3. Parinari excelsa, a large evergreen tree of the Zambezian dry evergreen forest. The abundance of climbers should be noted (photograph L. Lemaire).

nized in Upper Shaba, the south-eastern part of Zaire (Fig. 5).

The Luiswishi dry evergreen forest

Several aspects of D.E.F. ecology were obtained from studies carried out at Luiswishi (27° 36′ 10″ E., 11° 29′ 05″ S., 1208 m a.s.l.), in the surroundings of Lubumbashi (Zaire). Several D.E.F. patches exist at this locality; they are scattered in a 25 km^2 area, the largest one covering some 4 ha. This seasonal tropical forest corresponds to the *Parinari-Syzygium* forest type of Fanshawe (1960) and belongs to the *Entandrophragmeto-Diospyretum hoyleanae* association (Schmitz 1988). It is subjected to a rainy season (November to March), a dry season (May to September) and two transition months (April and October). As already stated (Malaisse 1974), the dry season must be divided, from a phenological point of view, into a cold dry season (mid-May to mid-July) and a warm dry season (August to early October). This heavy seasonality is reflected in air and soil temperatures as well as in soil humidity (Malaisse and Kapinga 1987; Dikumbwa 1991).

Stem density of the Luiswishi D.E.F. is of some 50 500 stems ha^{-1}, of 15 880 stems over 1.3 m height, 1463 stems over 5 cm DBH, 567 stems over 10 cm and 231 stems over 20 cm DBH. These data coincide with values published for rain forests. The basal area is about 33.7 m^2 ha^{-1}, a somewhat low value if compared to the 30 to 47 m^2 ha^{-1} observed in tropical rain forests (Malaisse 1984).

Phanerophytes are the dominant biological type (Fig. 6). They account respectively for 87 and 99% in the raw and balanced spectrum (Malaisse 1984). Leaf area is mostly of the micro- or mesophyllous type (Fig. 7), thus of medium size, and smaller than in tropical rain forests. The leaf anatomy of the Luiswishi D.E.F. presents a tendency towards sclerophylly, i.e. a relatively thick lamina, frequent trichome, various sclerous tissues, but relatively rare hypoderms (Malaisse and Conlonval-Elenkov 1981). The L.A.I. varies from 5.0 (mid-December) to 2.6 (beginning of October) (Fig. 8); the ratio between the maximum value and that at the time of maximum defoliation is 1.91. This value agrees very well to the 2.11 ratio observed in Panama by Golley et al. (1975) for a tropical rain forest.

The relative importance of the dissemination agents varies according to strata. The dominant lignous species are, with the exception of *Parinari excelsa*, either passive ballistic autochores (*Brachystegia* spp.) or anemochores of the heavy type (*Entandrophragma delevoyi, Combretum* spp.) provided with wings (van der Pijl 1982), whilst zoochory prevail in the shrub layer (notably *Diospyros hoyleana, Tabernaemontana pachystigma, Artabotrys* spp.).

Termite mounds are also to be found in the forest. Their density is of 4.7 termite mounds per hectare (Malaisse 1985). Other studies deal with litter fall (about 6 T leaves per year), litter breakdown (Malaisse et al. 1975) and soil microorganisms (Malaisse-Mousset 1974).

Fig. 4. Profile diagram illustrating one strip of the Luiswishi dry evergreen forest (50 × 5 m) (after Malaisse 1984).

From a dynamic point of view, Schmitz (1971) describes the regressive succession for "muhulu" in the surroundings of Lubumbashi. At first, the *Brachystegieto-Marquesetium* takes first the place of the *Entandrophragmeto-Diospyretum hoyleanae* climax. It is a semi-evergreen forest characterized by *Marquesia macroura, Brachystegia taxifolia* and *Salacia rhodesiaca.* It holds a transitional status between Zambezian D.E.F. and the substitutional open forests. It frequently looks like an even-aged stand. The first two trees do not regenerate under closed forest and do not survive their overcrowding by the D.E.F. woody climbers. They perfectly withstand early fires. The normal lifetime of this transition forest, when submitted to annual fires, is 150–250 years. Afterwards, the two species progressively disappear for the benefit of diverse *Brachystegia* and *Julbernardia* woodlands. Abundance and thickness of *Marquesia* are, in fact, an indication of the antiquity of D.E.F. deterioration.

Method

Starting from 1980, a wide collection of seeds and fruits has been assembled. It supplies a first set of informations dealing with nature, fresh weight and viability of D.E.F. diaspores. Seeds were introduced into the University nursery and followed up. Percentages of germination were noted. Voucher specimens and drawings of seedlings and saplings of about fifty dominant woody species were performed (Fig. 9) in order to provide a determination key. At the same time, the yearly height performances of the woody species were noted.

During the same period, weekly visits were paid to the experimental forest site in order to note any kind of injury. First of all, date, location and species of fallen trees were noted. The initial canopy gap opening was recorded on a map, as were the location of the trunks of the surrounding trees and the spatial coverage of their crowns. A profile diagram was drawn showing shoot position, height; species' names were recorded. A horizontal survey of ground vegetation, the "herb layer", was conducted on the basis of permanent plots (1 m^2). Changes were deduced from inventories realized in 1984, 1985, 1986, 1987, 1990 and 1991. Three fires occurred during our survey: in August 1983, October 1984 and September 1990; each time the burned area was mapped (Fig. 10).

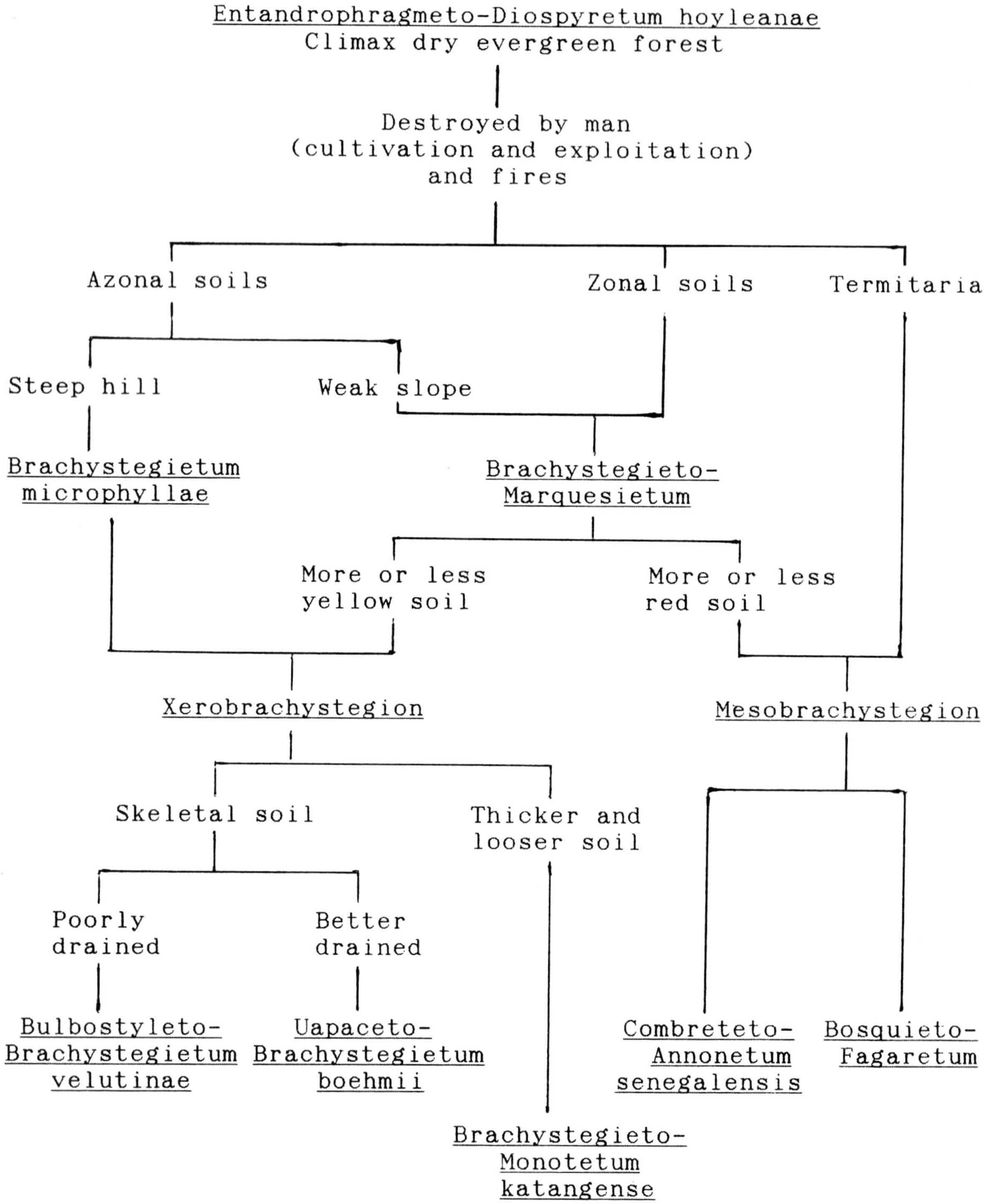

Fig. 5. The different terrestrial phytosociological units in Upper Shaba and their connections (after Schmitz 1971, modified).

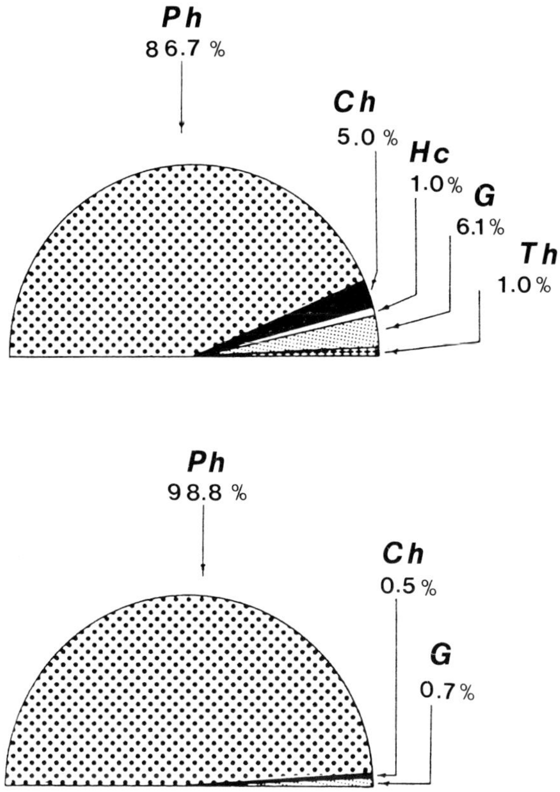

Fig. 6. Raw and balanced spectra of the biological types for the Luiswishi dry evergreen forest: Ph = phanerophyte, Ch = chamephyte, Hc = hemicryptophyte, G = geophyte and Th = therophyte (after Malaisse 1984).

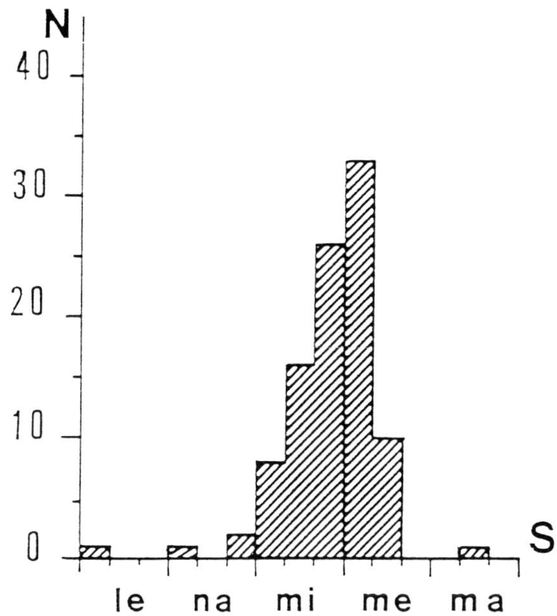

Fig. 7. Raw spectrum of leaf area for a Zambezian dry evergreen forest (classes according to Raunkiaer (1934); subclasses to Taylor (1975)): N = % age, S = surface, le = leptophylle, na = nanophylle, mi = microphylle, me = mesophylle and Ma = macrophylle (after Malaisse 1984).

Results

Regeneration potentials

A first set of data relating to dispersal units of Zambezian D.E.F. is compiled in Table 1. Some species appear to suffer heavy consumption. Prime examples are *Grewia schmitzii, Bridelia duvigneaudii, Rourea orienalis* and *Entandrophragma delevoyi*. The seed basis of the last species is appreciated largely by rodents.

Counting of seedling density leads to high local variability (40 ± 30 individuals m^2 at the beginning of the dry season) of which 14 to 72% disappear before or during the second year of life.

Height values of seedlings and saplings are presented in Table 2. The generally low values should be noted. On an average, seedlings measure 5 cm at the end of the first year and 9 cm at the end of the second year. It is only after six years that some saplings (mainly woody climbers) reach heights of more than 1 m.

Unfortunately, no values are available for seed banks.

Gap dynamics

Observations carried out at the Luiswishi site can be condensed into the following general remarks.

1. Gap dynamics vary strongly according to the type of injury considered (Fig. 11). On the one hand, gaps may result from mechanical action. As with Riera et al. (1990) we distinguish a gap formed by the fall of a single branch ("volis" in French) or a broken crown or trunk, the fall of a single tree (with or without a heaving of its basis) or of a dead snag, from windfallen multiple trees and windfallen tree complexes shifted in time. At Luiswishi, we never observed zones

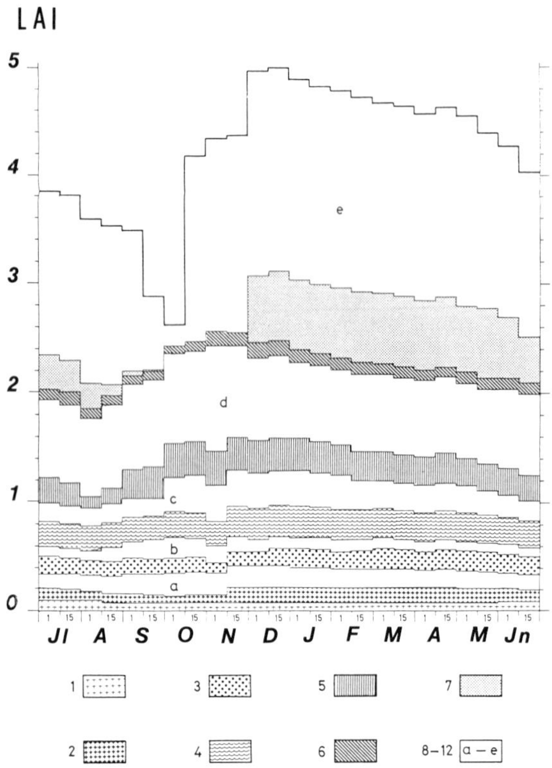

Fig. 8. Fortnightly leaf area index in the Luiswishi dry evergreen forest (climatic year 1982–83). 1: *Apodytes dimidiata* subsp. *dimidiata*, 2: *Diospyros hoyleana* subsp. *hoyleana*, 3: *Tabernaemontana pachysiphon*, 4: Other species, 5: *Entandrophragma delevoyi*, 6: *Brachystegia taxifolia*, 7: *Combretum acutifolium*, 8(a): *Parinari excelsa*, 9(b): *Combretum gossweileri*, 10(c): *Syzygium guineense* subsp. *afromontanum*, 11(d): *Aidia micrantha* var. *msonju*, 12(e): *Brachystegia spiciformis* var. *schmitzii* (original, data from Malaisse 1984).

where roots were torn from the soil leaving a portion of bare soil or mineral soil exposed. Gaps were mostly caused by falling broken trunks and the remaining basis measuring between 0.8 and 3.5 m height. Falls resulted (a) from the weight of a large number of vines that grow over the canopy of bended trees, (b) from the development of a bees' nest (mainly *Apis mellifica adansonii*, accessorily *Meliponula bocandei* or *Trigona erythra interposita*) or (c) from lightning.

On the other hand, fire also induces strong perturbations. The devastating effects of recurrent fires have frequently been underlined. In fact, only late fires have to be considered. D.E.F. patches are most frequently surrounded by a strand of *Aframomum stipulatum*, ten or twenty meters in width, dominating a loosely scattered carpet of *Setaria thermitaria* clumps. This herbaceous layer differs from the surrounding savannah (or woodland) grass layer not only by a smaller biomass (68%), but also by a higher hydration. In mid-June 1985, hydration amounted to 20.1 and 244.6% for the savanna grass layer and the *Aframomum* belt, respectively. Therefore, early fires generally do not approach D.E.F. Late fires, poked by south-east trade-winds, are, however, able to burn the *Aframomum* belt and to reach the forest skirt. Fire rarely seems to penetrate deeply into D.E.F. The effect of fire is mainly confined to the edges, where, in the dry season, the earlier herbaceous vegetation and the subsequent thicket-like regeneration provide ample inflammable material which allows the fire to reach a short distance into the "muhulu". Inside the forest, the absence of a herbaceous layer and the shelter from wind provided by the canopy presumably stop the spreading of fires (Cottrell and Loveridge 1966). Figure 8 supports these views as does the frequency of fire-scarred trees occurring at the forest edges. Besides surface fires, we once noted a ground fire — fire in organic layer (Trollope 1978) — progressing few cm per hour in the humus layer. It was restricted to a small area of 250 m^2. The passage of such a fire reduces root competition and increases the availability of mineralized soil nutrients.

From our observations it appears that forest resilience still exists after one or two severe fires, but three or more annual repeated fires lead to the total destruction of the D.E.F.

2. Gap dynamics should be differentiated according to the areas involved since gap size affects germination and survival through its influence on physical factors.

For small gaps — of up to 250 m^2 — resulting from the fall of branches or trees, closing of canopy will take about 45 years. Regeneration results more from coppices and root-suckers than from seedlings and saplings. Medium flexible shrubs may continue to grow despite having been bent to the ground. For instance, the abundance of orthotropic shoots sent up from curved stems of *Diospyros hoyleana* has been noted. In those small clearings, the mat of roots and superficial mycorrhizae of the mature community is

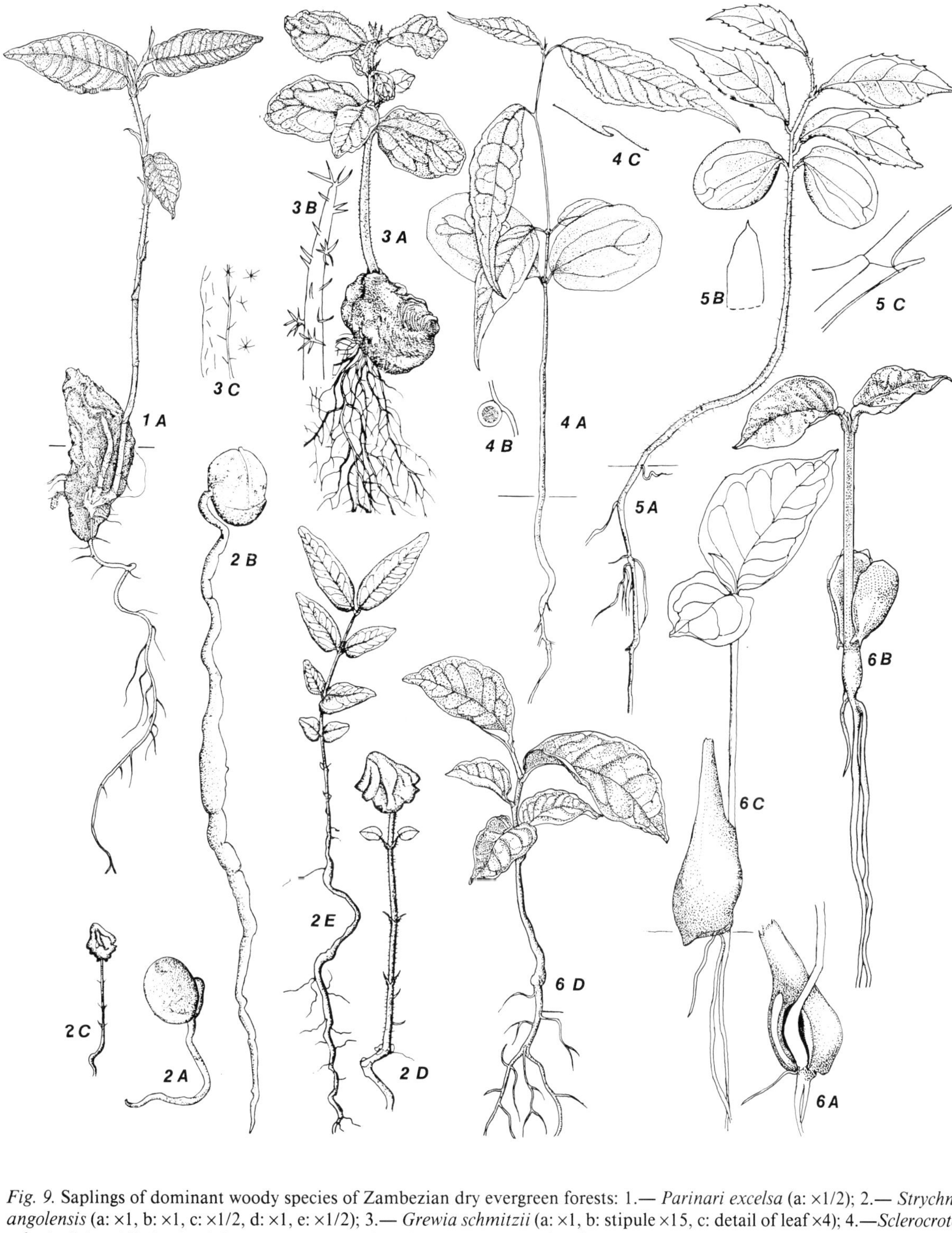

Fig. 9. Saplings of dominant woody species of Zambezian dry evergreen forests: 1.— *Parinari excelsa* (a: ×1/2); 2.— *Strychnos angolensis* (a: ×1, b: ×1, c: ×1/2, d: ×1, e: ×1/2); 3.— *Grewia schmitzii* (a: ×1, b: stipule ×15, c: detail of leaf ×4); 4.—*Sclerocroton schmitzii* (a: ×1/2, b: glandular spot ×11, c: tooth ×12); 5.—*Rawsonia lucida* (a: ×1, b: stipule ×12, c: tooth ×7); 6.—*Entandrophragma delevoyi* (a: ×1/2, b: ×1/2, c: ×1/2, d: ×1/2). (Voucher references Malaisse 12952 (1), 13943 (2), 12292 (3), 13961 (4), 13944 (5) and Blancke 53 (6); drawings by Ngoy Lunda Abidja.)

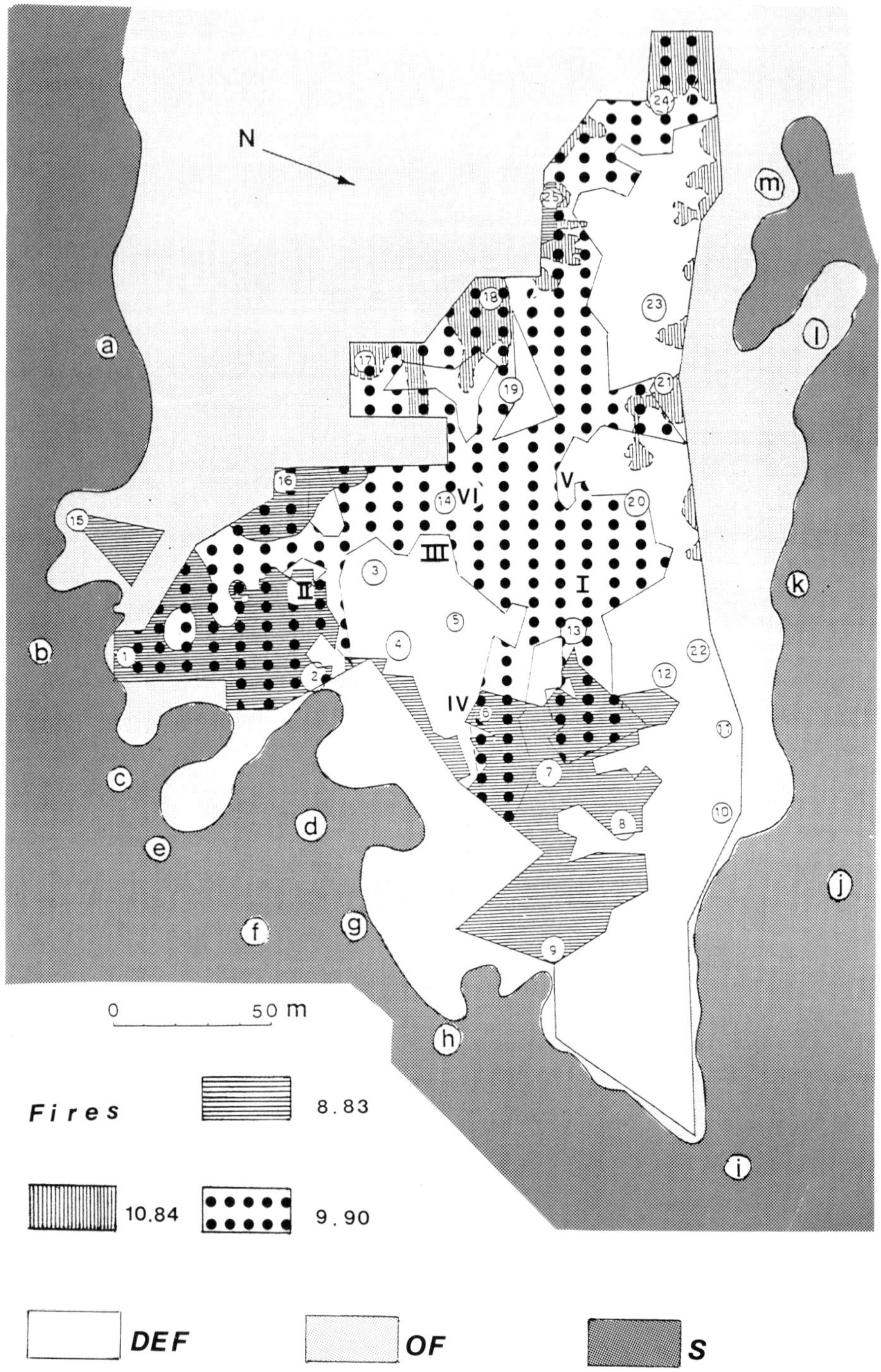

Fig. 10. Map of a dry evergreen forest patch at Luiswishi (surroundings of Lubumbashi, Zaire): Position of gaps (I—VI), termite mounds in (1—25) and outside (a—m) the forest and successive burned areas; DEF = dry evergreen forest, OF = open forest, S = savanna.

Table 1. Characteristics of dispersal units for some Zambezian dry evergreen forests species

	(a)	(b)	(c)	(d)	(e)
Adenia lobata	XI	1.5	0.0	0.0	100.0
Apodytes dimidiata subsp. *dimidiata*	XI	10.8	0.0	0.0	100.0
Bridelia duvigneaudii	X	6.6	0.0	60.0	40.0
Daturicarpa elliptica	V	7.5	3.6	3.6	92.9
Dichapetalum bangui	X	20.9	0.0	0.0	100.0
Diospyros hoyleana subsp. *hoyleana*	XII	28.6	3.4	0.0	96.6
Diospyros mespiliformis	VIII	29.1	1.9	0.6	97.5
Diospyros undabunda	X	35.1	–	–	–
Entandrophragma delevoyi	IX	18.8	3.7	41.2	55.1
Grewia schmitzii	VII	5.3	7.4	55.5	37.0
Leptoderris nobilis	X	12.9	8.9	8.2	82.9
Marquesia macroura	IX	1.0	4.5	7.5	88.0
Opilia celtidifolia	XI	141.8	21.8	1.8	76.4
Rhynchosia hirta	IX	6.3	–	–	–
Rothmannia whitfieldii	XI	8.2	18.4	0.0	81.6
Rourea orientalis	XI	72.0	42.1	9.1	48.8
Sclerocroton schmitzii	VI	3.8	0.0	26.8	63.2
Sorindeia katangensis	XII	26.8	15.4	14.1	70.5
Strophanthus welwitschii	X	1.6	3.3	1.1	95.6
Strychnos angolensis	XI	38.0	0.0	0.0	100.0
Strychnos lucens	X	18.9	0.0	0.0	100.0
Syzygium guineense subsp. *afromontanum*	—	29.1	–	–	99.5

(a) Collecting month.
(b) Fresh weight (g) of 100 seeds or fruits.
(c) Imperfectly developed diaspores (%).
(d) Diaspores attacked (%).
(e) Germinative faculty (%).

so closed that only seedlings with roots integrated into the mat can survive (Gomez-Pompa and Vasquez-Yanes 1976). Restoration of the forest will take much longer, at least 90 years.

When canopy opening is followed by surface fire, which corresponds to "slash-and-burn" agriculture, D.E.F. species diversity is decreasing with the loss of fire-sensitive species. Trapnell (1959) recognized an ecological group of fire-sensitive species, usually evergreen, present in mateshi and unable to withstand dry season fires. For north-eastern Zambia, the list of plants belonging to this group was completed by Lawton (1978) to a total of 17 canopy trees, 14 understorey trees, 20 shrubs and scandent shrubs, 12 climbers, 2 suffrutices and 1 herb. Most of these plants are found in Shaba.

On the other hand, herbaceous aliens take over in the grass layer of the gap. Our surveys allow to recognize several ecological groups according to their particular strategies.

A first lot of species appears immediately after fire during the next rainy season and quickly occupies the surface of the gap. The respective species are easily dispersed by wind. Their germination is induced by increases in light, temperature and ash. Examples are *Pteridium aquilinum*, the bracken, and *Crassocephalum rubens* (Fig. 12). A second group develops later, but is still typical of post fire herbaceous vegetation. The plants included in this group belong to the soil-seed bank and are dispersed by zoochory, mainly rodents as is the case of *Aframomum stipulatum*. Other plants, able to survive at

Table 2. Mean height (cm) of 32 seedlings and saplings of Zambezian dry evergreen forests (* new stems, ° "in situ" measurements, ** values for vines in absence of any support, S_1 = seedling one year old, S_{12} = sapling twelve years old)

Species	S_1	S_2	S_3	S_4	S_5	S_6	S_7	S_{11}	S_{12}
Albizia adianthifolia	5.6 ± 1.0°	–	–	–	–	–	–	–	–
Artabotrys monteiroae	6.1 ± 0.3°	–	–	9.2 ± 2.0	–	–	–	–	–
Baphia bangweolensis	–	–	14.8 ± 6.2	51.3 ± 23.0	90.4 ± 56.0	101.3 ± 57.8	–	356.4 ± 72.6	–
Bequaertiodendron magalismontanum	–	7.8 ± 1.8	–	–	–	–	–	–	–
Brachystegia taxifolia	–	–	7.3 ± 2.5*	20.6 ± 10.6	29.9 ± 16.9	49.6 ± 27.9	–	196.0 ± 31.4	–
Cissus petiolata	5.2 ± 0.5°	–	8.3 ± 3.1	12.2 ± 4.3	12.9 ± 5.8	–	–	–	–
Combretum acutifolium	–	9.3 ± 2.6	–	–	–	–	–	–	–
Craterosiphon schmitzii	–	12.7 ± 3.9	14.9 ± 3.1	–	–	–	–	–	–
Dichtyophleba lucida	–	–	6.1 ± 2.0	–	–	–	–	–	–
Dichapetalum bangii	5.3 ± 0.9	8.2 ± 1.2	8.7 ± 2.9	–	–	–	–	–	–
Diospyros hoyleana subsp. *hoyleana*	–	–	7.2 ± 0.9	–	–	–	–	–	–
Diplocyclos decipiens	5.6 ± 2.9	–	–	–	–	–	–	–	–
Entandrophragma delevoyi	6.1 ± 1.4	7.3 ± 1.6	9.2 ± 2.1	–	–	–	–	–	–
Erythrophleum suaveolens	–	11.5 ± 2.7	–	–	–	120.7 ± 60.4	180.0 ± 79.5	–	666.7 ± 101.8
Grewia schmitzii	3.3 ± 0.1	–	–	–	–	–	–	–	–
Guibourtia coleosperma	4.6 ± 0.5	10.7 ± 1.5	–	–	–	–	–	–	–
Leptoderris nobilis	4.4 ± 1.5	5.2 ± 1.8	6.9 ± 2.6	15.9 ± 9.5	21.4 ± 28.4	110.9 ± 124.8	–	229.0 ± 25.5**	–
Landolphia eminiana	9.7 ± 2.5	–	10.3 ± 0.7	–	22.0 ± 9.1	31.4 ± 13.6	–	–	–
Landolphia parvifolia var. *parvifolia*	–	–	–	–	9.4 ± 2.2	12.0 ± 3.4	–	–	–
Ochna puberula	5.9 ± 1.1	–	–	–	–	–	–	–	–
Opilia celtidifolia	–	10.5 ± 1.7	–	–	–	–	–	–	–
Parinari excelsa	15.1 ± 3.3°	–	–	–	–	–	–	–	–
Rhynchosia hirta	–	20.5 ± 22.8	23.8 ± 17.7	–	144.5 ± 41.0	288.0 ± 46.0	398.0 ± 75.2	–	650.0 ± 32.6
Ritchiea quarrei	–	–	5.0 ± 1.0	–	–	–	–	–	–
Rothmannia whitfieldii	–	5.2 ± 1.5	–	–	–	–	–	–	–
Rourea orientalis	4.2 ± 1.1	7.1 ± 1.5	8.7 ± 2.4	–	–	–	–	–	–
Sclerocroton schmitzii	9.4 ± 1.9	11.3 ± 2.5	13.7 ± 2.6	–	–	–	–	–	–
Sorindeia katangensis	5.2 ± 1.3	–	11.0 ± 2.4	15.9 ± 3.8	–	26.8 ± 8.4	–	170.0 ± 23.0	–
Strophanthus welwitschii	–	–	3.0 ± 0.9	–	–	–	–	–	–
Strychnos angolensis	–	–	9.8 ± 2.3	–	–	–	–	–	–
Tabernaemontana pachysiphon	3.8 ± 0.2	–	–	–	–	–	–	–	–
Uvaria angolensis var. *angolensis*	5.2 ± 0.5	–	–	–	–	–	–	–	–

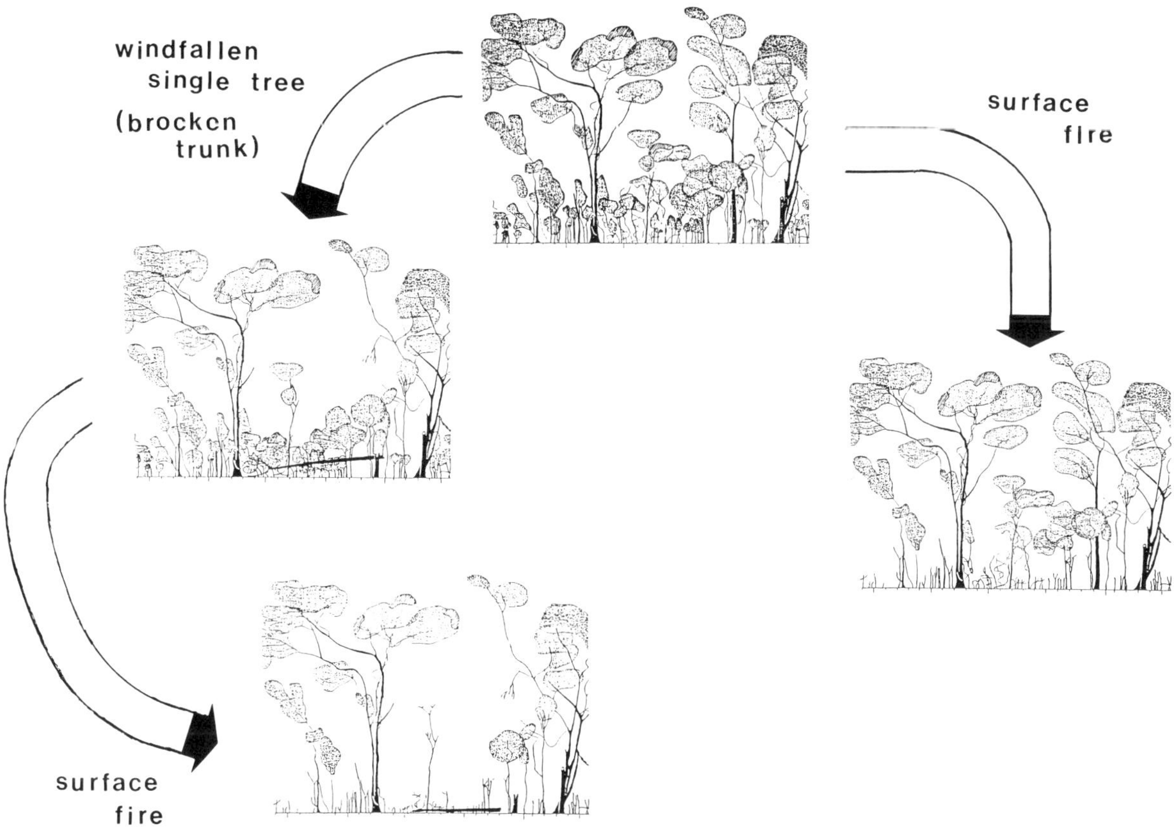

Fig. 11. Gap dynamic in Zambezian dry evergreen forests according to the type of injury.

very low densities under thick cover, explode with the opening of the canopy in places where sunlight penetrates to the ground. Examples are *Panicum lineatum* and *Bromuniola gossweileri.* According to Lawton (1964), the last species is a survivor from a forest that existed under a cooler, wetter climate. A later stage may be dominated by secondary shrubs or trees of low stature. At Luiswishi, only *Harungana madagascariensis, Trema orientalis* and *Kotschya strigosa* were recorded as representatives of this pioneer group.

Species belonging to the climax vegetation depend more on shoots than on seedlings for colonization and take over after 3 to 5 years, in absence of new fires. Species dynamics may vary according to the type of injury concerned. *Diospyros hoyleana*, the dominant shrub, can be studied as a good example. It should also be noted that the basic dynamic patterns described above are repeated throughout the region.

Patterns of restoration

From a sylvicultural point of view, the following recommendations can be derived from our observations:

1. As it seems impossible to suppress bush fires, each remnant D.E.F. must be protected from fire. This can be done by burning a fire break around it early in the dry season, in any case before the end of June. This will prevent the very destructive late fires, often started by hunters or honey gatherers.

2. Selective logging should be preferred to clear-cutting. This strategy ensures that canopy gap sizes are minimized and that suitable site-conditions are maintained. Moreover, it is in accordance with villager's mentality. It avoids the development of large carpets of dense herbaceous vegetation (*Aframomum stipulatum, Pteridium aquilinum, Smilax kraussiana* sociations) which provides abundant ignition mate-

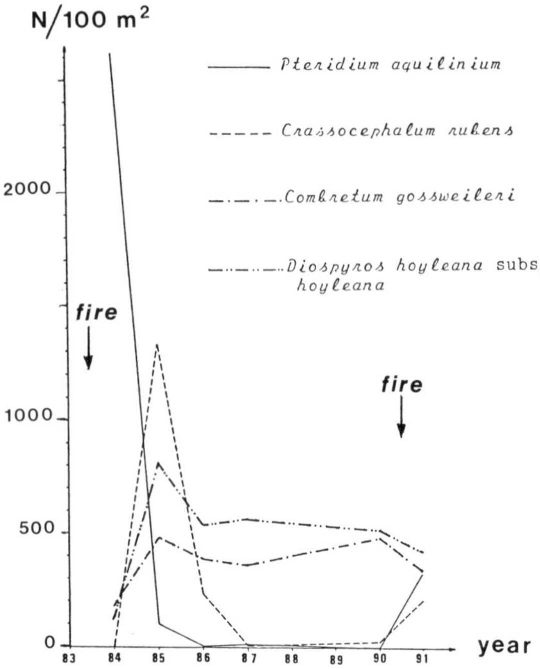

Fig. 12. Post fire densities (number per 100 m^2) of four species in a Zambezian dry evergreen forest gap.

rial and enhance the danger of repeated severe fires.

3. Gap enrichment with local trees should be recommended and preferred to planting of exotic trees. Saplings of *Erythrophleum guineense* and *Parinari excelsa*, belonging to the D.E.F. community are the most indicated. The first species is reported to prosper exceedingly well under openings of the canopy in the vicinity of Ndola and Kitwe in Western Province of Zambia (Fanshawe 1960). *Maesopsis eminii, Marquesia acuminata* and *Parkia filicoidea* are also highly promising species. The first has occasionally been observed in Zambian riparian forests and germinates well. *Marquesia acuminata* is a tall evergreen tree often reaching a height over 30 m, usually fluted for about 2–3 m up the bole; it may occur in dense pure stands and is able to regenerate and maintain a forest community under existing conditions (Lawton 1964). *P. filicoidea* is a fast-growing tree. *Entandrophragma delevoyi* is another possible tree for enrichment plantings. Lawton (1964) reports its presence on open areas in "chipya vegetation", which proves its heliophily. The tree grows slowly, but its wood quality is remarkable. The same holds true for the ten other species of this major timber genus distributed throughout the tropical lowlands of Africa. This species is threatened with extinction in its areas of natural distribution (Hedberg 1979) as a result of exploitation. It is easily killed by fierce fires, and there is evidence that it was a more important component of the canopy in the past (Lawton 1964).

4. Another pathway of restoration consists of the use of woody climbers, mostly with edible fleshy fruits such as berries and drupes. They are good pioneers and should be used in order to initiate the progressive succession. They produce a dense surface cover competing with success against the grass layer. This cover depresses grass necromass, reducing the danger of fire passage. Moreover, woody climbers improve the amount of litter, inducing a higher carbon content of the soil. In normal situations, their seedlings (*Landolphia* spp., *Uvaria angolensis, Combretum gossweileri*) form a particularly noticeable part of the herb layer (Lawton 1964). Climbers play an important role in recolonization as well as in reinforcing young regrowths. Their quick growth and their dark foliage close skirts and the upper level of gaps (Schmitz 1971).

Furthermore, the pleasant taste of their fruits, their appetency, as well as their nutritional value (Malaisse and Parent 1985) make these plants attractive to the inhabitants. Moreover these evergreen thickets will serve as refuge for a lot of animals, including monkeys, rodents (Malaisse and Parent 1982) and frugivorous birds, which are also observed on termite hill thickets and are active in zoochory (Malaisse and Anastassiou-Socquet 1983). Fanshawe (1960) already noted that in the copper belt of Zambia, any site with a dense, evergreen thicket is a potential D.E.F.

Discussion

The fact that the Zambezian D.E.F. relics are extremely interesting to ecologists and phytogeographers has already been underlined by Fanshawe (1960). From a forestry point of view their restoration and protection are urgent. It has been possible to provide a framework for the interpretation of the gap dynamic as well as to derive the main patterns of restoration.

In absence of agriculture, our ecological unit, the wetter northern part of the Zambezian region, presents an environment that permits a density of one man per 7 to 10 km^2 in a hunter-gatherer subsistence economy (Malaisse 1991). The actual human density is about 6 inhabitants per km^2. With the foundation, and development of the old colonial mining towns large cities have developed, for example, Lubumbashi, Kolwezi and Likasi on the Zairian side with 700 000, 300 000, and 210 000 inhabitants, respectively. The same applies to the Zambian copper belt with Ndola, Kitwe, Chingola and Chililabombwe. As most of the citizens are either not attached to electricity or unable to pay for it, they are more and more dependent on wood (firewood and charcoal) for cooking all the year round and warming during the cold dry season. For instance, the daily input of Lubumbashi town amounts to 1700 charcoal bags of 48 kg! (Malaisse and Binzangi 1985). Therefore, deforestation progresses. The surrounding woodlands are replaced by savannas. The mean radius of savannization, "the white belt of *Imperata cylindrica*", scattered with patches of *Pennisetum polystachyon* on charcoal oven places, progresses about one km each year. At present, this radius extends about 40–50 km around most of the Zambezian towns (Binzangi 1989). To prevent further exploitation, the protection and restoration of Zambezian D.E.F. and woodlands are urgently needed. In the absence of any forestry service, as far as Zaire is concerned, the only way of action goes through local village inhabitants and rulers. Unfortunately, more and more of them are involved in charcoal production (Lootens-De Muynck et al. 1982).

Therefore, we believe that only a well-supported political willingness coupled with a real improvement of the local inhabitants' way of living will provide a suitable framework for a sustainable restoration of these forests.

Acknowledgements

We are indebted to Kisimba Kibuye and Muzinga Yumba for assistance in field observations at the Luiswishi site, to Ngoy Longwa and Ngoy Lunda for drawings of seedlings. Ir. V. Blancke realized the 1990 survey of burned area as well as gap survey in March 1991. Mrs. V. Bouyer-Renard and Mrs. A. Bodenghien provided strong support and logistic facilities. Their kind assistance is acknowledged.

References

Benson, C. W. & Stuart Irwin, M. P. 1965. The birds of Marquesia thickets in Northern Mwinilunga district, Zambia. Arnoldia (Rhodesia) 1(30): 1–4.

Binzangi, K. 1989 Contribution à l'étude du déboisement en Afrique tropicale: le cas du Shaba méridional. Thèse Doct., Univ. Lubumbashi, Lubumbashi, 236 pp.

Chapman, J. & White, F. 1970. The evergreen forests of Malawi. Oxford, Commonw. For. Inst., 190 pp.

Cottrell, C. B. & Loveridge, J. P. 1966. Observations on the Cryptosepalum forest of the Mwinilunga district of Zambia. Proc. Trans. Rhod. Sci. Ass. 51: 79–120.

Dikumbwa N'Landu. 1991. L'impact des facteurs écoclimatiques sur les cycles biogéochimiques en forêt dense sèche à *Entandrophragma delevoyi* De Wild. au Shaba méridional (Zaïre). Thèse Doct., Univ. Liège, Liège, 210 pp.

Duvigneaud, P. 1958. La végétation du Katanga et de ses sols métallifères. Bull. Soc. Roy. Bot. Belg. 90: 127–186.

Fanshawe, D. B. 1960. Evergreen forest relics in Northern Rhodesia. Kirkia 1: 20–24.

Fanshawe, D. B. 1969. The vegetation of Zambia. Kitwe, Min. Rural Dev., For. Res. Bull. 7: 67 pp.

Golley, F. B., McGinnis, J. T., Clements, R. G., Child, G. I. & Duever, M. J. 1975. Mineral Cycling in a Tropical Moist Forest Ecosystem. Univ. Georgia Press, Athens, 248 pp.

Gomez-Pompa, A. & Vasquez-Yanes, C. 1976. Estudia sobre sucesion secundaria en los Tropicas Calido-Humedos: el ciclo de vida de las especies secundarias. In: Gomez-Pompa, A., Vasquez-Yanes, C., del Amo Rodriguez, S., Butanda Cervera, A. (Eds.), Regeneracion de Selvas. Compania Editorial Contineneta1 S.A., Mexico: 579–593.

Hedberg, I. 1979. Preliminary list of rare and threatened species in Zaïre. In: Hedberg, I. (Ed.), Systematic Botany, Plant Utilization and Biosphere Conservation. Almqvist & Wiksell, Stockholm.

Lawton, R. M. 1963. Palaeoecological and ecological studies in Northern Province of Northern Rhodesia. Kirkia 3: 46–77.

Lawton, R. M. 1964. The ecology of the Marquesia acuminata (Gilg) R.E.Fr. evergreen forests and the related chipya vegetation types of North-Eastern Rhodesia. J. Ecol. 52 (3): 467–479.

Lawton, R. M. 1978. A study of the dynamic ecology of Zambian vegetation. J. Ecol. 66 (1): 175–198.

Lootens-De Muynck, M. T., Mbuyi, B. D. & Binzangi, K. 1982. Typologie et aspects socio-économiques de quelques villages de bûcherons et charbonniers des environs de Lubumbashi (Shaba, Zaïre). Geo-Eco-Trop 6 (1): 45–63.

Malaisse, F. 1974. Phenology of the zambezian woodland area with emphasis on the miombo ecosystem. In: Lieth, H. (Ed.), Phenology and Seasonality Modeling. Springer, Berlin-Heidelberg-New York.

Malaisse, F. 1984. Structure d'une forêt dense sèche zambézienne des environs de Lubumbashi (Zaïre). Bull. Soc. Roy. Bot. Belg. 117 (2): 428–458.

Malaisse, F. 1985. Comparison of the woody structure in a regressive Zambezian succession with emphasis on high termitaria vegetation (Luiswishi, Shaba, Zaire). Bull. Soc. Roy. Bot. Belg. 118 (2): 244–265.

Malaisse, F. 1991. Apport et avenir des produits sauvages comestibles dans l'alimentation des villageois en région zambézienne. In: Les petits paysans dans les pays en voie de développement. Conseil Interuniversitaire de la Communauté française de Belgique: 165–175.

Malaisse, F. & Anastassiou-Socquet, F. 1983. Spectres biologiques, surfaces foliaires, types de fruit et modes de dissémination des diaspores des plantes des hautes termitières établies en forêts claires du Shaba méridional. Ann. Fac. Sc. Lubumbashi 3: 13–39.

Malaisse, F. & Binzangi, K. 1985. Wood as a source of fuel in Upper Shaba (Zaire). Commonw. For. Rev. 64 (3): 227–239.

Malaisse, F. & Colonval-Elenkov, E. 1981. Ecomorphologie et anatomie des feuilles des forêts denses sèches zambéziennes. Bull. Soc. Roy. Bot. Belg. 114 (2): 209–228.

Malaisse, F., Freson, R., Goffinet, G. & Malaisse-Mousset, M. 1975. Litterfall and litter breakdown in miombo. In: Golley, F. & Medina, E. (Eds.), Tropical Ecological Systems; Trends in Terrestrial and Aquatic Research. Springer, New York-Heidelberg-Berlin.

Malaisse, F. & Kapinga, I. 1987. The influence of deforestation on the hydric balance of soils in the Lubumbashi environment (Shaba, Zaïre). Bull. Soc. Roy. Bot. Belg. 119 (2): 179–191.

Malaisse, F. & Parent, G. 1980. Les chenilles comestibles du Shaba méridional (Zaïre). Les Naturalistes Belges 61 (1): 2–24.

Malaisse, F. & Parent, G. 1982. Rodents of the miombo woodland area: a nutritional and ecological approach. Ecol. Food Nutrition 11 (4): 211–216.

Malaisse, F. & Parent, G. 1985. Edible wild vegetable products in the Zambezian woodland area: a nutritional and ecological approach. Ecol. Food Nutrition 18 (1): 43–82.

Malaisse-Mousset, M. 1974. Premier aperçu sur la microflore de quelques sols du Haut-Shaba (Zaïre). Pédologie (Gand) 24 (1): 64–70.

Oatley, T. B. 1969. Bird ecology in the evergreen forests of North-Western Zambia. Puku 5: 141–180.

Raunkiaer, C. 1934. The Life Forms of Plants and Statistical Plant Geography. Clarendon Press, Oxford.

Riera, B., Puig, H. & Lescure, J.-P. 1990. La dynamique de la forêt naturelle. Bois et Forêts des Tropiques 219: 69–78.

Schmitz, A. 1962. Les muhulu du Haut-Katanga méridional. Bull. Jard. Bot. Etat Brux. 32: 221–299.

Schmitz, A. 1971. La végétation de la plaine de Lubumbashi (Haut-Katanga). Publ. I.N.E.A.C., Sér. Sc. 113: 388 pp.

Schmitz, A. 1988. Revision des groupements végétaux décrits du Zaïre du Rwanda et du Burundi. Musée royal de l'Afrique centrale, Tervuren (Belgique). Ann. Sc. Econ. 17: 318 pp.

Taylor, S. 1975. Optimal leaf form. In: Gates, D. & Schmerl, R. (Eds.), Perspectives of Biophysical Ecology. Ecol. Stud. 12. Springer, New York-Heidelberg-Berlin.

Trapnell, C. G. 1953. The Soils, Vegetation and Agriculture of North-Eastern Rhodesia. Lusaka.

Trapnell, C. G. 1959. Ecological results of woodlands burning experiments in Northern Rhodesia. J. Ecol. 47 (1): 129–168.

Trollope, W. S. W. 1978. Fire behaviour — a preliminary study. Proc. of the Grassland Soc. of Southern Africa 13: 123–128.

van der Pijl, L. 1982. Principles of Dispersal in Higher Plants. 3rd ed. Springer, Berlin.

White, F. 1981. Vegetation map of Africa (scale 1:5 000 000). UNESCO/AETFAT/UNSO. Oxford Univ. Press, Oxford.

White, F. 1983. The vegetation of Africa. A descriptive memoir to accompany the UNESCO/AETFAT/UNSO vegetation map of Africa.

8. Problems associated with raising trees from seeds
The Ethiopian experience

D. TEKETAY
Swedish University of Agricultural Sciences, Faculty of Forestry, Department of Forest Site Research, S-901 83 Umeå, Sweden; Permanent address: Alemaya University of Agriculture, Faculty of Forestry, P.O. Box 138, Dire Dawa, Ethiopia

Key words: Afforestation/reforestation, artificial regeneration, nursery problems, seed biology/ecology

Abstract. Ethiopia is one of the countries in which plant resources are vanishing at an alarming rate. This contributes to the improverishment of ecosystems, affecting agricultural production, wildlife and people's welfare. The remaining small patches of relict forests will disappear in the very near future, if the current trend of their utilization is not changed. One of the alternatives which would also help to restore the devastated forest areas is artificial regeneration of forests. Despite various difficulties, efforts are being made to enrich the remaining natural forests, to reforest devastated areas and to establish community forests and plantations. Most afforestation employs exotic species, because they are easy to raise in nurseries and that they are fast growing once they are established in the field while indigenous species are difficult to raise and relatively slow growing. Despite these advantages, exotic species may have some risks, and problems have developed in some exotic plantations. Habits, habitats/ecology, means of propagation, uses, distribution and remarks on the germination of woody species, mostly trees, are presented for many indigenous species. Prevention of deforestation, restoration of degraded forest lands, and the protection and conservation of the remaining forests, require goodwill and sacrifices in terms of labor, time and money from every citizen, dedication from concerned bodies in the government and support from the international society.

Introduction

Tropical and adjacent regions house about two-thirds of all angiospermic species known (Heywood, 1985). However, the pressure on these plant resources, especially forests, has reached a very critical stage leading to their destruction, even before we have a chance to collect, describe and study them thoroughly. The unabated transformation and devastation of tropical wildlands has become one of the most pressing issues of our times. Not only are the rates of deforestation very high, but also approximately 40% of the existing forest areas have been degraded in recent times (Bawa et al. 1990). The reasons for the destruction of tropical plant resources are associated with population growth coupled with clearing/burning for more land to cultivate and settlement as well as to obtain wood for fuel, construction, household and farm equipment, cutting for timber, pulp, food and medicine, economic difficulties and inflation leading to dependence on wood as source of income, overgrazing and road construction.

The plant resources in the tropics are important not only to the countries which own them but also they are economically, socially and environmentally important to the world as a whole (FAO 1982). Hence, there is an urgent need for harmonizing national and international efforts to conserve and rationalize the management and utilization of these resources. To do this, there are three crucial interrelated issues one has to address in the management of indigenous forests: degradation and depletion of resources, regeneration and restoration of forest ecosystems and conservation of genetic resources (Bawa et al. 1990). The challenges generated by the reduction and degradation of forest cover can be adequately met only if serious attempts are made to restore forest ecosystems. Restoration inevitably must involve improved reforestation of degraded lands through plantation of native/exotic species and the extension of forest boundaries by artificial and natural regeneration.

Restoration of degraded lands through plantation of native species and extension of bound-

H. Lieth and M. Lohmann (eds.), Restoration of Tropical Forest Ecosystems, 91–100.

aries by artificial regeneration depends upon the possibilities of raising seedlings either from seeds or vegetative propagules. Raising seedlings from seeds poses some problems as it is governed by a number of factors. Our knowledge about these factors is very limited. Basic information about seed biology of native plants in Ethiopia is very scant or absent. At a time when many of these species or their habitats are being threatened, it is our urgent task to learn more about them and use the information thus gained to propagate, utilize and conserve such species.

In this paper, I would like to share my experience on difficulties being encountered in raising trees in nurseries and establishing them in the

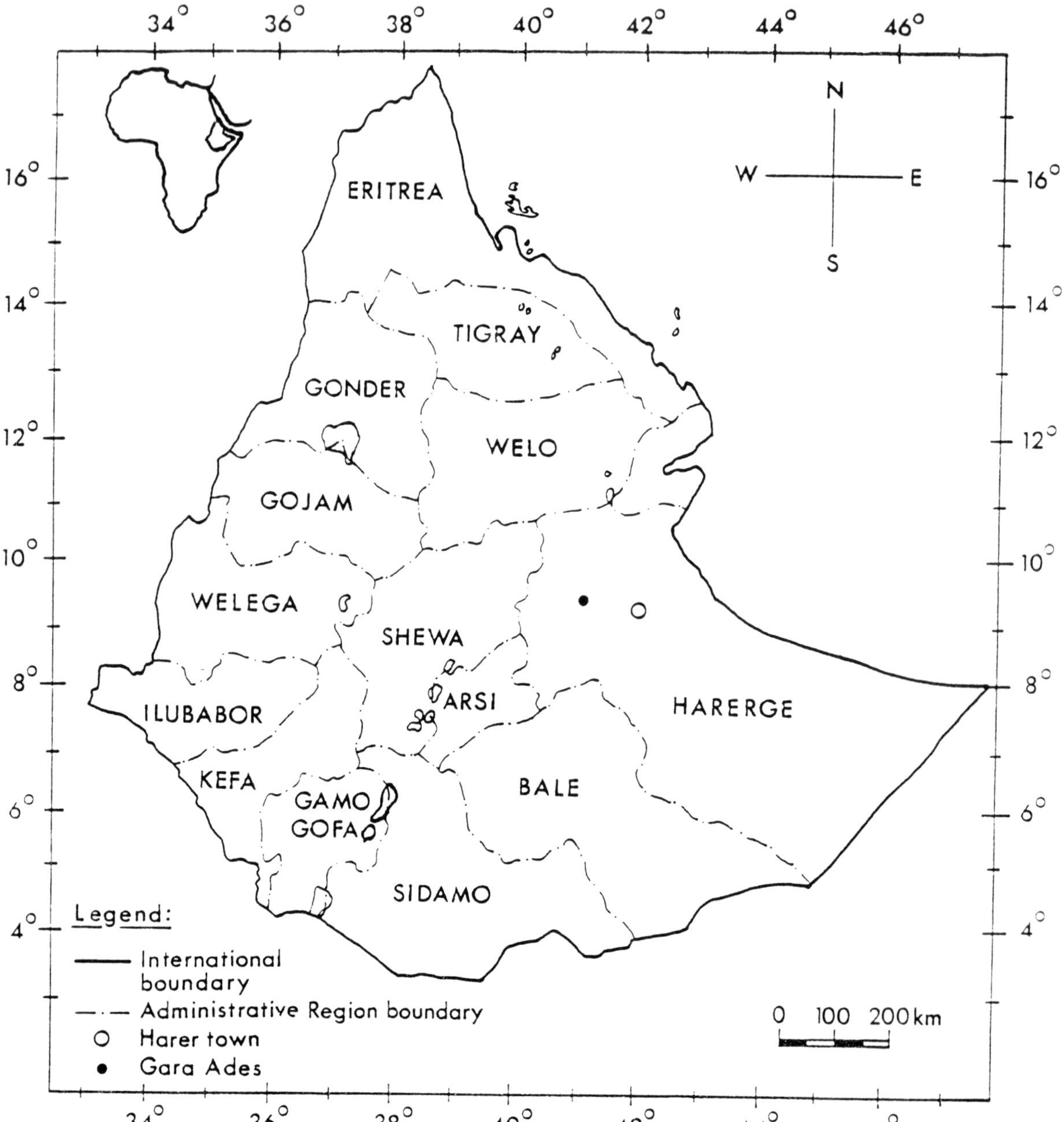

Fig. 1. The study area: Gara Ades, Southeastern Ethiopia.

field in Ethiopia, with special reference to Harerge, southeastern Ethiopia (Figure 1).

General conditions of forests in Ethiopia

There are four types of forest/tree holdings in Ethiopia at present: state owned forests, community forests owned by peasant and urban dwellers associations, forests owned by producers cooperatives and privately owned trees (trees growing or planted in farms or home gardens).

Ethiopia is one of the countries in which the forest resources are vanishing at an alarming rate. The natural forests of Ethiopia have been estimated at 27 150 000 ha, i.e. 22.2% of the country, out of which closed forests are only 4 350 000 ha (3.6%) and the rest 22 800 000 ha (18.6%) being savanna woodland (Lanly 1981; Grainger 1986). On the other hand, average annual deforestation in closed forests was estimated at 10 000 ha (Lanly 1981; Grainger 1986). The situation is similar in the open savanna woodlands. The devastation of the vegetation cover contributes to the impoverishment of the ecosystem, affecting agricultural production, wildlife and people's welfare. Possible causes of this crisis are discussed in detail by Tewolde B. G. Egziabher (1988, 1989). Unless the present trend of forest utilization is changed, every piece of relict forest remaining will be gone in the very near future. This brings us to the most important question, i.e. "How can this change be brought about?".

Some measures which could help achieve the change

The following are some of the measures that should be taken if the remaining few natural forests are to be saved.

1. Deforestation must be stopped, and the only way it can be stopped is to provide realistic alternatives for the people, especially for farmers who are totally dependent on the forests for various services. Introduction and improvement of existing agroforestry practices, establishment of community woodlots, encouraging farmers to plant trees, and promotion of multipurpose trees, are just some possible alternatives. Alternative energy, building material and forage sources should also be investigated.
2. Management plans for the natural forests should be prepared and implemented, with full community participation, to provide wood for different purposes to the people, allow natural regeneration of trees and generate income to cover expenses.
3. Representative areas of the different forest types should be set aside in protected reserves where the genetic diversity of both plants and wild animals will be conserved.
4. The biology and ecology of indigenous tree species should be studied. Particular attention should be devoted to germination strategies.
5. Wherever possible, plantations should employ multipurpose trees.
6. Better communication and education is necessary to inform the public of issues, problems and solutions in forestry and land use.
7. New afforestation programs should be initiated and existing programs strengthened.

The present paper addresses the question of afforestation.

Artificial regeneration of forests in Ethiopia

Efforts are being made to enrich the remaining natural forests, to reforest devastated areas and to establish community forests and plantations in Ethiopia. Some 30 000 ha of non-industrial plantations were established in Ethiopia between 1976 and 1980 (Grainger 1986) but this has not compensated for deforestation. These plantations have been established by the Ministry of Agriculture, cooperatives and aid agencies (e.g. CARE), using mainly exotic species, especially *Eucalyptus* and *Cupressus*. Some factors contributing to the popularity of exotics is their freedom from pests and diseases, their fast growth, and their ease of nursery propagation. However, the use of exotic species poses some risks. One example is the cracking of trees in one of the largest cypress plantations at the Munessa-Shashemane State Forestry Project in Ethiopia. In 1988 cracking of stems was observed in some 7- to 9-years-old trees (A. L. Raunio, unpublished report). Some weeks later many affected trees died. In Northern Tanzania, some 200–300 ha of cypress has been lost because of the cracks and deaths. The problem is also known in Kenya where it is associated with drought years. The exact cause has not been detected, but may be due to water stress coupled with some deficiency of micronutrients (Raunio, pers. comm.). Another example is the explained death of *Acacia saligna* trees after establishment and good early growth at Dengego hills, in the Harerge region, southeastern Ethiopia.

Problems in nurseries

The activities and achievements in nurseries have direct bearing on artificial regeneration of trees. Hence, it is appropriate to consider some of the problems encountered in nurseries, and to suggest some possible solutions. The following are some of the problems observed in Ethiopian nurseries.

Lack of appropriate technology

The seeds of trees are collected either by traditional ways of climbing them or cutting their branches and at times felling them. Climbing is in the first place a very risky and difficult task while in the second place it does not work for all the trees, especially for huge trees with their trunk diameter reaching more than a meter and their heights as high as 45 m. These features may prevent good mother trees from being collected while favoring the poor ones.

After the fruits are collected extraction and cleaning of seeds are performed by hand, sometimes in crude ways such as the use of pieces of hides to crash cones and extract seeds of *Juniperus procera* and remove the pericarp from dried fruits of *Olea africana.*

After seeds have been extracted and cleaned in laborious ways, the seeds are stored in various containers simply at room temperature whether they can retain their viability or not, for the simple reason that there are no storage facilities which help to keep seeds viable for longer periods. Consequently, considerable quantity of seed which had been obtained with great efforts may be wasted.

The shortage of transportation facilities limits the ability to collect seeds from large areas and many provenances, and to transport seedlings at the right time and to the right place.

Protection of seedlings after establishment

Millions of seedlings are reported to have been planted every year by peasant and urban dwellers associations, producer cooperatives and the state. However, many of these seedlings succumb immediately after establishment or in due time for lack of proper protection, especially in plantations owned by associations and road-side plantations. As a result, resources, labor and time are wasted for nothing.

Shortage of trained manpower

Ethiopia has insufficient qualified staff to provide competent nursery management. Few individuals have training in seed biology/ecology and plantation forestry at the professional and technical level. Consequently, little is known of the behavior of indigenous forest seed, and their requirement for germination. Thus the successful propagation of indigenous trees from seeds remains elusive.

Planatation management is also deficient. For instance, most of the successful cypress plantations in the Harerge region, some of which more than 1000 ha, have not been pruned or thinned. The species are not self-cleaning making manual pruning a necessity. Thus, though the trees may provide conservation and other values, they will lose much of their timber value.

Lack of knowledge about indigenous trees

Knowledge about the phenology of indigenous trees, germination requirements of their seeds, dispersal of propagules, natural regeneration of forests and growth characteristics of trees, etc. is either scant or lacking.

In an ongoing research, I am studying the germination ecology of woody, mostly tree, species (Table 1) from Gara Ades montane forest, 9° 20′ N and 41° 15′ E (Figure 1), in southeastern Ethiopia. Every species tested so far seems to have dormancy imposed in one or the other way, though the degree varies from species to species. The dormancy ranged from 20–100% when the

Table 1. Species being studied with their habit, habitat/ecology, means of propagation, uses distribution and remarks[a]

No.	Species/ family	Habit	Habitat/ecology	Means of propagation	Uses	Distribution	Dormancy/ germination
	A. Gymnosperms						
1	*Juniperus procera* Hochst. ex Endl. Cupressaceae	— evergreen tree up to 45 m — endangered sp.	— in highland and mountain areas, on rocks and on soils with good drainage, tolerant of widely varying soils except soils heavier than sandy clay, restricted to sites with mean annual temperature of below about 19 °C, lower mean annual temperature about 7–10 °C, optimum mean annual rainfall 400–1200 mm, drier parts of mountains with a pronounced dry period, in pure montane *Juniperus* forest, in montane vegetation and in mixed evergreen montane forests — 1100–3500 m	— seeds — seedlings — 35 000–42 750 seeds/kg (47 300[b])	— timber, indoor and outdoor construction — transmission and other poles, pencils, furniture, medicine, soil and water conservation, planted around churches, fuel, cedar wood oil for cosmetic industry	— from Ethiopia and Sudan to Zimbabwe–Namibia	— dormancy[c] medium — improved germination with pretreated seeds
2	*Podocarpus gracilior* pilger Podocarpaceae	— evergreen tree up to 45 m	— in highlands, semi-humid lower highland forest, pure *Podocarpus* forest, montane vegetation and in mixed evergreen montane forests — 1700–2500 m	— seeds — seedings — 1 900 seeds/kg[b] — raised in nurseries for afforestation	— timber, interior carpentry, flooring, furniture, boxes, veneers, shuter, outdoor construction when treated with preservatives, fuel, edible oil,[b] medicine	— Ethiopia, Kenya Tanzania, Transvaal and Uganda	— dormancy very high — presowing treatment required to improve germination
	B. Angiosperms						
3	*Abutilon longicuspe* Hochst. ex A. Rich Malvaceae	— shrub about 4 m	— in *Juniperus–Podocarpus* forest at openings[b] — 2350 m[b]	— no report of cultivation	— bark used as rope	— Angola, Ethiopia, Kenya, Malawi, Mozambique, Nubia, Saudi Arabia, Sudan, Tanzania, Uganda	— dormancy very high — presowing treatment required to improve germination
4	*Acacia abyssinica* Hochst. ex Benth. Fabaceae	— tree up to 20 m	— woodland, wooded grassland, forest margins and along sides of streams and rivers — 1500–2800 m	— no report of cultivation — 13 300 seeds/kg[b]	— feed, fuel, charcoal, bark as rope, local ploughs, building houses, fences, apiculture, soil and water conservation, nitrogen fixation, shade, bark eaten during scarcity of food[b]	— Ethiopia, Kenya, Malawi, Mozambique, Sudan, Tanzania, Uganda, Zaire and Zimbabwe	— dormancy very high — presowing treatment required to improve germination
5	*Bersama abyssinica* Fresen. ssp. *abyssinica* Verdcourt Melianthaceae	— shrub or small sized tree up to 18 m	— essentially in *Juniperus–Podocarpus* forest and degraded remnants of it, in thickets and copses, in grassland and open woodland and even cultivation, on slopes and hills, in gallery forest and montane scrub 1700–2715 m	— no report of cultivation — 3 650 seeds/kg[b]	— shade in coffee, medicine	— from East Africa to Malawi and Zambia	— dormancy very high — no report on how to improve germination

Table 1. (Continued)

No.	Species/ family	Habit	Habitat/ecology	Means of propagation	Uses	Distribution	Dormancy/ germination
6	*Calpurnia aurea* (Ait.) Benth. Fabaceae (Papilionoideae)	— shrub or small tree up to 10 m	— forest margins, bushland or grassland, favored by overgrazing — 1650–3000 m	— seeds — seedlings — cuttings — cultivated — 38 460 seeds/kg	— live hedge, sticks for torch-lights, medicine, insecticide, leaf-sap used as an aromatic on the head by women (Ethiopia), for handles of hammers and axes, house building, roots for charcoal, ornamental, shade in coffee, seeds to stun fish, cleaning teeth[b]	— westwards to Central African Republic and south to South Africa (Cape Province); also in India	— dormancy high — presowing treatment required to improve germination
7	*Clematis hirsuta* Perr. & Guill. var. *hirsuta* Ranunculaceae	— Woody climber, 5 m or more	— rich forest soils, edges and remnants of forests, wooded grassland, dissected plateau lava beds, on open slopes, along roadsides, on mountains and in valleys — 890–3100 m	— no report of cultivation — 520 800 seeds/kg[b]	— stem and branches used as ropes to tie things, medicine	— widespread in many parts of tropical Africa and Arabia	— dormancy medium — presowing treatment required to improve germination
8	*Clematis simensis* Fresen. Ranunculaceae	— woody climber, 10 m or more	— forest margins, bushland, grassy or rocky mountain slopes with dense vegetation, old lava beds with shrubby vegetation, in cultivated land — 1500–3350 m	— no report of cultivation — 512 800 seeds/kg[b]	— medicine, for making tattoos and engraving iron	— Angola, Cameroon, Congo, Ethiopia, Fernando Po, Kenya, Madagascar, Malawi, Nigeria, Sudan, Tanzania, Uganda, Zimbabwe and Arabia	— dormancy low
9	*Croton macrostachyus* Hochst. Euphorbiaceae	— tree up to 30 m	— forest, edges of thickets, savanna, border of cultivated fields, waste ground, along rivers — 700–3400 m	— no report of cultivation — 11 670 seeds/kg[b]	— shade in coffee, indoor carpentry, ordinary furniture, veneers, inner layers of plywood, boxes, cartes, apiculture, medicine	— widespread in Africa from Guinea to Ethiopia and from Ethiopia to Angola and Mozambique	— dormancy very high — presowing treatment required to improve germination
10	*Dodonea anguistifolia* L.f. Sapindaceae	— shrub up to 3 m	— edges of upland forest, upland bushland and grassland, secondary forest and scrub, invading areas recently cleared of forest, also in cultivated areas — 500–2900 m	— no report of cultivation — 96 430 seeds/kg[b]	— hut construction, walking stick, medicine	— pantropical	— dormancy high — presowing treatment required to improve germination
11	*Ekebergia capensis* Sparrm. Meliaceae	— tree up to 30 m	— in montane forest — 1680–3000 m	— no report of cultivation — 3 950 seed/kg[b]	— interior carpentry, joinery and veneers, mortar, household and farm[b] tools	— widespread in tropical Africa from Senegal to Ethiopia and southwards to South Africa	— dormancy medium

Table 1. (Continued)

No.	Species/ family	Habit	Habitat/ecology	Means of propagation	Uses	Distribution	Dormancy/ germination
12	*Ficus sur* Forssk. Moraceae	— tree up to 30 m	— upland rain forest and riverine forest, mountain grassland or secondary scrub — 1400–2500 m	— no report of cultivation — 3 125 000 seeds/kg[b]	— Shade in coffee, fruits and young arial shoots edible, leaves and fruits stimulate milk production in cows, medicine, wood is suitable for several applications	— widespread in tropical Africa, west to Senegal and Cape Verde Islands, south to Cape Province of South Africa and Yemen	— dormancy very high
13	*Indigofera rothii* Baker Fabaceae	— shrub or small tree up to 4 m	— upland scrub, forest margins — 2150–2800 m	— no report of cultivation — 242 130 seeds/kg[b]	— for fencing houses and farms,[b] farmers claim that land previously covered by this plant gives better yield[b]	— endemic to Ethiopia	— dormancy very high — presowing treatment required to improve germination
14	*Maesa lanceolata* Forssk. Myrsinaceae	— tree/shrub up to 10 m	— woodland, thicket, preferably at forest edges and in secondary scrub — 1800–1400 mm of annual rainfall — 1500–2870 m	— no report of cultivation — 91 240 seeds/kg[b]	— Oil from fruit used as grease for pottery and medicine, fruit also used to stun fish, wood used as fuel and extract of it is used to tan leather red	— all over tropical Africa including the islands, and in the hotter parts of subtropical Africa and in the Arabian Peninsula	— dormancy very high
15	*Maytenus arbutifolia* (Hochst. ex A. Rich.) Wil. Celastraceae	— tree about 8 m	— forest forest margins, degraded forest, river banks, grassland and scattered thickets and along roads — 1200–3000 m	— no report of cultivation	— fuel, feed, for household utensils, used against skin disease of cattle[b]	— Burundi, Ethiopia, Kenya, Rwanda, Saudi Arabia, Somalia, Tanzania, Uganda, Yemen and Zaire	— dormancy very high
16	*Mysine africana* L. Myrsinaceae	— evergreen shrub up to 3 m	— steep mountain slopes, forest borders, along mountain rivers or gullies or in shrubby grassland, rocky areas — 1500–3800 m	— seeds — cultivated elsewhere but not in Ethiopia — 56 180 seeds/kg[b]	— feed, edible fruits, hardwood resistant to termite, house construction and for cleaning teeth, medicine	— Angola, Ethiopia, Kenya, Malawi, Mozambique, Namibia, Somalia, South Africa, Sudan, Zaire, Zambia, Zimbabwe, Afghanistan, Arabia, The Azores, Belutchistan China, The Himalayas, Kashmir, Socotra	— dormancy very high — presowing treatment required to improve germination

Table 1. (Continued)

No.	Species/ family	Habit	Habitat/ecology	Means of propagation	Uses	Distribution	Dormancy/ germination
17	*Olea africana* Miller Oleaceae	— tree up to 10 m	— woodlands, thickets, highland forests *Podocarpus* and *Juniperus* forests, very common in secondary scrub — 1500–2500 m (up to 2775 m[b])	— seeds — seedlings — 6 700 seeds/kg[b]	— edible fruit, edible oil, medicine, condiment, fancy furniture, cabinet work, tannery and veneering, excellent for fuel and charcoal, one of the hardest commerical timbers in Africa, for cleaning teeth[b]	— widespread in Africa, Afghanistan, Arabia, Belutchistan China, The Himalayas, India, Madagascar and Mauritius	— dormancy very high — presowing treatment required to improve germination
18	*Rhus glutinosa* A. Rich. spp. *neoglutinosa* (M. Gilbert) M. Gilbert Anacardiaceae	— shrub or tree up to 10 m	— forest margins, often conspicuous after clearance of *Podocarpus* or *Juniperus* forest, less often in more open bushland on rocky slopes — 1500–2700 m	— no report of cultivation — 263 850 seeds/kg[b]	— fuel, construction	— endemic to Ethiopia	— dormancy high — presowing treatment required to improve germination

[a]Sources of information: Bolza & Keating (1972), Breitenbach (1963), Cufodontis (1953–1972), Demel Teketay (1987), Demel Teketay & Assefa Tegineh (1991), FAO (1986), Getachew Aweke (1979), Hall (1981), Hedberg & Edwards (1989), Jansen (1981), Jones (1989), Von Maydell (1986), Poluha (1990) and Seegeler (1983).
[b]Own observation.
[c]Observation from ongoing experiments on dormancy: low = 0–40%; medium = 41–50%; high = 51–80%; very high = 81–100%.

seeds were exposed to light from fluorescent tubes under fluctuating temperatures, 20 °C (day) and 10–12 °C (night). Some reasons for seed dormany include hard seed coats and rudimentary embryos surrounded by copious endosperm. The most difficult species to germinate, even with different pre-sowing treatments, are *Podocarpus gracilior, Bersama abyssinica, Maesa lanceolata* and *Ficus sur.* One of the dominant and most important species, *Juniperus procera,* had medium dormany (about 50%). It did not respond well so far to the pre-sowing treatments applied on the seeds, such as acid and hot water scarification. The species which responded very well to pre-sowing treatments include *Acacia abyssinica, Calpurnia aurea* and *Indigofera rothii,* all of which are legumes.

Until more is known about the propagation of indigenous species, it is likely that exotic species will be preferred by nurseries because of their predictable propagation.

Competition between forestry and agriculture

Enrichment and reforestation/afforestation programs may involve, especially in a country like Ethiopia, areas which had already been changed into a farm land. This inevitably creates conflicts between those who are responsible for the programs and farmers. For example, some reforested areas in the Chercher Highlands of south- eastern Ethiopia have been ploughed and grazed. Seedlings have also been uprooted in some areas. Clearly, successful reforestation requires the voluntary and active participation of farmers. Greater consultation prior to afforestation, community participation in planning and implementation, and agroforestry are all possible solutions.

Financial problems

All activities in nurseries require money. As in all other tropical countries, funds allocated to nurseries in Ethiopia are very limited, thereby restricting the successful implementation of plans. Most of the activities are performed manually. In the Harerge region and also in several other regions in Ethiopia, laborers are paid in kind, i.e. grains and oil, in a program known as "food for work". This program is supported by foreign aid from different countries. It attracts many individuals, including farmers. However, in spite of its advantages, it seems to have a drawback in that some of the farmers are tempted to abandon their farms and to be fully engaged in the nursery activities. This has been observed to play a role in the decline of agricultural production, making farmers partially or totally dependent on the grains that they obtain from the "food for work" program. It has also a direct bearing on the urban population which depends on the farmers for agricultural products.

Conclusions

The public awareness about the consequences of destruction of forests and vegetation has increased enormously in the last two decades in Ethiopia. However, being aware of the problems is not enough. Efforts should be geared, on the one hand towards overcoming the problems mentioned above and on the other, towards the restoration of degraded forest lands as well as protection and conservation of the remaining forests. This requires goodwill and sacrifices in terms of labor, time and money from every citizen, dedication from concerned bodies in the government and support from the international society. It also requires competent and adequately trained staff, and knowledge of the ecological characteristics of the species involved.

Acknowledgements

I would like to thank Dr. Anders Granström for various important discussions, for reading the paper and for his comments and suggestions; Ms. Anna Liisa Raunio for providing me with her unpublished reports, for important discussions about the topic and for reading the manuscript; Dr. Anders Persson for reading the manuscript and Mrs. Ingrid Englund for typing Table 1.

References

Bawa, K. S., Ashton, P. S. & Mohd. Nor, S. 1990. Reproductive ecology of tropical forest plants: management issues. In Bawa, K. S. & Hadley, M. (Eds.), Reproductive Ecology of Tropical Forest Plants. UNESCO, Paris.

Bolza, E. & Keating, W. G. 1972. African timbers — the properties, uses and characteristics of 700 species. Division of Building Research, Commonwealth Scientific & Industrial Research Organization, Melbourne, Australia.

Breitenbach, F. V. 1963. The indigenous trees of Ethiopia. Ethiopian Forestry Association, Addis Ababa.

Cufodontis, G. 1953–1972. Enumeratio plantarum Aethiopiae Spermatophyta. Bulletin du Jardin de l'Etat Bruxelles, Bruxelles.

Demel Teketay. 1987. Revision of the genus *Clematis* L. in Ethiopia. Belmontia 19 (108): 1–42.

Demel Teketay & Assefa Tegineh. 1991. Shade trees of coffee in Harerge, Eastern Ethiopia. The International Tree Crops Journal 7: 17–26.

FAO. 1982. Conservation and development of tropical forest resources. FAO Forestry Paper 37. FAO, Rome.

FAO. 1986. Databook on endangered tree and shrub species and provenances. Forestry Resource Division, FAO Forestry Department.

Getachew Aweke. 1979. Revision of the genus *Ficus* (Moraceae) in Ethiopia. Laboratory of Plant Taxonomy and Geography, Agricultural University, Wageningen.

Grainger, A. 1986. Deforestation and progress in afforestation in Africa. The International Tree Crops Journal 4: 33–48.

Hall, J. B. 1981. Data sheets on species undergoing genetic impoverishment, *Juniperus procera* Hochst. ex Endl. Forest Genetic Information 10: 25–29.

Hedberg, I. & Edwards, S. 1989. Flora of Ethiopia. The National Herbarium, Addis Ababa.

Heywood, V. H. 1985. Flowering Plants of the World. Croom Helm, London & Sydney.

Jansen, P. C. M. 1981. Spices, condiments and medicinal plants in Ethiopia, their taxonomy and agricultural significance. PUDOC, Wageningen.

Jones, S. 1989. The influence of stratification, scarification, hot water and maternal plant on the germination of *Juniperus excelsa* seeds from Eritrea. The International Tree Crops Journal 5: 221–235.

Lanly, J. P. (Ed.). 1981. Tropical Forest Resources Assessment Project (GEMS): Tropical Africa, Tropical Asia, Tropical America. FAO/UNEP, Rome.

Poluha, E. 1990. Risks, trees and security: a baseline study of Beddedo, a peasant association in Wello, Ethiopia. Swedish University of Agricultural Sciences, Working Paper 111, Uppsala.

Seegeler, C. J. P. 1983. Oil plants in Ethiopia, their taxonomy and agricultural significance. PUDOC, Wageningen.

Tewolde B. G. Egziabher. 1988. Vegetation and environment of the mountains of Ethiopia: implications for utilization and conservation. Mountain Research and Development 8 (2/3): 211–216.

Tewolde B. G. Egziabher. 1989. The environmental variables which led to the ecological crisis in Ethiopia. COENOSES 4 (2): 61–67.

Von Maydell, H.-J. 1986. Trees and shrubs of the Sahel, their characteristics and use. GTZ, FRG.

9. Mixed species plantations using high value rainforest trees in Australia

D. LAMB and P. LAWRENCE
Botany Department, The University of Queensland, Brisbane, Australia

Keywords: *Araucaria, Flindersia, Hypsipyla,* polycultures, underplanting

Abstract. There have been a number of trials to assess the growth of high value rainforest tree species grown under plantation conditions (monocultures and mixtures) in the wet tropics of Australia. Many of these date back over 50 years or more. Only one species (*Araucaria cunninghamii*) has been subsequently used in an extensive plantation programme.

In the years since many of those trials were carried out plantation establishment techniques have improved together with our understanding of the ecology of some of these species. In addition, economic circumstances and the value of these timbers is changing. The paper reviews several trials involving young and old mixed species planting of rainforest trees and considers some of the ecological and silvicultural issues needing resolution.

Introduction

Rehabilitation of degraded tropical forest lands is expensive. In many cases this has meant that rehabilitation has been carried out with the intention of obtaining a commercial return of some kind. If the degraded lands are still occupied by some trees, one approach has been to carry out enrichment planting to increase the stocking of commercially valuable timber trees (e.g., Tang and Wadley 1976). If the degraded lands have too few trees then rehabilitation has often been achieved by establishing plantations.

During the last few decades plantations in the tropics of the Asia-Pacific region have frequently established using fast-growing species such as *Acacia, Pinus, Eucalyptus* or *Albizia.* These have been grown for general purpose structural timbers or pulpwood and in many areas excellent growth rates have been achieved. The argument we wish to advance is that further research is needed to investigate the plantation potential of some of the more valuable cabinetwood species in rainforests of the region. One reason is simply commercial — the *Pinus, Acacia* and *Eucalyptus* plantations mostly produce timbers for the lower value end of the market. Is there scope to supply the higher value section of the market as well? A second reason has to do with maintaining regional biological diversity. Can indigenous species be used to create new forests on presently degraded lands? Might commercially oriented research help us find ways to begin restoring some of the forests that once occupied what are now degraded lands?

The idea of using high value rainforest tree species in plantations is not new and there is a long history of trials with many species in different parts of the tropics (Streets 1962). But with few exceptions these earlier trials seem to have failed, or at least the results were not encouraging enough to have prevented the present preference for *Pinus, Acacia* and *Eucalyptus.* But there are several reasons for thinking that it is time to reassess the question once more. We will illustrate this by briefly describing the course of events in tropical Australia related to this issue.

The use of high value rainforest tree species in plantations in tropical Australia

The first planting trials using rainforest species were carried out in 1903 (Anon. 1983). These were underplantings or enrichment plantings in logged forests with few remaining commercial

H. Lieth and M. Lohmann (eds.), Restoration of Tropical Forest Ecosystems, 101–108.

Table 1. Optimal growth rates of various high value rainforest timber trees grown in Australia in comparison with those typically achieved in *Araucaria* and *Pinus* plantations

	Age (yr)	Diameter breast height (cm)	Mean annual increment (m^3 ha^{-1} yr^{-1})
**Flindersia brayleana*	18	49.9	3.9
**Grevillea robusta*	36	22.3	2.4
**Elaeocarpus grandis*	24	29.3	3.4
Cedrela odorata	23	32.2	4.4
Swietenia macrophylla	21	25.7	6.8
**Toona australis*	56	42.6	8.0
**Araucaria cunninghamii*	50	–	15.0
Pinus caribaea	30	–	18.0

*Indigenous to Australia.
Source for high value tree data: Cameron & Jermyn (1992).

species. At the time only 10 species were deemed to be commercial although this number reached 150 by 1980. In 1914 the first small-scale trial plantings were carried out using native species. This work continued until 1930 by which time over 100 plantation trials containing at least 35 rainforest species were established. Twelve of these were native hardwoods, three native conifers and the remainder were exotic species. The only species to be subsequently used in large-scale plantations was the native conifer *Araucaria cunninghamii.*

There were various reasons for not using the other species. One was that the growth rates achieved were generally modest in comparison to those that could be achieved with exotic pines (Table 1). Thus most of the hardwoods had mean annual increments of less than 8 m^3 ha^{-1} yr^{-1} in comparison with the 18 m^3 ha^{-1} yr^{-1} of *Pinus caribaea.* (It should be noted that this is the current average for *P. caribaea* and it has been achieved after considerable research into breeding and fertiliser technology.)

A second reason for not pursuing the plantation option was that comparisons were invariably made between plantation logs and logs from natural forests; as long as the latter were still available, plantation grown timber appeared less attractive to saw millers.

Finally, several species suffered severe insect attack when grown in plantation monocultures. The most well known of these was *Toona australis* (Red Cedar), a member of the Meliaceae. This was one of the most commercially attractive of all the native hardwoods and was highly prized by early loggers. Unfortunately it appeared to invariably attract a tip boring moth (*Hypsipyla robusta*) when grown in plantations. The moth caused repeated dieback of the growing shoots and eventually severe stem malformation.

Several factors have changed, however, since these early assessments were made.

i) Planting techniques: a good deal of research has been carried out investigating improved methods of plantation establishment for pines since most of the early rainforest species trials were established. This research has addressed site preparation, weed control and fertiliser requirements and the results have led to much faster earlier growth rates in plantations. There is little doubt that similar growth benefits would apply to new rainforest tree plantations.

ii) Markets: in Australia the local production of rainforest timbers ceased when all tropical rainforests (about 1 million hectares) became nature reserves following their registration on the World Heritage Register in 1989. Rainforest timbers can still be imported, but various projections suggest that the number of countries still able to export these timbers in ten years time will have been drastically reduced. That is, both the local and overseas sources of supply have declined and will continue to do so.

iii) Prices: as a result of the diminishing supply, prices for the higher quality timbers have been rising and show every sign of continuing to rise. By contrast, pulpwood prices and the value of structural grade timbers seem unlikely to rise in the same way given the very large supplies becoming available in the Asia-Pacific region from sources such as the pine plantations in New Zealand and Chile.

iv) Processing machinery: veneer and milling equipment are now able to accept much smaller logs than was once the case meaning plantation rotations can probably be reduced.

v) Ecological knowledge: more is known about rainforest ecology and the physiological requirements of many rainforest species than when the early trials were carried out. That is, there is greater scope for matching species to sites and for providing optimal growing conditions.

Taken together these factors suggest it is time to reassess the decisions made many years ago to abandon attempts at commercial plantation establishments using rainforest tree species.

Many of the earlier trials were plantation monocultures. The reason for this was that single species plantations are obviously easier to manage than mixed species plantations. *Araucaria cunninghamii* grew well in these monocultures but all the other species tested grew slowly or even failed. This raises the question of whether rainforest tree species should be grown in plantation monocultures or whether other silvicultural techniques must be used. The fact that a species grows naturally in a diverse forest ecosystem does not necessarily mean that commercial plantations should attempt to simulate these conditions. But might some kind of a mixture be beneficial? The surprising fact is that there are apparently very few good, carefully designed, trials that compare plantation monocultures with either a temporary mixture (e.g. crop trees grown under an early cover crop of another species) or a permanent plantation mixture. Where mixed species trials have been carried out they are commonly unreplicated and poorly designed. This means it is difficult to be unequivocal when assessing the results although the past trials do have value in suggesting species and trends worthy of future investigation.

A review of many of the trials carried out using rainforest species from the tropics and subtropics species of Australia has recently been compiled by Cameron and Jermyn (1992). This describes growth rates of trees in single species trials as well as in underplanting trials and in permanent species mixtures. Some of the growth rates in single species trials have been given earlier (Table 1). Results of some of the underplantings and mixed plantings are described below.

Experiences with underplanting trials in Australia

One of the most valuable rainforest timber species in Australia is *Toona australis* (Red Cedar), a member of the Meliaceae. As mentioned earlier plantations of this grown in the open are invariably attacked by the tip borer *Hypsipyla.* However trees grown in shade beneath a canopy cover usually grow better because the tip borer attack is reduced (Mitchell 1971; Anon. 1983). Cameron and Jermyn (1992) quote reports suggesting an overstorey should be maintained at 30–35 m^2 ha^{-1} until four to eight years when it could be reduced to 20 m^2 ha^{-1}.

The exotic *Swietenia macrophylla,* another member of the Meliaceae, also grows better when initially established under the shade of a nurse crop but the shade had to be removed at an early stage to allow rapid early growth to be sustained (Cameron and Jermyn 1992). By contrast, a third member of the Meliaceae, and also from outside Australia, *Cedrela odorata* (West Indian Cedar), did not appear to be badly attacked by the native Australian *Hypsipyla* nor to benefit from early shading. In fact, the experience from a number of trials at different sites showed that the best growth of this species commonly occurred in the open without any early shading (Cameron and Jermyn 1992). Quite different results for this species have been reported elsewhere. Streets (1962), for example, quotes the results of trials in Fiji which reported very fast early growth by *Cedrela odorata* grown under *Leucaena leucocephala.*

Most of the trials noted above were carried out in logged-over forest where the canopy of the residual trees provided the cover. A number of other trials have been carried out using *Pinus* plantations as nurse crops. One of the most interesting of these involved planting the native conifer *Araucaria cunninghamii* beneath a *Pinus elliottii* overstorey. The *Araucaria* produces a more valuable timber than *Pinus* but is more sensitive to soil nitrogen deficiencies than the pine (Bevege and Richards 1970). But experience has shown that *Araucaria* underplanted to pine can be grown without the nitrogen fertiliser otherwise needed when grown in the open at the same site. Such results might be expected from underplantings beneath a nitrogen fixing species

Fig. 1. Plantation mixture involving alternative rows of *Araucaria cunninghamii* (Hoop Pine) and *Flindersia brayleana* (Queensland Maple). The trees are 60 years old. Note figure in foreground.

but *Pinus* is not normally thought of as a nitrogen fixer. Bevege and Richards (1970) suggested several possible explanations for the observation and trials were carried out to explore using *Pinus* plantations as a means of establishing *Araucaria* plantations although this was never finally implemented on a large scale.

Other trials also explored using *Pinus* as a cover crop for various rainforest timber species. The most carefully designed of these used a 16-year-old *Pinus elliottii* plantation having a stocking of 740 trees ha^{-1} and a basal area of 35 m^2 ha^{-1}. This was underplanted with *Flindersia brayleana, Flindersia australis, Swietenia macrophylla, Grevillea robusta, Elaeocarpus grandis, Flindersia ifflaiana and Ceretapetalum apetaeum.* As in the case involving *Araucaria cunninghamii* the intention was to explore the possibility of using a *Pinus* cover crop to facilitate the establishment of the higher value hardwood species on an otherwise poor site. The *Flindersia brayleana* grew well and reached 18 m at age 20 years. *Swietenia macrophylla* reached 13 m at the same age. All the other species were less vigorous. There was no evidence that any of the underplantings had any adverse effect on the growth of the overstorey *Pinus* trees (Cotsell 1991). Again the result demonstrated that it was possible to use *Pinus* as a cover crop to establish valuable rainforest timber trees on an otherwise poor site although there was no direct comparison made showing how these various species might perform at this site when grown in the open.

In summary, some species appear to benefit from early cover and others do not. There are no specific details on what species should be used to provide cover, how much cover should be provided for different species or how long it should be retained. Nor is it clear just what the ecological processes are that foster better growth in those species benefiting from the initial cover and why they should grow faster.

Experiences with permanent mixed plantings

There are only two studies of permanent plantation mixtures where both species were planted at the same time reported by Cameron and Jermyn (1992). One of these involved *Flindersia brayleana* planted with *Araucaria cunninghamii* (Fig. 1). Both species grew well in the mixture and there is some evidence that the mixture improved the performance of the *Flindersia*. This will be discussed in a little more detail below. The other mixture trial involved *Cedrela odorata* and *Araucaria cunninghamii*. In this case the *Cedrela* grew faster than *Araucaria* in the early stages of the trial at one site but in most cases it appeared to grow best in single species open plantings.

Why might plantation mixtures be beneficial?

This field evidence is not large and does not include the various trials that presumably went unreported because they failed. If the evidence is so sparse why persist with cover crops or mixed plantations? What reasons might there be for believing they would offer any benefits?

The potential benefits fall into five broad categories:

a) Between-tree competition is reduced and hence production is increased because the phenologies of the species in the mixture are out of phase.

If competing species are growing at different times of the year interspecific competition for resources will be less than intraspecific competition. There is some evidence that such differences in growth phenologies may contribute to the maintenance of diversity in natural plant communities (Grubb 1977; Rogers and Westman 1981).

b) Between-tree competition is reduced and hence production is increased because differences in stem and root architecture lead to a partitioning of resources.

If one species is essentially deep-rooted while another has most of its roots in the upper soil profile then competition between the two for soil water and nutrients may be reduced. Similarly, if one species is light demanding and another is shade tolerant then the overall leaf area index of the plantation mixture may be greater than for one species alone. In both cases the site is more fully exploited.

c) Nutritional benefits

There are a number of examples of improved plantation growth on poor soils when nitrogen fixers are mixed with the commercially desirable tree species (e.g. Gordon and Dawson 1979; De Bell et al. 1989). Furthermore, even when non-nitrogen fixing species are involved there is the possibility that litter decomposition proceeds faster in mixed species plantations than in monocultures (Simmonds and Buckley 1990). Litter breakdown is usually not limiting in tropical hardwood forests but the mechanism may acquire significance when more slowly decomposing conifer species are used.

d) Reduced insect damage

There is some evidence that diverse plant communities lose a lower proportion of available leaf area to herbivores than simple plant communities (Brown and Ewel 1987). Further, polycultures of interplanted crops offer support to fewer insect pests at lower densities than monocultures (Root 1973; Risch 1981). That is, the biological diversity, structural complexity and microclimate heterogeneity of mixed forests appear to confer greater resistance to plant herbivores than do simple monospecific forests. Interestingly it appears that some of exotic species grown in trials and plantations in some isolated Pacific Islands have so far escaped many of the insect problems these species commonly experience when planted in continental situations.

e) Beneficial microclimate beneath a cover crop

Some species may require or benefit from shade in the earlier seedling stages when seedlings are becoming established. Certain members of the Meliaceae also appear to be subject to less tip borer attack when grown beneath a cover crop, although there are remarkably few quantitative studies that have attempted to quantify the relationship (but see Weaver and Bauer 1986; Grijpma 1976).

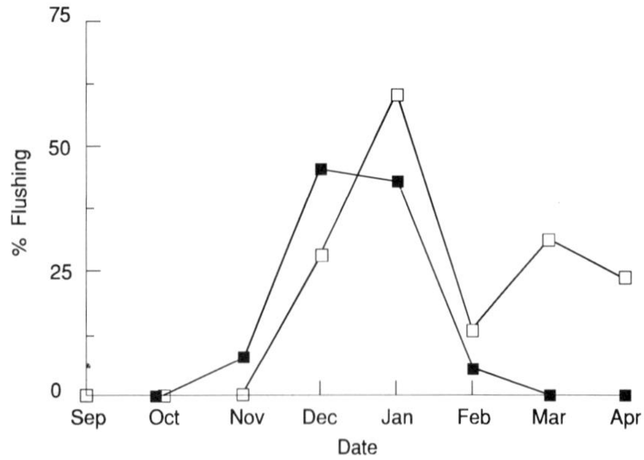

Fig. 2. Timing of leaf flush in *Araucaria cunninghamii* (Hoop Pine ■) and *Flindersia brayleana* (Queensland Maple □) showing both species reach a peak around December–January during the wet season. [*Source:* Lawrence 1991]

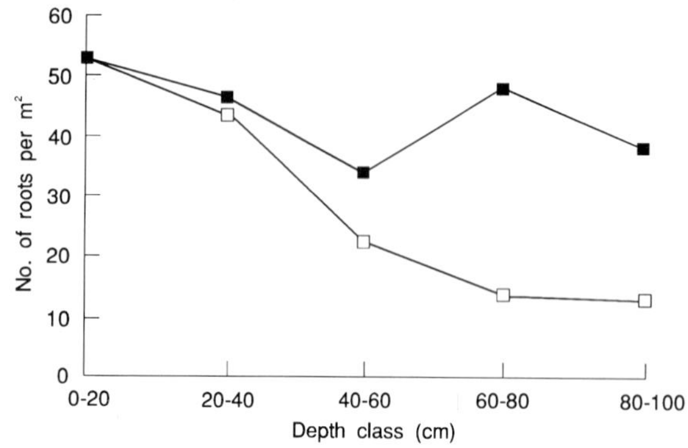

Fig. 3. Density of roots 1–5 mm diameter in the top 100 cm of soil in a mixed plantation of *Araucaria cunninghamii* (Hoop Pine ■) and *Flindersia brayleana* (Queensland Maple □). There was no difference between the two species in the density of other root size classes. [*Source:* Lawrence 1991]

A case study

In 1931 the Queensland Forest Service planted a trial plantation involving alternative rows of *Araucaria cunninghamii* and *Flindersia brayleana.* The trial was not replicated but consisted of a 2.9 ha block of mixed plantation, a 8.6 ha plantation of *Flindersia brayleana* and a 13.2 ha plantation of *Araucaria cunninghamii.* The mean annual rainfall at the site is 1190 mm and it adjoins an area of subtropical rainforest. Further details of the plantations and the site are given in Lawrence (1991).

The trial was used to explore two of the reasons why mixed plantations might be more successful than monocultures.

Hypothesis 1: That competition between the two species is reduced because of different growth phenologies.

This hypothesis was examined by assessing the timing of leaf flush activity, the timing of stem growth, the seasonality of litterfall and the timing of root activity (Lawrence 1991). All of these occurred at the same time in each species irrespective of whether the trees were in a pure or mixed plantation. The timing of leaf flushing is typical of these data and the results are shown in Fig. 2. These indicate that both species commenced leaf flush and reached a peak at similar times of the year suggesting that their phenologies were not greatly out of phase and that the two species might be directly competing for resources. That is, there were no particular advantages of this mixture over a monoculture of either species.

Hypothesis 2: That competition between the two species is reduced because of differences in root architecture and rooting depth.

This hypothesis was examined using trenches to directly assess rooting depth of the two species. Results are given in Fig. 3 which shows there were more small *Araucaria* roots at depth than *Flindersia* roots. That is, there was evidence suggesting vertical partitioning of the soil profile and hence a possible reduction in root competition between the two species.

The trial was not designed to allow a statistically valid comparison of productivity in the mixed and pine stands. However, there was some evidence that *Flindersia* trees in the mixed plantation were larger than *Flindersia* grown in a monoculture (Fig. 4). The cause of this is likely to have been a reallocation of carbon resources from low branches into greater stem growth caused by the surrounding tall *Araucaria* trees. By contrast the *Araucaria* trees appeared to be marginally smaller in the mixed plantation than in the pure stand.

It is not possible to draw definitive commercial conclusions from such a trial because the significance of such differences depend on the

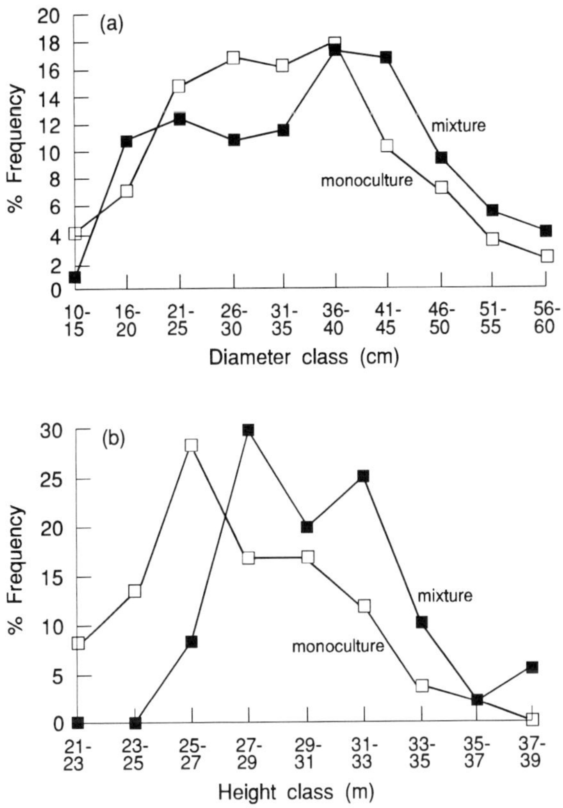

Fig. 4. The frequency of *Flindersia brayleana* (Queensland Maple) trees of different diameter (upper graph) and different height (lower graph) classes when grown in a monoculture ($n = 154$) or in a plantation mixture ($n = 191$) with *Araucaria cunninghamii*. [*Source:* Lawrence, 1991]

relative prices of the two timbers and the tree densities used. What the results do show, however, is that the topic is well worth researching further. As a result of these studies we have commenced work examining the growth responses of these two species plus other high value species in several carefully designed plantation configurations. One is an agroforestry trial involving *Araucaria* and *Flindersia* planted at variable densities using an adaptation of Nelder's (1962) variable density design. Another is a replicated plantation mixture involving 16 species in which the trees have been provided with optimal growth conditions (e.g. weed control, fertilisers etc.). This trial will be used to test the predictive value of a growth model based on the physiological attributes of the species (D. Vieglais, pers. comm.).

Some questions needing answers

Plantations of high value trees deserve greater consideration as a means of rehabilitating degraded tropical lands. But for their potential to be evaluated a good deal more needs to be known about their silviculture and management. Some of the key issues deserving further research are:

a) Growth rates
Just what are the growth rates of the potential tree potential tree species if provided with optimal conditions (that is, better conditions than were used in many of the earlier evaluation trials). Is there scope for improving such growth and form by better seed provenance collections?

b) Timber quality
What is the relationship between growth rate and timber quality? Is the quality of timber from fast-grown plantations poorer than from natural forests? Is it still high enough to attract premium prices?

c) Nurse or cover crops
Which species benefit from cover crops? How much cover is needed and for how long? Are there short-lived nurse crop species that might be suitable or must the nurse crop be physically removed? What are the nutritional benefits?

d) Permanent mixtures or monocultures
Are mixtures beneficial for most rainforest tree species when grown in plantations? If so how many species are necessary to confer the benefits of a mixture (two? five? ten?). What are the ecological mechanisms conferring advantages in a mixture?

e) Complimentary species
How can one judge which species might combine best in a mixed plantation? What are the critical species traits? In the *Araucaria–Flindersia* example above, it seemed the difference in rooting depth might be important. Are there other factors to guide the choice of species for mixtures?

f) Silvicultural management and harvesting
If mixtures are used how should they be managed? Should such plantations be thinned or be left to grow undisturbed? How might species growing at different rates be

harvested? (For example, would planting mixtures in alternative rows enable row harvesting to overcome this problem?)

Conclusion

Rehabilitation of degraded tropical lands will have to be approached in a variety of ways and the socio-economic conditions at a particular site might be more critical in determining the most appropriate approach than any ecological or silvicultural factors. Where socio-economic conditions are not limiting there are already well-established techniques for establishing plantations of fast-growing tree species on most types of degraded forest land. We argue here that some of the sites should be evaluated to determine whether high value rainforest tree species can be used instead of the more common fast-growing exotic species. That is, it is time to assess plantation value increment per hectare and not just volume increment. We do not wish to overstate the role of plantation mixtures in any such evaluation because there is evidence that some valuable rainforest species will grow well in monocultures. We do wish to emphasise however, that plantation mixtures deserve a more comprehensive evaluation than they seem to have received to date.

References

Anonymous. 1983. Rainforest research in north Queensland, Position Paper, Queensland Department of Forestry, Brisbane.

Bevege, I. & Richards, B. N. 1970. Nitrogen in the growth of *Araucaria cunninghamii* Ait. underplanted in *Pinus* stands. Ecology 51: 134–142.

Brown, B. J. & Ewel, J. J. 1987. Herbivory in complex and simple tropical successional ecosystems. Ecology 68: 108–116.

Cameron, D. & Jermyn. 1992. Review of plantation performance of high value rainforest species. Australian Centre for International Agricultural Research, Canberra.

Cotsell, N. 1991. Hardwood underplanting — Tripconys logging area, Interim Report 1966–1989, Queensland Forest Service.

De Bell, D. S., Whitesell, C. D. & Schubert, T. H. 1989. Using N-fixing Albizia to increase growth of Eucalyptus plantations in Hawaii. Forest Science 35: 64–75.

Gordon, J. C. & Dawson, J. O. 1979. Potential uses of nitrogen fixing trees and their shrubs in commercial forestry. Botanical Gazette (Suppl.) 140: 102–107.

Grijpma, P. 1979. Resistance of Meliaceae against the shoot borer. *Hypsypila* with particular reference to *Toona ciliata* M. J. Roem. var. *australis* (F.V. Muell.) C.D.C., pp. 69–79. In: Burley, J. & Styles, B. T. (Eds.), Tropical trees, variation breeding and conservation. Linnean Society Symposium, Series No. 2.

Grubb, P. 1977. The maintenance of species richness in plant communities. The importance of the regeneration niche. Biological Review 52: 107–145.

Lawrence, P. 1991. In search of a successful plantation mixture. Unpublished B.Sc. (Hons) thesis, Botany Department, The University of Queensland.

Mitchel, A. L. 1971. Planting trials with Red Cedar. Australian Forestry 35: 8–16.

Nelder, J. 1962. New kinds of systematic designs for spacing experiments. Biometrics 18: 283–307.

Risch, S. J. 1981. Insect herbivore abundance in tropical monocultures and polycultures: an experimental test of two hypotheses. Ecology 62: 1325–1340.

Rogers, R. & Westman, W. 1981. Growth rhythms and productivity of a coastal subtropical eucalypt forest. Australian Journal of Ecology 6: 85–98.

Root, R. B. 1973. Organisation of a plant-anthropod association in simple and diverse habitats: the fauna of collards (*Brassica oleracea*). Ecological Monographs 43: 95–124.

Simmonds, E. & Buckley, P. 1990. The ground vegetation underplanted mixtures of trees. Paper presented at 'Ecology of Mixed Species Stands of Trees', British Ecological Society & IUFRO Division 52.01. Heriot-Watt University, Edinburgh, 2–6 Sept. 1990.

Streets, R. J. 1962. Exotic Forest Trees in the British Commonwealth. Clarendon Press, Oxford.

Tang Hon Tat & Wadley, D. 1976. Report on the survival and development survey of areas reforested by line-planting in Selangor. Forest Research Institute Malaysia, Research Pamphlet No. 67.

Weaver, P. L. & Bauer, G. P. 1986. Growth, survival and shoot borer damage in mahogany plantings in the Lugvillo forest in Puerto Rico. Turrialba 36: 509–522.

10. The rehabilitation of the tropical rainforests' ecosystems

Cooperative research between UPM and YNU

ABAS BIN SAID
Centre of Applied Sciences, University of Agriculture Malaysia, Bintulu Campus, 97008 Bintulu, Sarawak, Malaysia

Keywords: Rainforest depletion, ecological balance, reforestation, Malaysian forest research

Abstract. The concern over depletion of the tropical rainforests has resulted in an increasing emphasis toward the programmes in rehabilitation in an effort to maintain the ecological balance within the ecosystem. In Malaysia, the loss of the natural tropical forest is due to harvesting of its timbers and conversion into lands for the purpose of agriculture, mining, industries and urban development. In view of the loss, an extensive reforestation programme is necessary both to sustain the forest resources and to rebuild the deteriorating ecosystem. Malaysian forest research has included planting of native species on barren land and in secondary vegetation since the twenties and much experience is available. This cooperative research between University of Agriculture Malaysia (UPM) and Yokohama National University (YNU) is conducted to study effective techniques to rehabilitate the disturbed forest areas on the campus. Indigenous species of mainly commercial species are selected for the planting. The methods used in seedlings and seeds collection, seed germination and planting are discussed in this paper. Early results of planted seedlings performance are also mentioned.

Introduction

One major issue in the global concern over depletion of the tropical rainforest is that the developing nations, especially in the tropical regions, have allegedly been over-exploiting their forest resources at the expense of future generations. It has been argued that tropical rainforest depletion has detrimental effect on the ecosystem. Malaysia, especially the state of Sarawak, has been a central focus of this issue.

Among the many proposals to remedy the situation has been the suggestion to establish near-natural forest by planting native tree species on barren land or into degraded secondary vegetation caused by prolonged slash-and-burn agriculture. Much experience with planting exotic and native species on abandoned vegetable gardens and tin-tailings is available from many decades of experimental research by the Forest Research Institute, now FRIM, at Kepong, Malaysia peninsular. In Sarawak, native tree species have been planted by the Forest Department since the twenties. *Casuarina nobilis* (syn. *C. sumatrana*) has been planted in secondary vegetation on sandy podzol soils to produce firewood, several species of illipe-nut producing *Shorea* were planted on good riparian soils, and durian (*Durio zibethinus*) in taungya on loamy upland soils. A whole range of native timber tree species were experimentally planted on marginal soils near the present Kuching airport and reported in successive annual reports of the *Forest Department Sarawak*. The current taungya scheme at Sabal Forest Reserve also includes planting of native multiple-use trees, such as durian. As a result, our research could draw on a substantial body of local and natural experience. As a new aspect to the present research programme was the phytosociological analysis of relevant habitats for the area to be reforested prior to the selection of the species and the control of the early growth and stand development.

A joint research project entitled "Rehabilitation of Tropical Rainforest Ecosystems" was initiated and implemented by Universiti Pertanian Malaysia and Yokohama National University. The purpose of the project is to conduct experimental planting of indigenous tree species on degraded sites such as abundant shifting cultivation areas, secondary forests, and degraded industrial and urban lands. The four-year project is sponsored by Mitsubishi Corporation with a funding of $4 million ringgit. A 50-hectare

H. Lieth and M. Lohmann (eds.), Restoration of Tropical Forest Ecosystems, 109–117.

research site has been set aside for the joint project.

The first meeting on the project was held in Bintulu on 2 May 1990 among officials of Universiti Pertanian Malaysia, Yokohama National University and Mitsubishi Corporation. The agreement between Universiti Pertanian Malaysia and Yokohama National University was signed on 27 December 1990.

The project was officially launched by the honourable Datuk Abang Zohari bin Tun Abang Haji Openg, Minister of Industrial Development, Sarawak on 15 July 1991 in Universiti Pertanian Malaysia, Bintulu Campus. In conjunction with the ceremony, a two-day seminar was held with the theme "Rehabilitation of Tropical Rainforest Ecosystems: Present Research and Future Directions".

This report highlights the status of the joint research project, presenting achievements in the last two years.

Objectives

The joint research project is conducted to achieve the following objectives:

a) to establish indigenous tree species on degraded sites which include shifting cultivation areas, secondary forests, etc.;
b) to develop appropriate techniques to establish indigenous tree species within varying site conditions;
c) to better understand ecosystem development of sites after being planted with indigenous tree species;
d) to build scientific knowledge and understand better the species selected for planting;
e) to create productive environment within the degraded sites;
f) to rebuild the landscape of the degraded sites.

Scope of study

This joint research is an attempt to explore technology in nursery practice, planting and site preparation required for indigenous tree species. The study involves trial planting of different indigenous tree species on varying site conditions, experimenting with different site preparation techniques, developing appropriate nursery practices for producing quality planting stocks. This study is also aimed to serve as a model of rehabilitation on degraded and using potential indigenous species. The primary goal of this rehabilitation project is to take degraded land and make it productive.

Description of study area

The study is conducted on a 47.5-ha site within the branch campus of Universiti Pertanian Malaysia in Bintulu, in the state of Sarawak, about 600 km north-east of Kuching, latitude 3° 12′ N, longitude 113° 05′E (Figure 1). Meteorological records obtained from Weather Service Station at Bintulu Airport approximately 9 kilometers from campus indicate a 12-year average rainfall of 3643.4 mm, with a standard deviation of 502.8 mm. The rainy season occurs in the months of November to January, during "Landas Season" of north-east monsoon (Figure 2a and b). In 1990–1991 there was a prominent drought from May until August. The mean annual rainfall readings for 1990 and 1991, of 2993.6 mm and 3221.7 mm, respectively, are low compared to the average of annual rainfall of the previous twelve years (1980–1992). The mean daily temperature is 26.7 °C and relatively consistent above 80 percent throughout the year (Figures 3 and 4) but very low values of saturation pressure deficit occur regularly around noon and may become severe during episodic drought events (Bruenig 1971).

The location of the research site is shown in Figure 5.

The soil in the Bintulu campus of Universiti Pertanian Malaysia belongs to *Nyalau* and *Bekenu* series, which are well drained. The Nyalau series is characterized by coarse loam, yellowish brown topsoil 9 cm deep, with a brownish yellow subsoil. The Bekenu series is characterized by mixed fine loam, light yellowish brown topsoil 4–15 cm deep and brownish yellow subsoil (Peli et al. 1984).

The project will be conducted at four (4) phases.

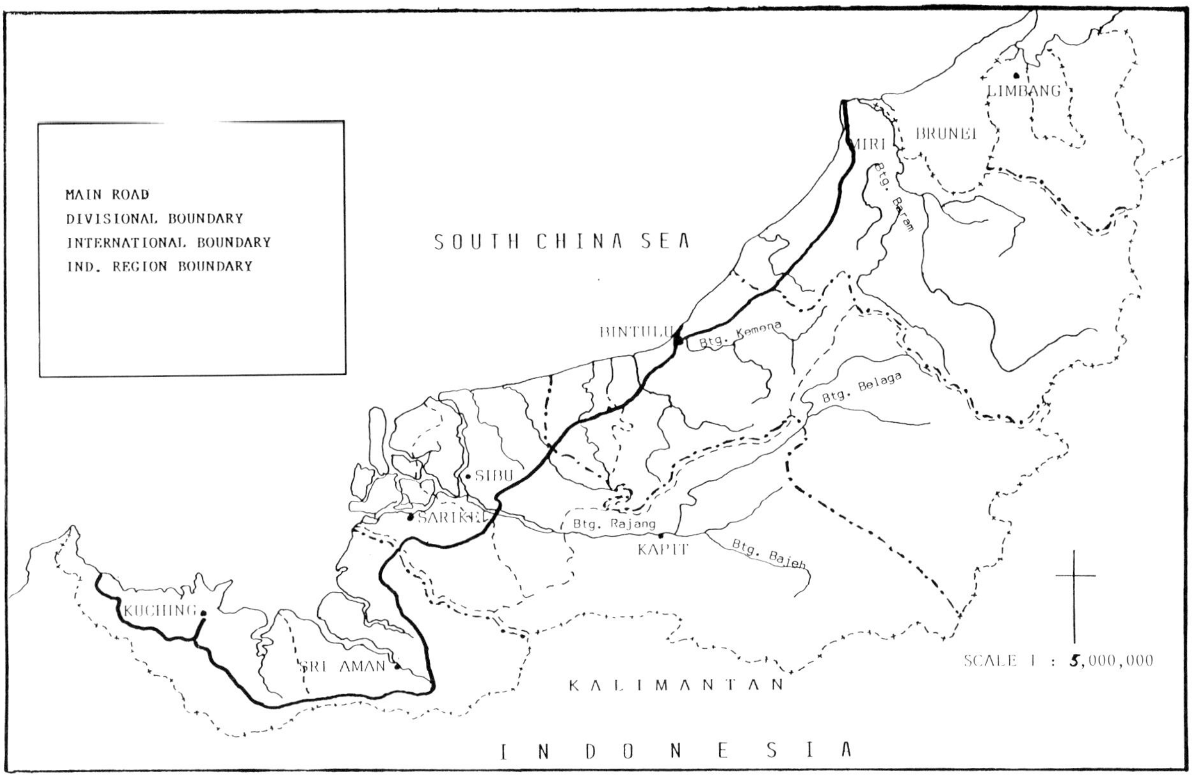

Fig. 1. Map of Sarawak.

Phase	Year	Area
1	1990–1991	3.4 ha
2	1991–1992	3.7 ha
3	1992–1993	7.0 ha
4	1993–1994	33.4 ha

Project activities

First year activities

The following activities were carried out during the first year.

Activity 1: Production of planting stocks
Planting stocks were obtained from two sources, i.e. germinated seeds or collection of seedlings from the forest floor. Seed and seedlings were collected mainly from forested areas around Bintulu. Some seeds were purchased from the people living within the fringe of the primary forest. Purchased seed were *Shorea* spp., *Dipterocarpus* spp., *Dryobalanops* spp., and *Eusideroxylon zwageri.* This ongoing activity started in October 1990. Seeds and seedlings collected are germinated and raised in the nursery. Germination rates of collected seeds have been encouraging, but survival of seedlings collected from the forest floor (wildings) were not satisfactory.

Two problems encountered are the irregular fruiting of Dipterocarp tree species and increasingly smaller areas of undisturbed forests. Fresh seeds could only be collected during the fruiting season. However, the fruiting of Dipterocarpaceae is irregular and has no definite seasons. In addition, accessibility to tree seed source is becoming more difficult since most lowland forested land have been harvested or converted into agricultural plantation and other land development. Most undisturbed forests are located very far and difficult to reach. Therefore, the choice of Dipterocarp species for research is indeed limited.

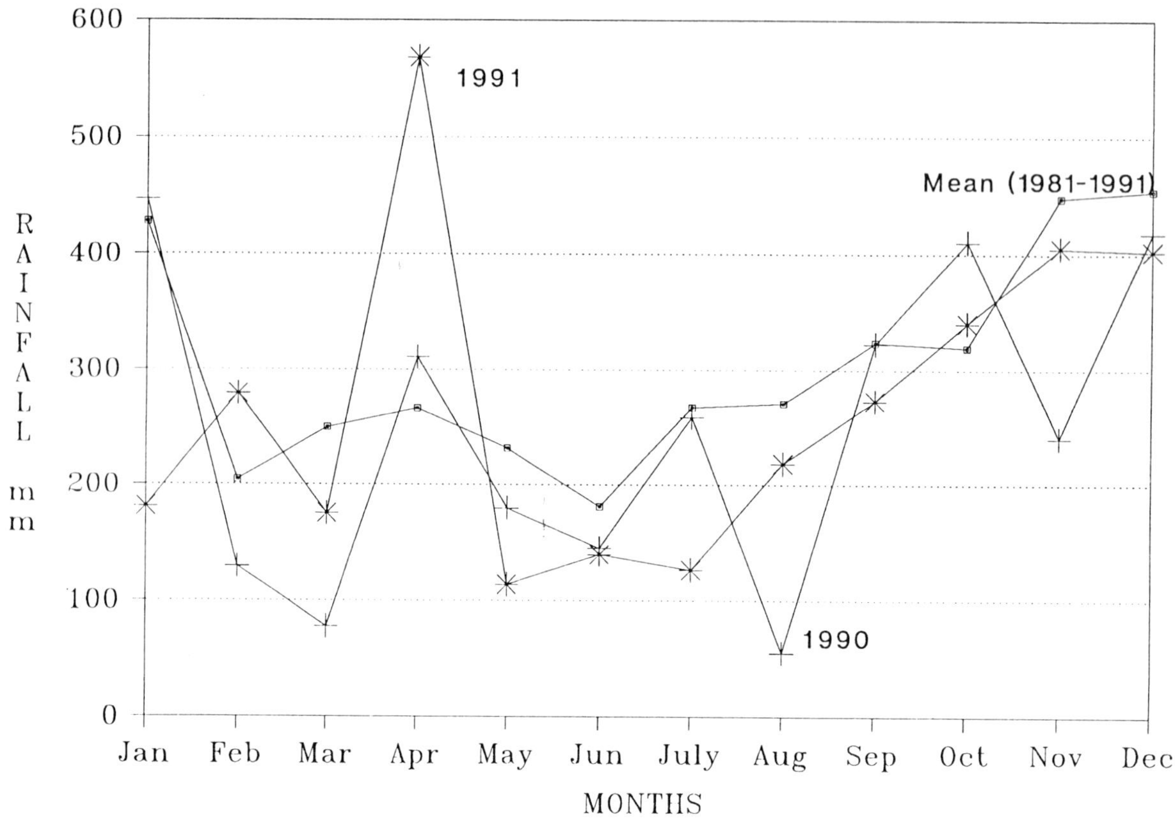

Fig. 2a. Bintulu rainfall records.

Activity 2: First plot site preparation

The first phase activities were conducted on a 3.4-ha grassland area. The existing vegetations were mainly grass species of *Ischaemum magnum* and *Miscantus floridulus* and woody species of *Trema orientalis.* Two different techniques of site preparation were adopted. **Method A**, a soil reconstruction technique, is for areas of severe soil erosion and compaction. In this method, site preparation involves several processes: (i) removal of existing vegetations, (ii) construction of wooden lattices to minimize soil erosion, (iii) ploughing to reduce soil compaction and improve aeration, and (iv) soil mulching using removed vegetation to further reduce erosion and enrich the soil (Figure 6). **Method B** is for areas where the condition of erosion and soil compaction is less severe. This method does not involve mechanical soil restoration but requires only removal of vegetation in line strips of one meter width, retaining half-meter strips of existing vegetation to provide shade to the planted seedlings (Figure 7).

Activity 3: First plot planting

Planting in the first phase began in March 1991 and extend to the end of 1991. Plantings were carried out carefully to ensure higher survival rate. Seedlings selected for planting were those with well-developed root systems. On the sites prepared with Method A, three seedlings were planted in one square meter. On site prepared using Method B, planting was carried out with 1×1 m spacing. Two major problems were

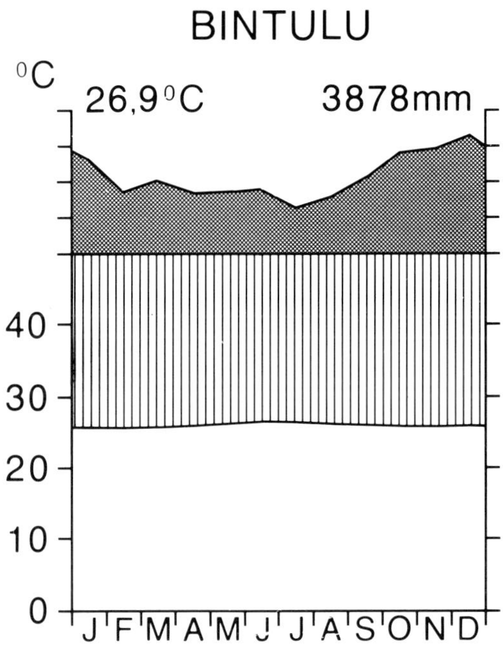

Fig. 2b. Climate diagram of Bintulu.

encountered in the early stage of planting. Firstly, planting coincided with the draught, which resulted in stress to the planted seedlings. Secondly, inadequate supply of seedlings which are ready to be planted interrupted the programme of planting and resulted in unavoidable delays. Data on planted seedlings are shown in Table 1.

Activity 4: Maintenance of the planted plot
Maintenance is required in the first two years after planting in an attempt to minimize competition from weeds and to enhance growth. A survey conducted at the sites showed that the most common weeds competing with seedlings were *Ischaemum magnum, Ludwigia hyssopifolia, Trema orientalis* and *Eleome rutidosperma* (Isa and Tawan 1992). The control measures experimented include manual weeding and spraying with weedicides. The planted seedlings were applied with fertilizer once every three months to facilitate growth.

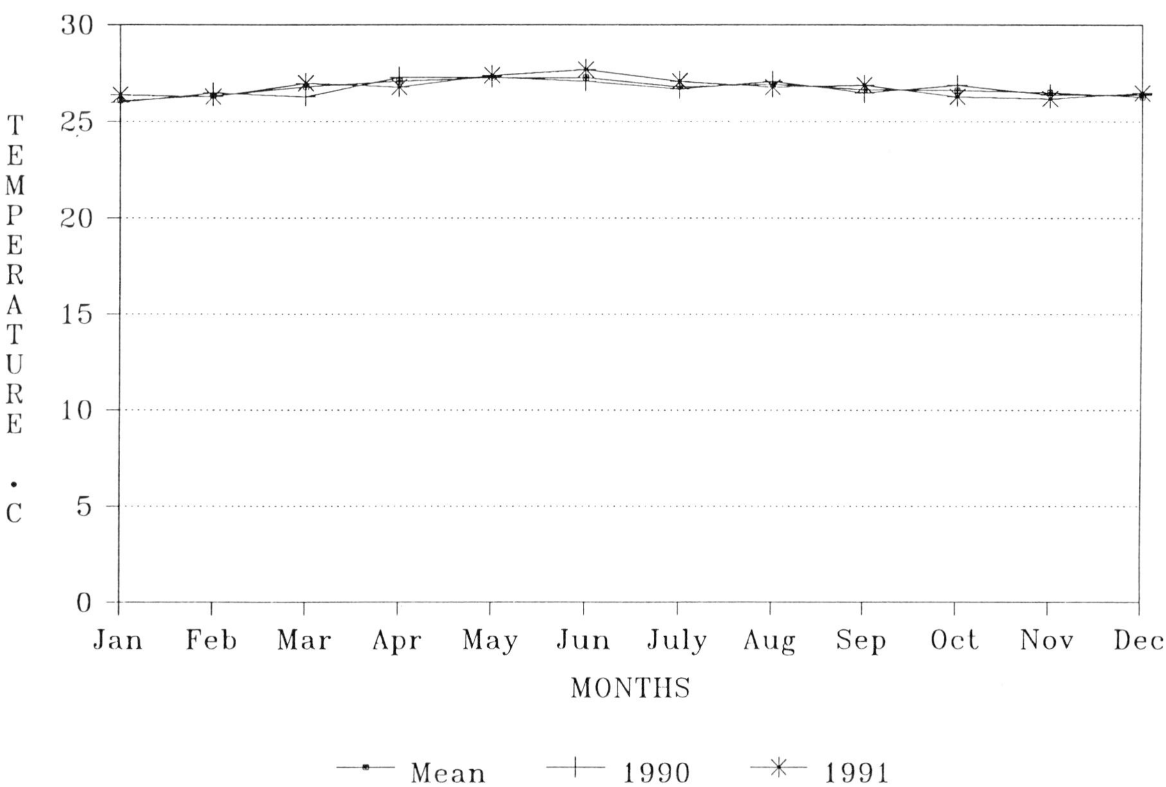

Fig. 3. Bintulu mean temperature range.

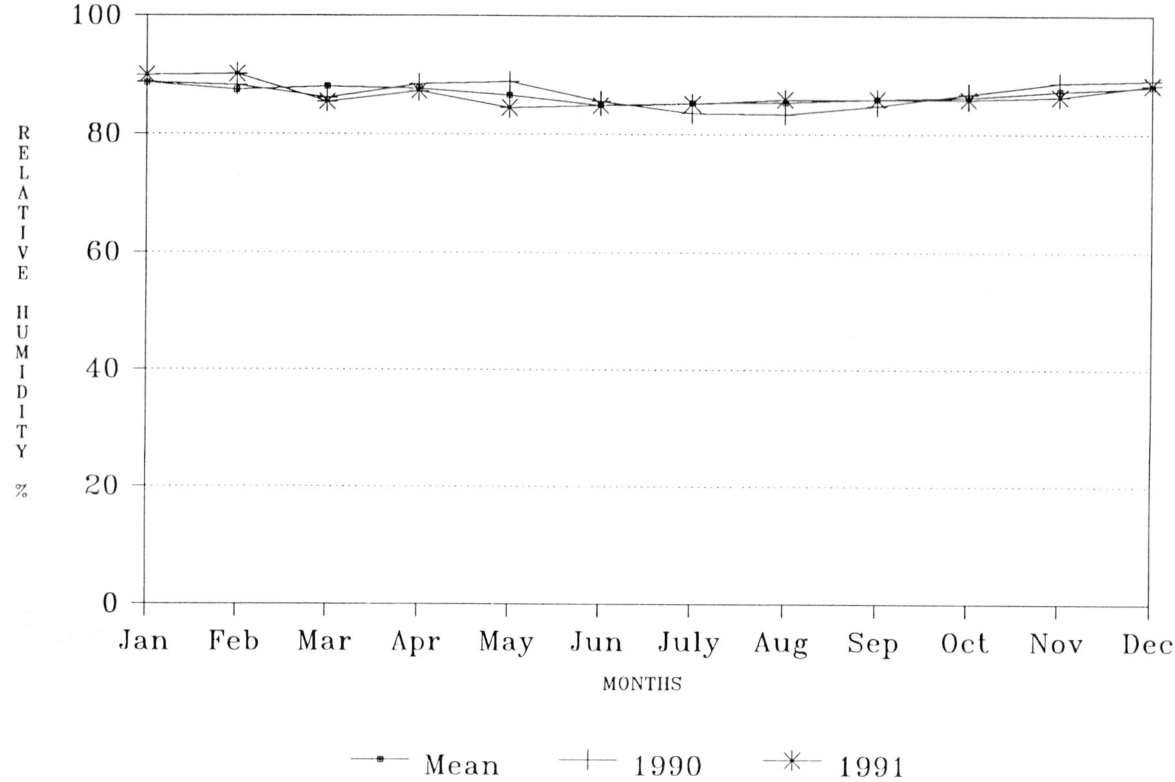

Fig. 4. Bintulu mean relative humidity records.

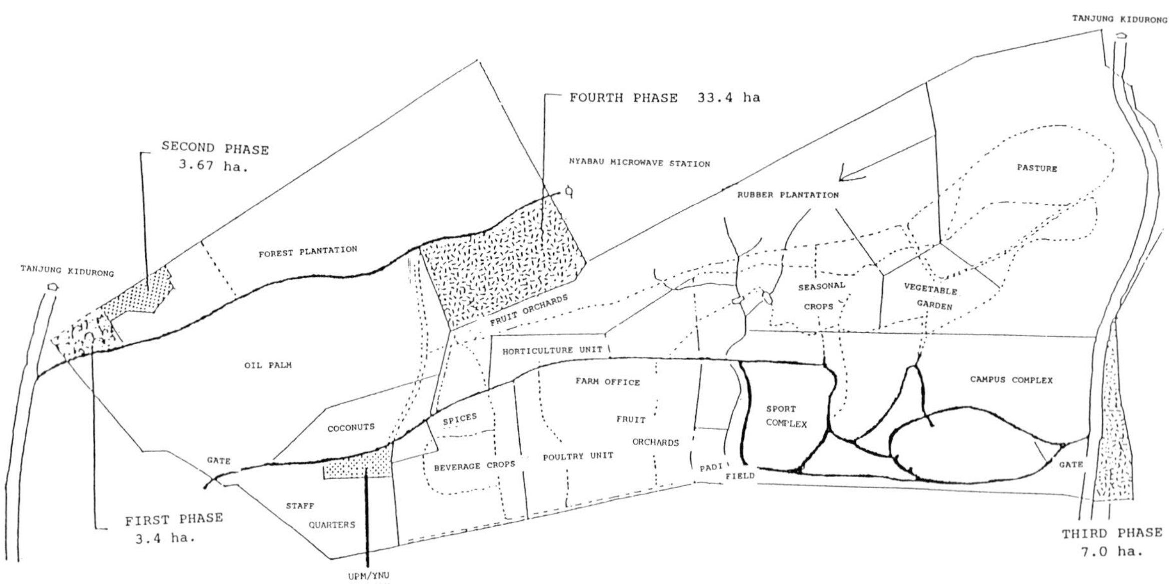

Fig. 5. Research project sites.

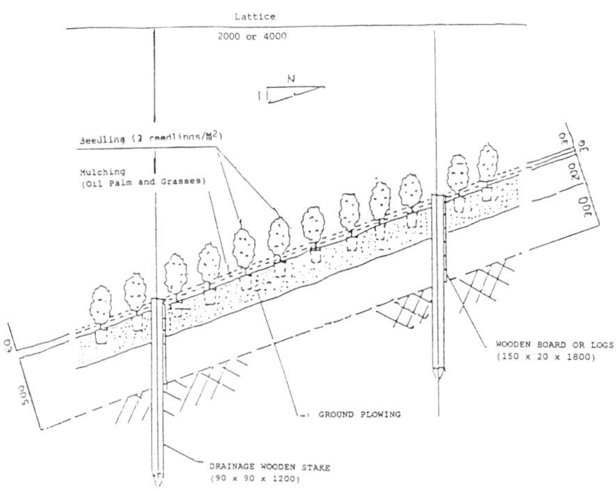

Fig. 6a. Cross-section through a slope plantation.

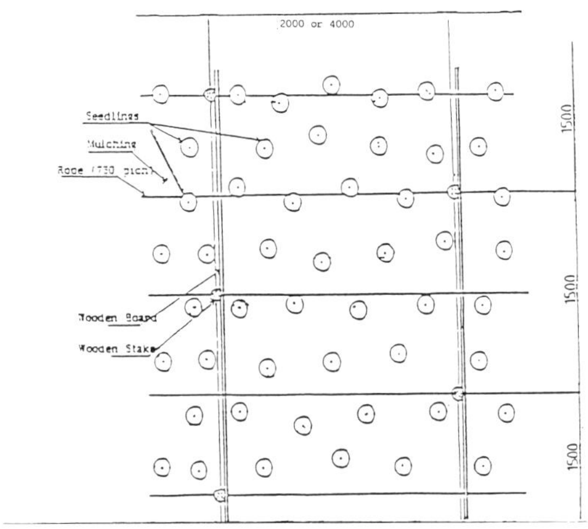

Fig.6. Planting pattern of seedlings.

Second year activities

Activity 1: Production of planting stocks
Collection of seeds and seedlings continues in the second year to ensure sufficient supply of materials for planting in the later stages. In 1991, seedling collection was emphasized as not many Dipterocarp trees were found flowering and fruiting. However, seeds were collected and germinated from some Dipterocarp species which were fruiting.

Activity 2: Second phase plot preparation
The second phase is a 3.7-ha site located adjacent to the first plot. Vegetation consisted of germinated family, mainly *Ischaemum magnum* and *Miscantus floridulus* and secondary species trees dominated by *Melastoma* spp. and *Macaranga* spp. As in the first phase, Method A site preparation was adopted in areas with several eroded and compacted soil.

On sites covered with secondary forest species, Method C was adopted. It involves retaining the canopy of the secondary forest species to provide shade and removal of all the undergrowth to minimize competition.

Activity 3: Second phase plot planting
In areas involving Method A, the dense planting of three seedlings per meter square was carried out. In areas involving Method C, line planting under the forest canopy with spacing of 3×3 m was conducted. Planting in both areas are still in progress and expected to be completed in May 1992.

Progress of project

Despite the prolonged drought and limited knowledge on indigenous species, the project has been progressing as planned.

a) Seedlings have been successful raised in the nursery. The number of seedlings available in the nursery as shown in Table 3.
b) The number of seedlings planted in the first and second phase site are 42 232 and 11 971, respectively. The survival of planted seedlings have been encouraging.

Problems and recommendations

During the project implementation, several problems have been encountered which include:

i) In 1990–91, the prolonged drought has been the major cause of concern. Drought has affected the planting programme and mortality of the seedlings, both in the field and in the nursery.
ii) Unpredictable flowering and fruiting of Dipterocarpaceae has made it difficult to plan the production of sufficient planting stocks.
iii) Wildings have been unreliable as collected wildings have shown higher rates of mortality.
iv) Lack of information and understanding on the requirements of indigenous species has

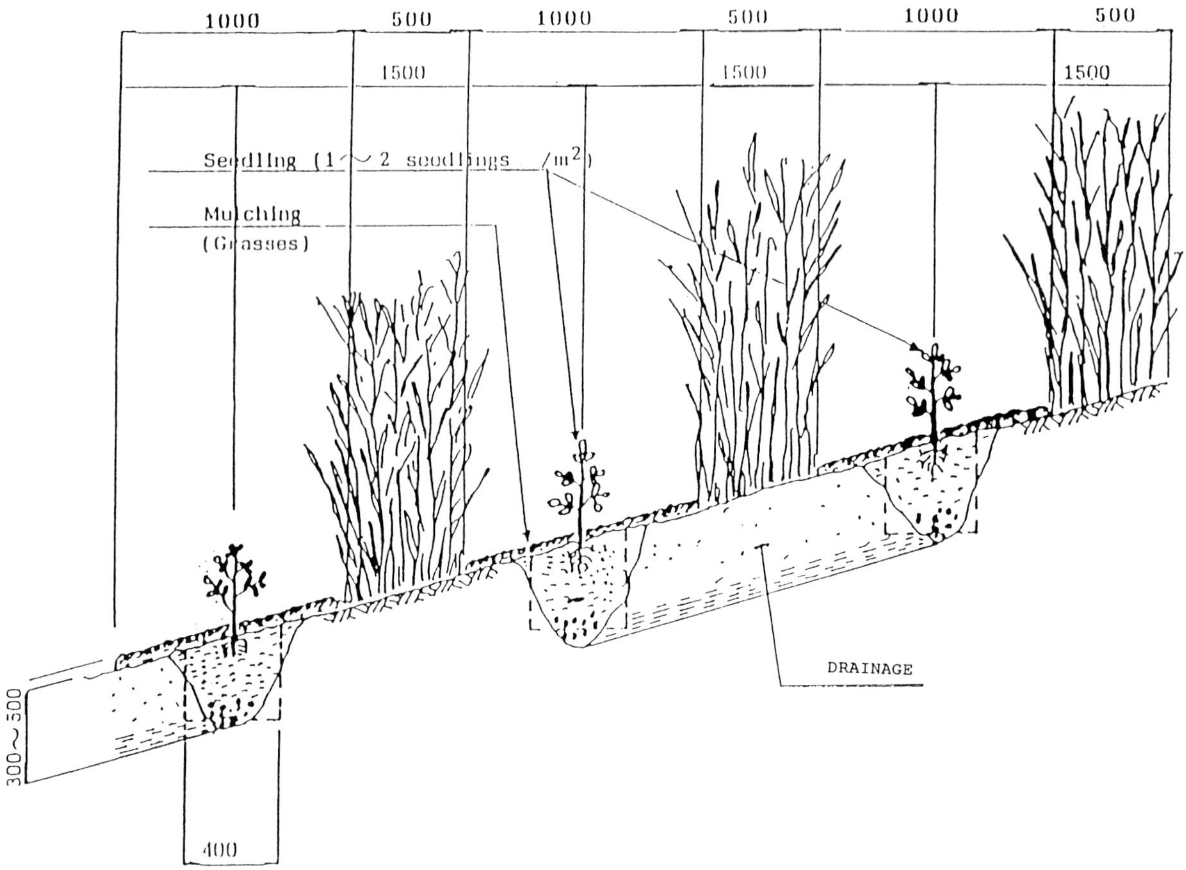

Fig. 7. Method B site preparation and plantings.

Table 1. Status of the first plot. Sites prepared using Method A and Method B

	Planted	Mortality	% Survival	Replanted
(i) Method A Dense planting (3 seedlings per 1 m²)				
Sub-plot A	2 757	882	68.0%	715
Sub-plot B	1 073	618	42.4%	430
Sub-plot C	1 797	744	58.6%	689
Sub-plot D	5 273	2 312	56.2%	1 247
Sub-plot E	26 363	6 595	75.0%	863
Total	37 263	11 151	–	3 944
(ii) Method B Pot hole planting	1 276	138	89.2%	–
Total A + B	42 232	12 284	70.9%	3 874

Table 2. First assessment during planting of the second plot. Sites prepared according to Method A and Method C

	Planted	Mortality	% Survival
Method A Dense planting (3 seedlings/m²)			
Sub-plot A2	10 357	na	na
Sub-plot B2	1 252	na	na
Sub-Total	11 609	na	na
Method C Line planting (3m × 3m)	362	20	94.4%
Total A + C	11 971	–	–

na — Data not available.

Table 3. Total number of seedlings in nursery

Species		Number
Scientific name	Common name	
1. *Shorea ovata*	Meranti Pitis	48 511
2. *Shorea mecistopteryx*	Meranti Kawang Burung	36 254
3. *Hopea kerangachsensis*	Luis	11 773
4. *Shorea leprosula*	Meranti Tembaga	10 174
5. *Hopea beccariana*	Merawan	3 545
6. *Eugenia* sp.	Ubah	2 907
7. *Calophyllum* sp.	Bintagor	2 458
8. *Drybalanops beccarii*	Kapur	1 855
9. *Pentaspadon motleyi*	Pelajo	1 760
10. *Palaquium* sp.	Nyatoh	1 709
11. *Vatica nitens*	Resak	1 281
12. *Parashorea parvifolia*	Urat Mata	1 072
13. *Durio zibethinus*	Durian	774
14. *Shorea materialis*	Selangan Batu	545
15. *Shorea macrophylla*	Engkabang Jantung	462
16. *Eugenia* sp.	Kelat	401
17. *Litsea* sp.	Medang	141
18. *Dracontomelon dao*	Sengkuang	82
19. *Whiteodendron moultonianum*	Kawi	81
20. *Artocarpus rigidus*	Terap	57
21. *Dacryodes* sp.	Seladah	30
22. *Garcinia* sp.	Kandis	18
23. *Elasteriospermum tapos*	Tapos	4
24. Other species	-	1 978
Total		127 872

caused planting to be conducted on a trial and error basis.

v) Control of weeds after planting requires a lot of effort as the climate is conducive to weed growth.

The following recommendations need to be considered to address the above problems:

i) Studies using alternative methods of producing planting stock are essential to meet the need for larger-scale replanting projects.

ii) Studies need to extend beyond the technology of rehabilitation. There is a need to encompass more scientific components such as understanding the varying requirements of indigenous species. This approach would be more beneficial for such a rehabilitation programme.

iii) Further training and exposure need to be granted to the local researchers involved in this project. Knowledge and experience are vital not only for personnel development but also to help handle future research in rehabilitation.

Acknowledgement

I would like to express my cordial appreciation to Mitsubishi Corporation for funding this project; Prof. Dr. Akira Miyawaki (Director, Institute of Environmental Science and Technology) and his colleagues for their continuous effort to lead and guide the implementation of this project; research members from Bintulu, Mr. Ismail Adnan Abdul Malek, Mr. Mohd. Zaki Hamzah and Mr. Othman Bojo for their commitment and cooperation to ensure the success of this project; Mr. Mohamad Azani Alias (Project Manager) for his effort in the field; and Associate Professor Dr. Ibrahim Mohd. Yusof (Director, UPM Campus Bintulu) for his support and permission for the use of the experimental sites.

References

Bruenig, E. F. 1971. On the ecological significance of drought in the equatorial wet evergreen (rain) forest of Sarawak (Borneo). Trans. First Aberdeen-Hull Symposium on Malaysian Ecology. Univ. Hull, Dept. Geogr., Misc. Ser. No. 11: 66–97.

Ipor, I. B. & Tawan, C. S. 1992. Some notes of common weeds in rehabilitation sites of YNU–UPM Joint Project (Unpubl.), 5 pp.

Peli, M., Ahmad Husni, M. H. & Ibrahim, M. Y. 1984. Report and map of the detailed soil survey of UPM farm, Bintulu Campus. Sarawak Bulletin Teknikal No. 1, 73 pp.

11. Rehabilitation of tropical forests from countryside to urban areas

KAZUE FUJIWARA
Yokohama National University, Japan

Keywords: Corridor, dense plantation, eco-technology, green patch, landscape, native species, rehabilitation, tropical forest

Abstract. Tropical forests have been reduced drastically in the last ten years. Tropical forests in the Philippines, Thailand and the Malay Peninsula have disappeared particularly rapidly. The Philippines, Indonesia and Thailand stopped exporting tropical timber in the 1980s. Some nearly natural tropical forests remain in some parts of the national park area. Even national parks do not have real virgin forests. Most are younger natural forests and/or substitute vegetation such as dry dipterocarp forests, or *Macaranga* secondary forests, *Trema orientalis* scrub, etc. Montane tropical forests still barely remain above 800 or 1000 m. In the lowland the original landscape has completely changed from mangrove and lowland tropical forests to urban and rural landscapes.

Most urban landscapes are similar, with trees planted along roadsides, residences and/or in buffer zones of commercial space. They are classified into two types of landscaping, as dominated by areas planted with either *Pterocarpus indicus* or *Tabebuia*. *Pterocarpus* plantation areas indicate areas of potential tropical rainforest, while *Tabebuia* plantings indicate areas of seasonal tropical forests.

The theory for rehabilitation of tropical forests should include three key points: 1. Rehabilitation of tropical forests in their original places, such as areas of abandoned shifting cultivation, areas eroded after cutting, etc. (mainly countryside); 2. Creation of harmonious landscapes, combined with rehabilitation of tropical forests in urban areas; and 3. Integrated rehabilitation of tropical forests and regional planning in rural areas.

1. Introduction

There are three so-called "laws of ecology": 1. Everything is connected to everything else; 2. Everything has to go somewhere; and 3. There is no such thing as a free lunch. But this is our way of life, and terrestrial environmental problems threaten its very existence. This includes tropical forests and Japanese forests. Tropical forests include not only countryside but also urban areas. Rehabilitation of tropical forests is needed in both country and urban areas, with corridors of tropical forest connecting these two types of forest area.

In general, people use exotic, beautiful blooming plants everywhere, not only in urban areas but also in the countryside. As a means of combatting the crisis of natural forests and diversity in ecosystems, we propose to combine the rehabilitation of natural forests, including tropical forests, with general landscaping. We published several results of rehabilitation of natural forests in Japan (e.g. Miyawaki et al. 1984; Miyawaki 1984; Fujiwara 1984). We then began to rehabilitate tropical forests in Thailand and Malaysia (Fig. 1). Now it is proposed to transfer tropical forest rehabilitation methods from the countryside into urban areas.

2. Method: Short way to rehabilitate natural forests

In Europe people have experience and a history of rehabilitating and recreating natural forests, using pioneer trees. These forests are preserved as nature conservation areas. In Japan Prof. Miyawaki introduced this method in 1969 (Miyawaki and Fujiwara 1969; Miyawaki et al. 1974). Then he developed the so-called "Miyawaki Method", which is to restore the topsoil on sites and then plant seedlings of the canopy species of the local potential natural forest, based on phytosociological field surveys (Braun-

H. Lieth and M. Lohmann (eds.), Restoration of Tropical Forest Ecosystems, 119–131.

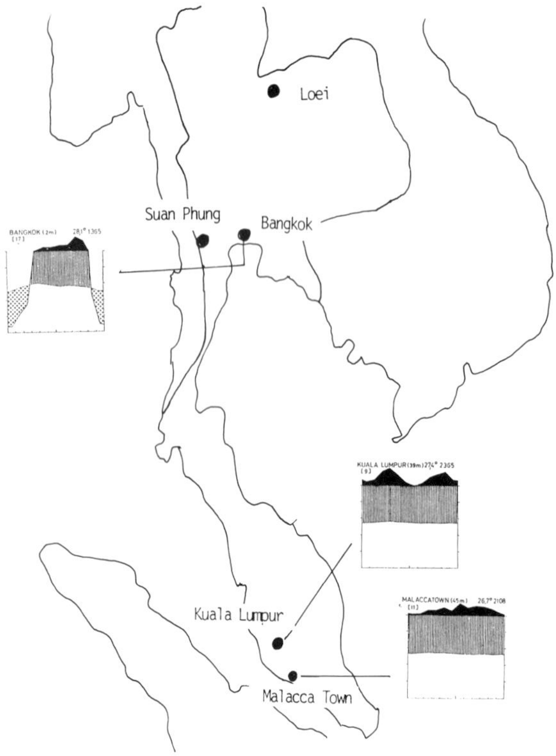

Fig. 1. Location of experimental sites for rehabilitation of tropical forests in the tropical Asia.

Blanquet 1928, 1932, 1951, 1964; Ellenberg 1968; Mueller-Dombois and Ellenberg 1956; Fujiwara 1987; Tüxen 1956). At first we planted one individual per square meter. Afterward two and/or three individuals per square meter were tried for normal sites and six to nine individuals per square meter in severe environments. Dense plantations give faster growth with less requirement for maintenance than plantations of individual trees. Also, recovery of the topsoil promotes the growth of seedlings without succession (succession develops topsoil). This lack of requirement for maintenance is important, since Japanese officers and/or company people change positions every two or three years and their plantation projects must show success within this period of time.

What is different between rehabilitation of native forests by Miyawaki's method and more common (forestry) plantation methods? The following can be cited: 1. Different methods of species selection from the forestry method. We select and use as many native canopy species as possible, based on the potential natural vegetation; 2. Mixed plantations, not single-species plantation; 3. Use of pot seedlings with well-developed root systems (with height of up to 80 cm); 4. Soil preparation, including provision of good drainage and use of organic fertilizers such as compost, weeds, dropped leaves, etc. In urban areas, trash, useless wood, broken blocks, concrete blocks, etc., are put into mounds. Mounds are useful in areas with bad drainage; 5. Dense plantation (3–9 individuals per square meter); 6. Mulching with rice straw, leaves, etc., for protection against soil dryness and loss of nutrients; 7. No management after two or three years from planting (Miyawaki 1984; Fujiwara 1984).

Why do Japanese people want natural forests so much? Generally people like exotic plants and/or beautiful blooming plants more than natural forests. In Japan environmental problems are related to Japanese social history after the Second World War. After World War II, people worked very hard to revive their economy; also they destroyed natural environments for construction of roads, buildings, new towns, reclaimed land, factories, etc. Then pollution followed and natural resources were lost. In the 1960s, people recognized the crisis of the destruction of natural resources and rose in an active movement to preserve natural resources. In the 1970s, pollution became a serious problem and pollution-related illnesses increased. Professor Dr. Akira Miyawaki proposed to rehabilitate natural forests as environmental protection forests for monitoring of pollution and restoration of natural environments. Shin Nippon Steel Company (Shin Nittetsu) tried the so-called "Miyawaki Method" at nine factories from Kyushu to Hokkaido, beginning in 1972. Then many companies made environmental protection forests at their sites, such as Tokyo Electric Company, Kansai Electric Company, Kyushu Electric Company, Honda Co. Ltd., Tore Co. Ltd., etc. Local governments, such as Kanagawa Prefecture, Yokohama City (Kanagawa Pref.), Fujisawa City (Kanagawa Pref.), Fuji City (Shizuoka Pref.), Chatan-cho (Okinawa), Okayama Prefecture, etc., did likewise, either at the same time or soon after (see this volume, p. 7, Fig. 1). I worked with Prof. Miyawaki, and I proposed

Table 1. The growth record of seedlings, from plantation in Malacca, Malaysia (surveyed on Jan. 21, 1992)

No.	Species	Height (cm)	Diameter (mm)	No.	Species	Height (cm)	Diameter (mm)
1	*Cinnamomum iners*	150	25.2	16	*Baphia nitida*	52	13.7
2	*Pterocarpus indicus*	200	27.4	17	*Eugenia grandis*	120	16.1
3	*Cinnamomum iners*	128	15.2	18	*Cinnamomum iners*	124	16.0
4	*Pterocarpus indicus*	270	33.1	19	*Shorea* sp.	30	3.2
5	*Cinnamomum iners*	156	21.0	20	*Pterocarpus indicus*	170	24.4
6	*Khaya grandiflora*	64	13.1	21	*Khaya grandiflora*	93	13.7
7	*Hopea odorata*	30	4.8	22	*Baphia nitida*	130	15.3
8	*Pterocarpus indicus*	235	32.9	23	*Eugenia grandis*	174	25.0
9	*Baphia nitida*	100	15.1	24	*Baphia nitida*	70	13.1
10	*Eugenia grandis*	170	19.5	25	*Pterocarpus indicus*	170	17.9
11	*Cinnamomum iners*	145	27.3	26	*Eugenia grandis*	190	23.5
12	*Baphia nitida*	93	13.3	27	*Khaya grandiflora*	97	12.0
13	*Hopea odorata*	53	8.8	28	*Eugenia grandis*	140	18.7
14	*Pterocarpus indicus*	260	37.2	29	*Baphia nitida*	110	15.0
15	*Hopea odorata*	80	10.0	30	*Khaya grandiflora*	75	14.2

the species for planting based on potential natural vegetation in each site by the phytosociological method.

3. Re-creation of environmental forests in urban areas

Why is rehabilitation of "native forests" needed in urban areas? During the Bonn Symposium one participant asked me. There is not so much space in urban areas. People in Europe, however, traditionally did preserve and create green patches, such as parks, gardens, allées (tree-lined avenues), etc., with native trees and some exotic species. We studied the restoration, rehabilitation and/or re-creation of native forests in urban areas and found it very valuable for city planning to keep functional ecosystems and living beings in urban areas. Parks, common spaces, and/or open spaces can become green patches. Green belts along rivers, roads, railroads, coastal areas, etc., and hedges can connect green patches in urban areas and make green networks.

Rehabilitation of tropical forests was done at JUSCO super shopping markets in Malacca and Kuala Lumpur (Malaysia) and in Lattanathibett (Thailand). Species were selected from the composition of the native forest for their rehabilitation and from characteristic landscaping species of these regions. All the sites are limited, with widths ranging from one to 16 m. Since all are on level terrain, mounds were constructed, from 0.8 to 8 m high depending on the site. Construction waste such as rocks, gravel, lumber, etc. from construction sites was put into mounds, and soil and topsoil, which have some nutrient content, were put on top of the mounds. Then pot seedlings with well-developed root systems were planted.

1. Malacca in Malaysia

The annual precipitation is 2210 mm and the mean annual temperature is 26.7 °C. Malacca has a rainy season from April to November, with a dry season in winter. Ayer Keroh Recreational Forest is located in Malacca. This forest is a young tropical forest. Probably the big trees were already cut down and used for timber. The remaining trees are preserved as a recreational forest. These are mostly *Artocarpus scortechinii*, *A. lanceifolius*, *Elaeocarpus* sp., *Palagium malagayi*, *Dipterocarpus* sp., *Dryobalanopsis aromatica*, *Shorea* sp., *Terminalia fedipisimas*, *Alangium* sp., *Garcinia arroviridis*, *Eugenia anisophylla* in the canopy layer.

Then seedlings were checked in nurseries. Seedling species, such as *Cinnamomum iners*, *Pterocarpus indicus* (angsana), *Dipterocarpus* spp., *Eugenia grandis*, *Milletia atropurpurea*, *Mimusops elengi* (bunga Tanjong), and *Khaya grandiflora*, were selected as the main trees to plant at the site of JAYA JUSCO in Malacca. Pot seedlings of 50 cm height were planted on the mound in August 1991, densely, three individuals per square meter. They grew to 150～270 cm by January 22, 1992 (Table 1). Plantation areas were behind and on the west side of the JAYA JUSCO building. The areas are mounds one to five meters wide and 0.5～1.5 meters high, as strips surrounding the site. Exotic tropical blooming-shrub species, such as Ixora sunkist (Malacca's symbol flower), *Durantha repens*, and *Muraya paniculata* were selected as the mantle community, and are planted as a 30 cm border around the mound. These are planted also especially at the front of the JAYA JUSCO Malacca site, along with tree species such as *Cassia fistra* (Malacca's symbol tree), *Samanea saman* (rain tree), Eugenia grandis, *Khaya* spp., *Peltophorum pterocarpum* (Yellow frame), *Ficus benjamina* (native), and shrub species such as *Bougainvillea* spp., *Hibiscus rosasinensis*, *Mussaenda* spp., *Allamanada cathartica*, *Ixora sunkist*, *Durantha repens*, *Murraya paniculata*, and *Crytostachys lakka* (red palm, native species). They are planted especially around parking areas. *Delonix regia* was not selected because it does not grow straight or as tall as in further north of tropical Asia, where there is a better developed dry season. After plantation, leaves of oil palm are used as mulching for protection against evaporation and soil erosion. In general rice straw is used for mulching. Since this is difficult to get in Malacca, leaves of oil palm are used instead.

2. Wangsa Maju in Kuala Lumpur

The annual precipitation is 240 mm and the mean annual temperature is 26.5 °C in Kuala Lumpur. Kuala Lumpur has two seasons of elevated rain, from March to May and from September to December with two drier seasons in-between. Potentially Kuala Lumpur is a tropical rain forest area. In the urban area we can see *Pterocarpus indicus* (local name: ansana) planted for allées, which grows very well and fast.

Wangsa Maju site is located next to a highway and is smaller than the Malacca site. Seedlings of native tropical species are planted on the slope next to the highway, which has a steep slope of 30～40 degrees. The Malay Peninsula has good experimental sites for planting tropical species. The Forest Research Institute Malaysia Arboretum (FRIM) was planted in 1932. It was green vegetable fields before plantation. Plantation areas in FRIM, with many species of *Shorea* (lauan), *Dryobalanopsis aromatica* (kapur), *Hopea odorata*, and *Dipterocarpus* spp., became real forests with several understorey species. Even in Kanching Recreation Forest, Selangor Province, *Dryobalanopsis aromatica* forest develops on the steep southeast slope of 60 degrees, from a plantation in 1935 (Table 2). The natural slope forest in Bukit Nanas is preserved in Kuala Lumpur. Its main canopy species are *Shorea bracteolata*, *Dryobalanopsis aromatica*, *Alstonia angustiloba*, *Garcinia arroviridis*, *Sapium baccatum*, *Knema laurine*, *Eugenia virens*, *Palaquim obovatum*, *Litsea costalis*, *Pometia pinnata*, *Arthocarpus clasticus*, etc. (Table 3). Possible seedling species to get from nurseries of UPM and FRIM were selected. These include *Shorea macroptera*,

Table 2. The vegetation composition of the plantation area in Kanching Recreation Forest in Selangor (June 29 in 1991, relevé by Fujiwara and Aschari)

Tree-1 layer: 40 m 80%, Tree-2 layer: 18 m 40%, Shrub layer: 4 m 40% Elevation: 100 m. Slope and exposition: SE 60°

Tree-1 layer			
3·4	*Dryobalanopsis aromatica*	2·2	*Dilenia reticulata* (30 m)
2·2	*Arthocarpus scortechinii*	(1·1)	*Shorea ovalis*
1·1	*Shorea bracteolata* (40 m)	1·1	*Campnosperma* sp. (33 m)
Tree-2 layer			
2·2	*Hopea odorata*	2·2	*Dryobalanopsis aromatica*
1·2	*Strobulus elongatus*		
1·2	*Pimeleodendron griffithianum*	1·1	*Pometia pinnata*
		1·1	*Litsea grandis*
1·1	*Dyera costulata*	1·1	*Arthocarpus scortechinii*
+	*Eugenia virens*	+	*Endospermum malaccensis*
Shrub layer			
2·2	*Peternandra* sp.	1·1	*Calamus* sp.
1·2	*Ochlanostachys amentacea*	+	*Randia scortechinii*
(+)	*Sapium baccatum*		etc.

Table 3. List of the main plants at Bukit Nanas in Kuala Lumpur

1. *Alstonia angustiloba*	17. *Baccaurea brevipes*
2. *Sapium baccatum*	18. *Atuna nannodes*
3. *Tristania whitiana*	19. *Dryobalanops aromatica*
4. *Garcinia atroviridis*	20. *Litsea costalis*
5. *Aquilaria malaccensis*	21. *Gironniera nervosa*
6. *Artocarpus elasticus*	22. *Aidia wallichiana*
7. *Scorodocarpus borneensis*	23. *Prunus polystachya*
8. *Albizia falcata*	24. *Eugenia virens*
9. *Elateriospermum tapos*	25. *Memecylon garciroides*
10. *Gironniera nervosa*	26. *Grewia antidesmaefolia*
11. *Shorea bracteolata*	27. *Litsea firma*
12. *Nephelium glabrum*	28. *Shorea* sp. (Balar)
13. *Ixonanthes icosandra*	29. *Dipterocarpus baudii*
14. *Ryparosa fasciculata*	30. *Knema scoriechinii*
15. *Actinodaphne sesquipedalis*	31. *Palaquium obovatum*
16. *Knema laurina*	32. *Gardenia tubifera*

S. parvifolia, S. acuminata, S. singrawang, Hopea odorata, H. lelferi, H. nervosa, Balanocarpus ieimii, Dipterocarpus kustleri, Anisoptera costata, Vitera paucifolia from FRIM (as 50 cm high pot seedlings) and *Calophyllum* spp., *Cinnamomum inerm*, etc. from UPM (as 140~300 cm high pot seedlings). They were planted mixed, four individuals per square meter.

Pot seedlings of 50 cm high were planted on October 5, 1991 and they grew satisfactorily (Table 4). The tallest seedling grew to 170 cm (*Hopea odorata*). Pot seedlings of 140~300 cm high were planted on January 14, 1992. They are not healthy, so additional plantation should be done with 30 cm high seedlings afterward. This experience showed that seedlings are the most important factor for re-creation of natural forest, especially tropical forest.

Exotic species and some native species were chosen for landscape planting, at the front of the building and around parking areas on the site. These include *Fagrea fragrans* (native species), *Michelia cempaka, Crytostachys lakka* (red palm, native species), *Roystonea* spp. (Royal palm), *Pisonia alba, Mussaenda* spp., *Bougainvillea* spp., *Phyllanthus myrtifolius, Ixora sunkist, Allamanda cathartica, Durantha repens, Rheo spatacea, Asplenium nidus, Ophiopogon* cult., *Acalypha siamensis, Hibiscus* cult., *Lantana camora*, and *Cytula* spp. After plantation, rice straw was used for mulching, providing better protection than oil palm leaves.

3. Lattanathibett in Nonthabri, Thailand

The lowland areas in Thailand were mostly developed and changed to urban cities, residences, villages, cultivated fields, etc. Semi-natural forests remained in very small areas or not at all. A village forest, with a canopy of *Dipterocarpus alatus*, was barely preserved in Nong Song Hong, Phakon Saqrakham in Prachin Buri. Its composition is shown in Table 5. Canopy layer is occupied by *Dipterocarpus alatus*. In the understorey many seedlings, three meters high, occurred densely, six to ten individuals per square meter. The following species were chosen: *Dipterocarpus alatus, Hopea odorata, Dalbergia cochinchinensis, Pterocarpus macrocarpus, Peltophorum pterocarpum, Eugenia cumni, Afzelia xylocarpus, Shorea roxburghii*, and *Tectona grandis*. Pot seedlings of various species were planted, four individuals per square meter, and others were planted six individuals per square meter. The growth of native species is shown in Tables 6 and 7.

Bangkok and the surroundings are flat and the ground water is very high. The area is frequently flooded after rain and during the rainy season. Therefore good drainage should be ensured before plantation. Higher mounds with sandy soil were constructed at the SIAM JUSCO site. The bases of these mounds were by construction wastes from the site, and were covered with sandy soil. Some places were especially very narrow (only 80 cm wide) for planting and not far enough above ground water. Therefore concrete walls of 80 cm high were constructed and construction waste, wood, concrete blocks, etc. were put in them (Figure 2). They were then covered by sandy soil mixed with some fertilizer. Originally this area was mud, which had bad drainage. It became hard during dry period and was not fit for plants. Therefore we changed the mud to sandy soil on the top of the mound.

The annual precipitation in Bangkok is 1420 mm and the mean annual temperature is 28.3 °C. Thailand has dry and rainy seasons, half a year each. Bangkok has a rainy season from May to October with a dry season from November to April. In the urban area we can see *Delonix regia, Tabebuia pentaphylla* (Thai cherry blossom), and *Cassia bareriana* as allée

Table 4. The growth record of seedlings planted at Wangsa Maju, Kuala Lumpur (surveyed on Jan. 21, 1992)

PQ 1 (Oct. 5, 1991 planted)				PQ 2 (Jan. 14, 1992 planted)			
No.	Species	Height (cm)	Diam. (mm)	No.	Species	Height (cm)	Diam. (mm)
13	*Hopea odorata*	70	11.8	1	*Calophyllum inophyllum*	140	10.0
14	*Shorea ovalis*	100	10.5	2	*Mesua ferrea*	140	14.8
15	*Hopea odorata*	60	12.5	3	*Calophyllum inophyllum*	300	26.8
16	*Hopea odorata*	90	11.0	4	*Cinnamomum iners*	70	4.0
17	*Alstonia angustiloba*	65	13.4	5	*Cinnamomum iners*	42	3.4
18	*Hopea odorata*	80	13.9	6	*Cinnamomum iners*	170	13.8
19	*Shorea platyclados*	100	11.3	7	dead	175	11.0
20	*Shorea platyclados*	95	9.1	8	*Cinnamomum iners*	180	15.2
21	*Shorea singrawang*	140	16.0	9	dead	210	11.8
22	*Hopea odorata*	120	15.8	10	*Swintonia ferrea*	180	16.4
23	*Hopea odorata*	120	15.5	11	*Mesua ferrea*	210	8.2
24	*Hopea odorata*	110	14.5	12	*Calophyllum inophyllum*	190	20.0
25	*Shorea ovalis*	70	12.3				
26	*Hopea odorata*	47	6.3				
27	*Hopea odorata*	110	6.0				
28	*Alstonia angustiloba*	50	6.9				
29	*Hopea odorata*	83	22.8				
30	*Hopea odorata*	153	17.5				
31	*Hopea odorata*	170	18.0				
32	*Anisoptera* sp.	60	7.8				
33	*Neobalanocarpus heimii*	100	13.8				

Table 5. The summarized vegetation composition of the village forest in Nong Song Hong, Plachinburi, Thailand (Feb. 12 in 1989)

Height and cover of tree layer (T): 20 m 75%, Height and cover of shrub layer (S): 4 m 80%, area: 30 × 50 m², elevation: 10 m (relevé by Tem Schmitnand and Fujiwara)

T	4·4	*Dipterocarpus alatus*		
S	3·3	*Strebulus* sp.	1·1	*Eriograsm* sp
	2·3	*Bambusa arundinacea*	1·1	*Grevia paniculata*
	(2·2)	*Dipterocarpus alata*	1·1	*Capparis micranta*
	1·1	*Artocarpus* sp.	1·1	*Randia bispinosa*
	1·2	*Helictres hirsuta*	+	*Miletia* sp.
	+	*Erythroxylon cambodianum*	+	*Milia javanica*
	+	*Casearea grewiaefolia*	+	*Alstonia scholaris*
	+	*Clerodendron* sp.	+	*Melastoma dilosa*
				etc.

trees and in gardens. These grow very well and more rapidly.

The rainy season was selected for planting. The area planted on July 15, 1991 (still rainy season) is growing better than that planted on October 7, 1991 (end of rainy season). Pot seedlings on the west side of the mound grew to 105 cm (*Hopea odorata*) or 185 cm (*Dalbergia cochinchinensis*) (PQ 1: south facing slope) and to 85 cm (*Hopea odorata*) or 125 cm (*Pterocarpus macrocarpus*) (Table 6). Pot seedlings planted in October on the east side of the mound grew to 57 cm (*Hopea odorata*) or 85 cm (*Pterocarpus macrocarpus*) (Table 7). Monitoring will be continued.

Two types of ornamental trees were selected, adult trees of 6~8 m high and under 10 cm DBH (diameter at breast height) and shrub species. The adult trees are *Delonix regia*, *Pterocarpus indicus*, *Tabebuia pentaphylla* (Thai cherry blossom), *Cassia bareriana*, *C. fistula*, and *Samanea samon* (rain tree). They are not evergreen and not native, but people can enjoy their flowers and shade. They are planted at the very front, next to the main road, and along the main entrance road, with shrub species. The shrub species are *Bougainvillea* spp., *Hibiscus mutabilis*, *Bauchinia integrifolia*, *B. purpurea*, *Ixora*

Table 6. The growth record of seedlings planted in permanent quadrat no. (PQ) 1 in Lattanathibet, Thailand (surveyed on Jan. 20, 1992)

No.	Species	Height (cm)	Diameter (mm)	No.	Species	Height (cm)	Diameter (mm)
1	*Dalbergia cochinchinensis*	185	30.20	16	*Hopea odorata*	80	10.03
2	*Pterocarpus macrocarpus*	110	10.88	17	*Dalbergia cochinchinensis*	130	20.65
3	*Tabebuia pentaphylla*	110	10.93	18	*Hopea odorata*	73	10.80
4	*Samanea saman*	110	10.10	19	*Pterocarpus macrocarpus*	80	10.70
5	*Dalbergia cochinchinensis*	130	30.04	20	*Hopea odorata*	85	10.12
6	*Hopea odorata*	70	10.32	21	*Pterocarpus macrocarpus*	83	10.00
7	*Hopea odorata*	90	10.02	22	*Hopea odorata*	105	10.18
8	*Dalbergia cochinchinensis*	105	30.06	23	*Pterocarpus macrocarpus*	167	10.54
9	*Pterocarpus macrocarpus*	155	10.90	24	*Hopea odorata*	105	10.21
10	*Dalbergia cochinchinensis*	47	20.39	25	*Tabebuia pentaphylla*	137	10.78
11	*Hopea odorata*	78	10.20	26	*Dalbergia cochinchinensis*	70	10.58
12	*Pterocarpus macrocarpus*	130	10.06	27	*Pterocarpus macrocarpus*	120	10.96
13	*Dalbergia cochinchinensis*	130	30.22	28	*Pterocarpus macrocarpus*	57	10.40
14	*Pterocarpus macrocarpus*	120	10.83	29	*Pterocarpus macrocarpus*	56	10.59
15	*Dalbergia cochinchinensis*	130	10.90				

Table 7. The growth record of seedlings planted in permanent quadrat no. (PQ) 2 in Lattanathibet, Thailand

No.	Species	Height (cm)	Diameter (mm)	No.	Species	Height (cm)	Diameter (mm)
1	*Pterocarpus macrocarpus*	125	10.25	9	*Dalbergia cochinchinensis*	140	20.70
2	*Dalbergia cochinchinensis*	120	20.73	10	*Dalbergia cochinchinensis*	130	13.39
3	*Hopea odorata*	85	10.57	11	*Pterocarpus macrocarpus*	67	10.50
4	*Dalbergia cochinchinensis*	150	30.10	12	*Pterocarpus macrocarpus*	155	10.21
5	*Hopea odorata*	70	10.05	13	*Pterocarpus macrocarpus*	170	20.50
6	*Pterocarpus macrocarpus*	20	10.73	14	*Pterocarpus macrocarpus*	160	10.41
7	*Hopea odorata*	60	10.05	15	*Afzelia xylocarpa*	87	10.69
8	*Hopea odorata*	65	10.98	16	*Dalbergia cochinchinensis*	130	20.95

spp., etc., planted under shading trees, and *Murraya paniculata*, *Wedelia trilobata*, and *Jusminum auriculatum* planted next to buildings, six individuals per square meter. They survived and spread branches well within one month after planting.

3. Rehabilitation of tropical forests in the countryside: Regreening Movement in Thailand

The Thai Government prohibits the export of tropical timber since 1986. Most natural tropical forests changed to secondary, dry-dipterocarp forests as a result of burning for shifting cultivation and hunting, and cutting for firewood and/or charcoal. The distribution of natural forests in central to northern Thailand is shown as Figure 3.

The forests in Thailand, which had occupied more than 70 percent of the country before World War II, are less than 30 percent today. The Regreening Movement (Chairman Prof. Dr. Shunji Murai, University of Tokyo) is cooperating with the so-called Princess Project on Conservation of Natural Resources (so called after Her Royal Highness the Princess Maha Chakrit Sirindhorn of Thailand).

Two test sites were selected, in Ratchaburi (western Thailand and near the border with Myanmar) and Loei (north Thailand near the border with Laos).

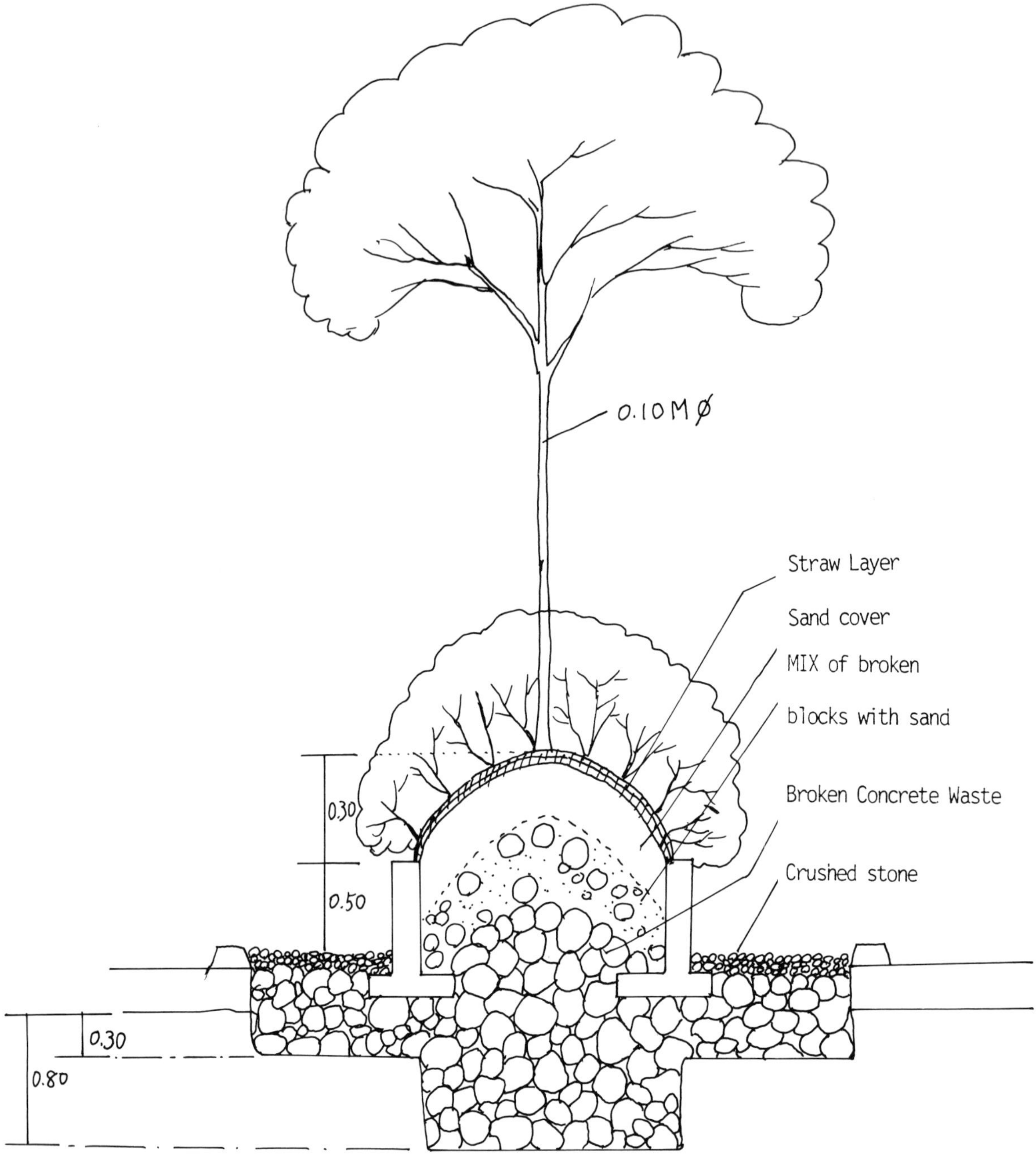

Fig. 2. The mound and planting method for islands between parking-lot sections in Lattanathibet, Thailand.

The actual vegetation in Ratchaburi is abandoned fields and dry dipterocarp forest. The sites have good soil, based on the survey by Dr. Pisoot Vijarnsorn (Figure 4). The potential natural vegetation will be Dipterocarpaceae forest. Pot seedlings from the Royal Forestry Nursery in Ratchaburi were *Sindora siamensis*, *Pterocarpus macrocarpus* and *Afzelia xylocarpa*, each 20∼50 cm high. The site was plowed and pot seedlings were planted three individuals per square meter by villagers, including children, under the leadership of the border police and village leaders. After the

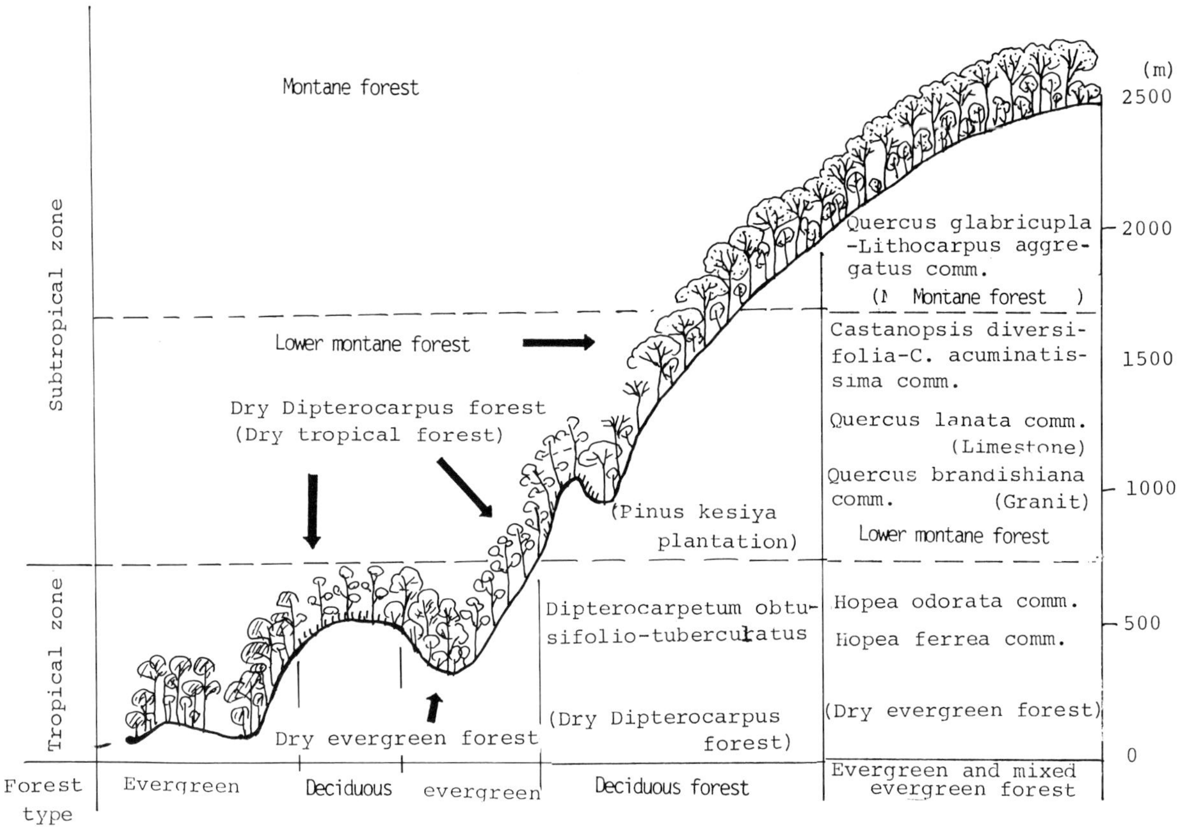

Fig. 3. Distribution of forest types in central to northern Thailand.

Table 8. Species list and stock of pot seedlings at Ban Bor Wee, Suan Phung District in Ratchaburi, Thailand

Pot seedling species		Ornamental species	
Species	Number	Species	Number
Sindora siamensis	6000	*Tabebuia pentaphylla*	3100
Pterocarpus macrocarpus	3500	*Cassia fistula*	1000
Afzelia xylocarpa	3000	*Peltophorum pterocarpum*	500
		Tecoma stans	150

plantation on June 25, 1991, the villagers got lunch from the Royal Department (based on lunch project from the Princess). Rice straw was used for mulching. After one month (June 17), the seedlings had grown to 20∼75 cm and with new leaves. Eighty-five percent of the plantation site was covered with weeds such as *Euphorbia* spp. (the dominant species), *Digitaria adscendens*, several Graminae species, Compositae, Cucurbitaceae, etc.

For weeding some money was needed, for example 40 Bt per day per person over several days. This is a big problem in a poor country. Ornamental species, such as *Tabebuia pentaphylla*, *Cassia fistula*, *Peltophorum pterocarpum* and *Tecoma stans*, were planted between the rehabilitation area of tropical forest and the road, as scattered plantations. These needed weeding more than those in the dense plantation area.

In Loei, natural species such as *Xylia xylocarpus*, *Pterocarpus macrocarpus*, *Intsia* sp., and *Dalbergia cochinchinensis*, were selected from the Royal Forest Nursery. They were planted on July 25, 1991.

In each case maintenance is a problem. Dense plantation is effective in reducing the maintenance time and work because pot seedlings cover the ground faster than plantation of scatter individuals.

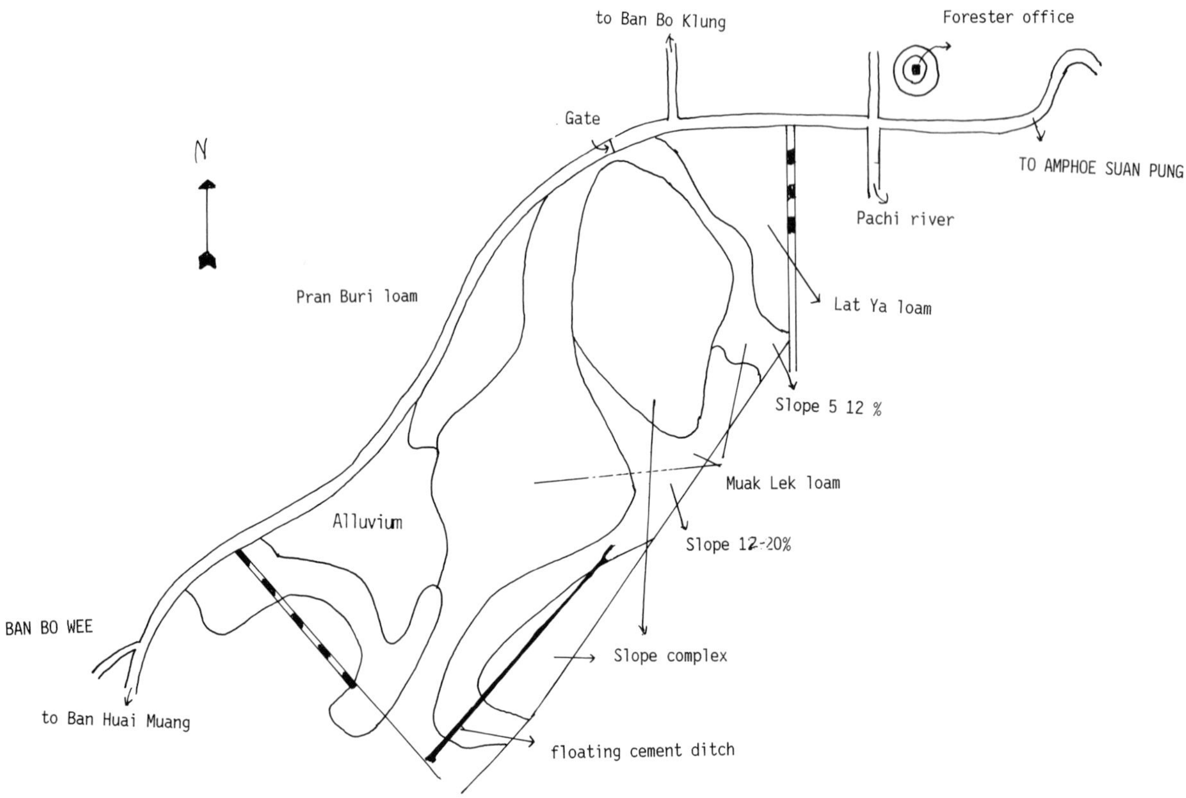

Fig. 4. Soil map of Ban Bor Wee, Suan Phung District in Ratchaburi, Thailand.

4. Local planning for sustainable rehabilitation of landscape

The rehabilitation of natural forests is very useful and valuable in both local and urban areas. In the countryside natural forests protect the soil from erosion, preserve animal habitats, maintain water balance, and provide timber and biological resources. Most people in tropical Asia are very poor. They do not have enough money to rehabilitate tropical forests or for their own living necessities or food. Therefore, in the past, national and local governments planted timber and/or natural species for forestation. Local people burned forests for hunting. They also need the firewood and charcoal. In order to meet these needs, it is proposed that dense plantation (for example six to nine individuals per square meter) of natural main-canopy species is very useful to restore the forests and utilize them as sustainable village forests. People should perform maintenance in the first two years. In this period government and/or some groups should help with labor costs. Village people can use young trees from thinning or plantation areas for firewood. When each plantation area has intervals of seven to nine years between selected cuttings, they grow faster and can also be used for timber. The areas surrounding the recovering natural forests can be used for agro-forestry, with fruit trees and medicine herbs or edible plants. Along with these, people can enjoy ornamental trees such as *Delonix*, *Tabebuia*, *Cassia*, etc., which fit the local climate and landscaping, and can be made as a continuation into urban areas as corridor of native forest.

In March 1992, I just finished cooperative work with the University of Philippines in Los Baños for restoring the water in Laguna Lake and the landscape of this region. This research was done under the project title "Control of agro-industrial pollution in Laguna Bay, Philip-

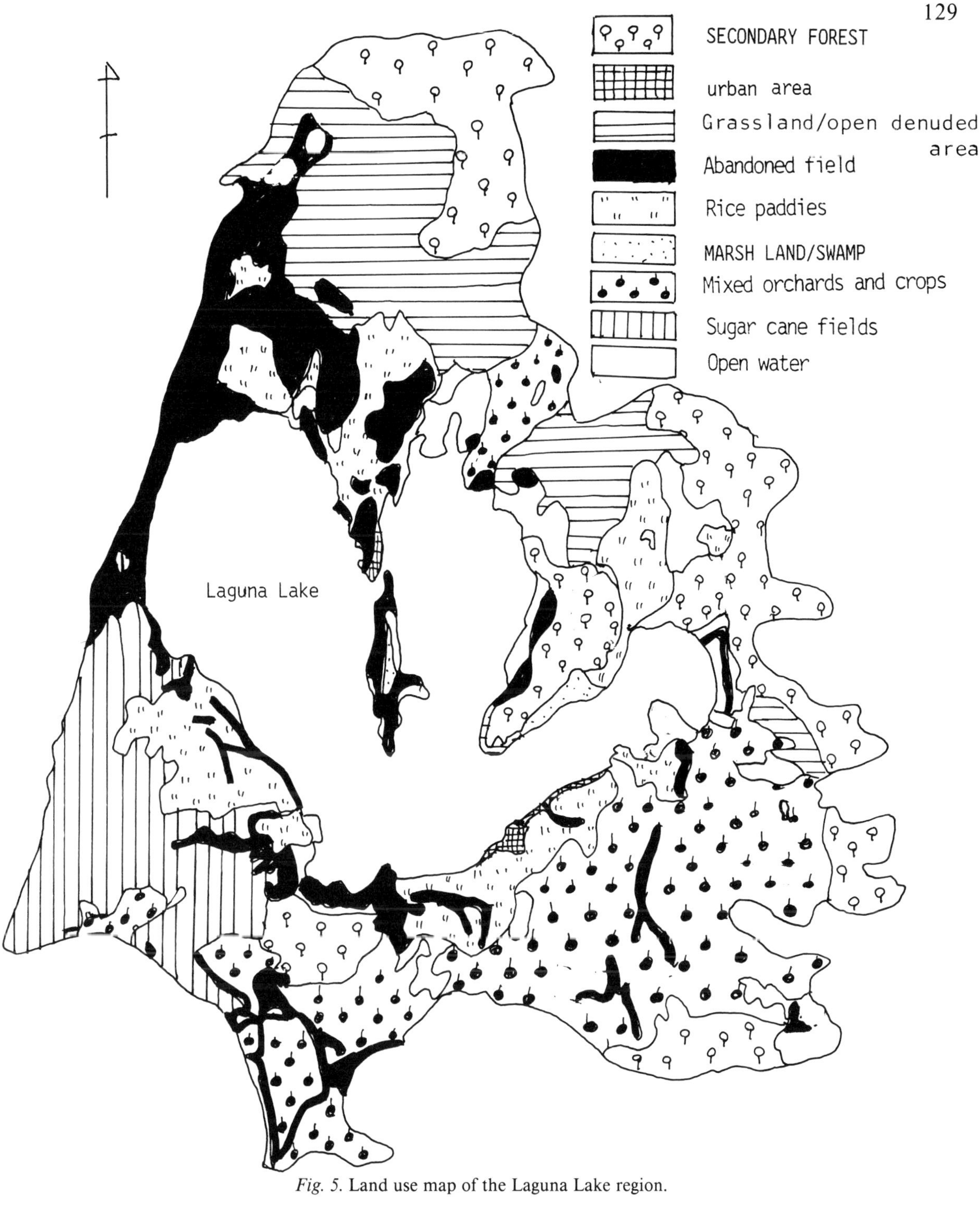

Fig. 5. Land use map of the Laguna Lake region.

pines", based on a grant from JSPS Cooperation Programmes with Southeast Asian Countries under the Core University System. The Philippines lost most of its natural forest already as exported timber. In Mt. Makiling area several types of forests barely remain. The lowland areas were developed for residences, factories (next to Manila), paddy fields, *Pan-*

danus and coconut plantations, secondary forests, secondary grassland, orchards, etc. (Figure 5). Water was polluted by factories, residences, and erosion of silt soil from paddy fields and Mt. Makiling. In such cases, rehabilitation plantation of natural forests as green patches in villages and on the eroded slope of Mt. Makiling will be proposed as a JAICA project, based on phytosociological field survey. Clearing the water in the canal from Manila Bay to Laguna Lake and rehabilitation of mangrove forests (plantation by Miyawaki Method) will also be proposed in factory and residence sites northwest of the lake site, with excavated silt soil which had accumulated one to two meters at the bottom of the lake. The rehabilitation of areas along rivers flowing into the lake is needed for purification of river water to stop the transportation of silt from paddy fields and waste water. On flooded areas strip forests on mounds will be constructed.

Rehabilitation of native forests — especially tropical forests — are the main key to local green planning and recovering the diversity of ecosystems from today, based on several experimental results.

5. Conclusion

Rehabilitation of tropical forests is a difficult problem in developing countries because of the cost for plantation and maintenance. Also the general population uses forests for firewood, charcoal, hunting, shifting cultivation, etc. Even governments in developing countries cut tropical forests for timber export, and used areas for rubber plantation (before, in tropical Asia) and now for oil palm plantation, pastures, etc. For these problems, local land use planning is most important where tropical forests are preserved. How many percent should be rehabilitated for local and global people still remains a question.

The forestry department succeeded at rehabilitation of several tropical forests in the Malay peninsula (p. 122). From 1972, we got ca 300 experimental sites in Japan, and in tropical Asia from 1990. This successful venture involves dense planting of mixed pot seedlings of canopy species which have well-developed root systems. The results showed that this is the best way to recover natural forests in a very short time with short, concentrated maintenance after plantation. They recorded the growth of one meter in a year, even in the warm temperate zone, and 50 cm to one meter in a half year in tropical Asia (Table 1, 4). In urban areas we can also rehabilitate tropical forests as green patches and corridors, in combination with ornamental species. These green patches and green corridors can then serve as connections between the countryside and urban area.

The above are summarized as follows:

1. Rehabilitation of tropical forests is needed especially in disturbed areas of their original places, such as sites of abandoned shifting cultivation, area eroded after cutting, which are mainly in the countryside.
2. Harmonious landscapes should be combined with rehabilitation of tropical forests and landscaping areas with traditional land use, excluding modern type.
3. Rehabilitation of tropical forests should be integrated into regional planning in rural to urban area.
4. It is possible to rehabilitate tropical forests even in narrow areas in cities and the countryside.

References

Braun-Blanquet, J. 1928. Pflanzensoziologie. Grundzüge der Vegetationskunde. 330 pp. J. Springer. Berlin.

Braun-Blanquet, J. 1932. Plant Sociology. 439 pp. New York.

Braun-Blanquet, J. 1951. Pflanzensoziologie. Grundzüge der Vegetationskunde. 2nd edition. 631 pp. Springer-Verlag, Vienna.

Braun-Blanquet, J. 1964. Pflanzensoziologie. Grundzüge der Vegetationskunde. 3rd edition. 865 pp. Springer-Verlag, Vienna/New York.

Ellenberg, H. 1956. Grundlagen der Vegetationsgliederung. Part 1: Aufgaben und Methoden der Vegetationskunde. "Einführung in die Phytosoziologie" (Walter, H. Ed.), vol. 4. 136 pp. Eugen Ulmer, Stuttgart.

Fujiwara, K. 1984. Investigation of the Growth Process of Environmental Forest in the Industrial Area in Camellietea japonicae. In: Abstracts, International Symposium on Urban Ecosystem and Environmental Science, pp. 20–22. Yokohama National University.

Fujiwara, K. 1987. Aims and Methods of Phytosociology of "Vegetation Science". Plant Ecology and Taxonomy to the Memory of Dr. Satoshi Nakanishi, pp. 607–628. The Kobe Geobotanical Society, Kobe.

Fujiwara, K., Aguiler, N., Hammavy, B. F., Barril, K. R. (in press): Vegetation of Laguna Lake Region in the Philippines.

Ng, F. S. P. (Ed.) 1978. Tree Flora of Malaya. Vol. 3, Forest Department, Ministry of Agriculture and Lands, Malaysia, 444 pp., Longman. Kuala Lumpur.

Ng, F. S. P. (Ed.) 1989. Tree Flora of Malaya. Vol. 4, Forest Department, Ministry of Agriculture and Lands, Malaysia, 549 pp., Longman. Kuala Lumpur.

Miyawaki, A. & Fujiwara, K. 1969. A Restoration Planning of the Green Environment and Landscape of "New Developing Area" in the Western Part of Fujisawa City. 38 pp. Fujisawa. (In Japanese with German summary, a vegetation map in color.)

Miyawaki, A., Fujiwara, K. & Kimura, M. 1984. Ecological and Vegetation Scientific Studies on Creation of Environmental Protection Forests in the Industrial Areas of Japan. Yokohama Phytosoc., Vol. 22. Part II. 151 pp. (In Japanese with German and English Summaries), Yokohama.

Miyawaki et al. 1974 The Phytosociological Investigation for the Creation of Environmental Protection Forests around Schools. 116 pp. Inst. Environ. Sci. & Techn., Yokohama National University. (In Japanese with German Summary.)

Mueller-Dombois, D. & Ellenberg, H. 1974. Aims and Methods of Vegetation Ecology. 547 pp. Wiley, New York.

Smitinand, Tem 1980. Thai Plant Names. Royal Forest Department. 379 pp.

Tüxen, R. 1956. Die heutige potentielle natürliche Vegetation als Gegenstand der Vegetationskartielung. Stolzenau/Weser. Angewandte Pflanzensoziologie, 13: 5–42.

Whitmore, T. C. (Ed.) 1972. Tree Flora of Malaya, Vol. 1, Forest Department, Ministry of Agriculture and Lands, Malaysia, 473 pp., Longman. Kuala Lumpur.

Whitmore, T. C. (Ed.) 1972. Tree Flora of Malaya. Vol. 2, Forest Department, Ministry of Agriculture and Lands, Malaysia, 444 pp., Longman. Kuala Lumpur.

PART TWO

Ecologically sound suggestions for management and aspects of nature conservation in tropical forestry

Introduction

As we already stated in the Preface, it seems unthinkable that the vast expanse of the natural tropical forest will remain untouched. Much of the land has already been logged and still more will be utilized. For that portion of the tropical forest which will remain in the hands of foresters, an ecologically sound procedure for sustainable management is needed. This topic was covered in six contributions during the symposium. Starting with the presentation of the ITTO guidelines for sustainable management of natural and planted tropical forests by Dr. Bruenig, the discussion of this topic brought contributions from East Asia, Africa and Australia. All contributions emphasized the multiple use of the forested landscape. Not only timber is wanted but other regionally important ecological benefits were taken into consideration. While the protection of the forest for environmental quality was the prime concern of the speakers of the six papers, some aspects of nature conservation for the benefit of the naturally evolved ecosystems were touched. Nature conservation of tropical forests was then the issue for the next two papers. We heard contributors from India and Thailand where these problems are especially pressing. The paper by Dr. Rheker had also a substantial section on tropical montane forest protection as an example from Africa. In the discussion of these papers the difficulty with maintaining an orderly, ecologically sound landscape management plan under the pressure of economic concern and in conflict with the immediate interests of the local population was stressed. The survival of all species now present in any given tropical ecosystem was questioned. In the discussion, land use management plans were proposed in which all aspects of an ecologically sound forestry concept would be achieved by a planned utilization pattern. Plantations for timber production should be mixed with natural areas from which rare species of wildlife and soil microorganisms could regenerate and reinvade the replanted areas. The need for long-term studies accompanying the restoration experiments was emphasized. The different goals between the foresters and the ecologists among the participants surfaced with both groups accusing the "greedy economists" as being the people to blame for the present miserable situation in the tropical forest ecosystems.

H. Lieth and M. Lohmann (eds.), Restoration of Tropical Forest Ecosystems, 135.

12. The ITTO guidelines for the sustainable management of natural and planted tropical forests

E. F. BRUENIG
University of Hamburg, Leuschner Strasse 91, D-2050 Hamburg 80, Germany

Keywords: Causal factors, criteria, failures, indicators, requirements, successes

Abstract. Rapid development of the global economy during the past 50 years has lead to non-sustainable exploitation of human and natural resources. The timber resources of the tropical and boreal forests are being recklessly and blatantly plundered. Tropical forests in particular are being cleared much in excess of the requirements of orderly and sustainable economic development.

In timber logging or slash-and-burn agriculture, social costs by far exceed the private and public benefits. Present practices of timber logging and trading are wasteful and destructive. Total costs substantially exceed values generated. The national social balances of the tropical timber business are negative. In response, the International Tropical Timber Organization (ITTO) prepared or is preparing the following guidelines:

— ITTO Guidelines for the Sustainable Management of Natural Tropical Forests, 1990;
— ITTO Guidelines for the Establishment and Sustainable Management of Planted Tropical Forests, 1991;
— Criteria and Indicators of Sustainable Forest Management;
— ITTO Guidelines for the Conservation of Biological Diversity in Tropical Production Forests (in prep.) in cooperation with the International Union for the Conservation of Nature, IUCN;
— Manuals to facilitate application in practice.

Successful application depends primarily on the political will and determination to change from predatory exploitation to wise use and sustainable utilization and conservation of natural resources.

1. Background

1.1. Evidence that sustainability is a possibility

Research, experimentation and practice of utilization, silviculture and management of natural forests and of forestry and agricultural tree plantations in the tropical zone have more than a century of history rich in experience of success and failure. The history of "agroforestry" in its many forms, especially in the form of the Southeast Asian Village and Home gardens and the Burmese taungya, is even longer and counts by thousands of years. The cradle of decidous tropical forest conservation and management have been India and Burma while that of evergreen tropical rainforest was Malaya (now West Malaysia or Malaysia Semenanjung). A great wealth of scientific knowledge from basic and applied research and experimentation in natural forest and in plantations of many kinds is now available in literature or in the data stores of research institutes.

In forestry practice, milestones of development in the evergreen rainforests were the sustainable management systems introduced in the lowland Dipterocarp forests of what was Malaya during the twenties and in the mixed peatswamp forests of Sarawak during the fifties and sixties. In Sarawak, working plans for each management unit regulated yield, harvesting procedures, inspection of exploited forest blocks and silvicultural treatment in the Permanent Forest Estate (PFE). The plans were integral parts of regional management plans to ensure coordination with general land use development. In tropical America, the development of procedures and systems of silviculture and management of natural and enriched natural forests has been impressively successful in Trinidad, Puerto Rico and Surinam. The history and state of knowledge is amply recorded in easily accessible litera-

H. Lieth and M. Lohmann (eds.), Restoration of Tropical Forest Ecosystems, 137–143.

ture (among others: Dawkins 1958, 1963; Figuera et al. 1978; Forest Department Sabah 1972; Francke 1942; Graaf 1986; Hamilton 1983; Leslie 1971, 1987; Williams and Hamilton 1982), but it has become fashionable to ignore this among non-professional interest groups and even by some professional foresters and scientists. In the case of planted forests, stories of success are much rarer. Some well recorded examples of agroforestry and forestry plantations of teak and other native or exotic hardwood species or confiers show that sustainability is possible if basic rules of good forestry are obeyed. Success stories exist especially in India, Indonesia, West and East Africa, Trinidad, Surinam, Puerto Rico and Brazil and are well covered in literature. A comprehensive review is available (Evans 1982).

These examples of success in natural and man-made forests have demonstrated the feasibility, but also the difficulties of sustainable management. The ecological, economic and societal risks in tropical forest conservation, utilization, management and development are at least as high as in temperate forestry. The high risks make continued basic and applied research essential to provide the needed reliable and sufficient data and information base for guiding and adapting management practice. This need has become particularly pressing in the present situation of accelerating global change, especially that of climate and atmospheric chemistry.

1.2. Contrary evidence that failures and misuse are the rule

In the practice of forestry, as in land use generally, something went very wrong in recent times in most tropical forest countries. The last 2 or 3 decades have seen a dramatically accelerating decline of the size and value of the tropical forest resource. The rapidly expanding population of indigeneous subsistence farmers claimed more and more forest land for the ancient traditional slash-and-burn shifting agriculture which is extremely wasteful of natural and human resources. Migrant settlers fail on foreign sites and escape into slash-and-burn cultivation. Public settlement schemes fail and add substantially to instability and hazards because they are not adapted to the natural and cultural environments.

New forms of industrial agriculture and husbandry, often purely speculative, clear large forest tracts, wasting valuable natural and human resources, producing benefits only to banks and speculators. The recent world-wide rush to supply the buoyant world timber market accelerates deforestation and the decline of the ecological, environmental and economic value of growing stock in the legally constituted Permanent Forest Estate (PFE). This timber rush has lead to dramatically deteriorating standards of timber harvesting and utilization. The manifold and complex linkages between forest management, licence policies, utilization, trading and marketing have so far proved resistant to improvement and control. The causes are largely socio-political. During the same period, the areas of annual reforestation and afforestation have, to some extent, expanded in the whole tropical zone, but remained below 10% of the deforestation which occurred during the same year. In the humid tropics, timber plantations under monocultural management have yet to prove their ecological and economic viability beyond the first or second rotation.

The forces driving the vicious circles of decline are often ill-conceived environmental, social, political or religious ideologies, mono-ideaisms, that is narrowly fixed sectarian ideas of all kinds, irrational policies of public subsidizing and unfeasible development and lending policies of donor country's aid agencies and banks. On a personal level, selfishness, reckless exploitation of privileges, ignorance and corruption contribute their share (Barnett 1989; Marshall 1990). Available scientific knowledge and local experience are being consistently ignored and not applied in the practice of forestry and agroforestry development projects. Disregard by the local developers of the local historic background and present socio-cultural conditions in the rural environment cause social conflicts and waste of natural and human resources. Insensitivity and ignorance of foreign aid "experts" of national and local needs and lack of expertise cause many projects to be wrongly oriented and finally to fail and fade into oblivion. The final outcome of initially often well-meant, but overly ambitious, and maladapted and ill-conceived projects is disillusionment among the needy people in the recipient country at high social and private costs.

2. The ITTO Guidelines

2.1. Introduction

In this rather desolate situation one of the many obstacles to improvement has been the lack of simple practical guidelines of best practice which would be accessible and useful for planners, managers, entrepreneurs, officials and politicians. ITTO in 1988 therefore decided to produce such guidelines. Guidelines for the management of natural forests were drafted in 1989, for planted forests in 1990, for sustainability generally and for the conservation of biological diversity in production forests in 1991. In principle, the guidelines are conceived as a structured check list of basic principles and approaches. Appropriate principal options are defined which should be adopted for action in forest management planning, execution and monitoring. They are not a manual or text book describing or prescribing techniques and procedures in detail. The guidelines for management of natural and planted forest and the criteria of sustainability put forestry in the wider ecological, environmental, technical, economic, sociocultural and political context in which it historically developed and traditionally operates. The major thrust is towards more environmentally and socially acceptable, multisectorally integrated multipurpose forest management. Accordingly, the instruments, procedures and substance of planning must meet certain criteria. This requires not only state-of-the-art techniques of harvesting, regeneration or afforestation, monitoring and enforcement, but also:

— consistent, rational and practical criteria and indicators of sustainability
— reliable and sufficient data bases and efficient information systems on which the realization and success of planning, implementation, monitoring and enforcement essentially depend.

Criteria and indicators of sustainability must be quantitatively definable and be suitable for continuous assessment of compliance and deviations. Table 1 gives an example of practical criteria and indicators, as well as methods of surveying the indicators which produce the information source which might be used when implementing the ITTO Guidelines.

The unifying core element of sustainable management is the integrated multisectoral spatial information systems (IMSIS). Many tropical countries possess adequate land capability surveys, land use and economic development plans and forestry and forest industries research and development plans, especially in connection with the Tropical Forest Action Plan of the World Bank and FAO. But nowhere are the plans, data bases and information systems of the various sectors and government departments efficiently linked and effectively integrated. Thus, such unifying IMSIS is an indispensible instrument to secure comprehensive, balanced and sustainable economic and social development (see the contribution on the Research and Development Programme for Forestry in Sarawak elsewhere in this volume).

2.2. The guidelines for natural forests

The guidelines for natural forest (ITTO 1990) define 41 principles to be observed and recommends a corresponding set of 36 possible actions with respect to:

— appropriateness of forest policy, asking for strong commitment "at the highest level"
— establishment of an adequate permanent forest estate (PFE)
— adequate power, strength and efficiency of the public forest service to ensure enforcement
— high quality of planning and execution of forest management
— state-of-the-art harvesting, proper roading, suitable choice and efficient use of appropriate equipment and machinery by skilled, well-trained personnel
— effective protection of the forest against encroachment, re-entry, fire and chemicals
— legal, economic and fiscal arrangements to provide adequate incentives to the private

Table 1. Field and office criteria and indicators of sustainability at management unit level: the control of compliance with ITTO management guidelines in logging and managing tropical rainforest

Criteria	Indicator	Information source
Utilization efficiency productivity (natural and commercial yields)	Waste of timber Canopy structure and closure, damage to soil and residual stand	Ground check (block inspections) Remote sensing Ground checks
Species preservation, biodiversity	Canopy structure and closure, proportion of nature reserves in management area	Remote sensing, Ground checks
Habitat preservation, biodiversity	Pattern of unlogged areas and nature reserves, pattern of distribution of classes of canopy structure and closure	Comparison of remote sensing images and management map prescriptions, ground checks
Prescribed sustainable yield	Amounts of actual harvest and wastage vs estimates	Field and loading checks, block inspections
Environmental compatibility	Sediment load and silting of water bodies	Continuous ground checking (monitoring) (difficult!)
Security of area and tenure	Encroachment of rights of permanent forest area	Remote sensing images
Integration	Economic development and land use plans	Comparison of plans with reality
Societal compatibility	(1) Settlements of customary rights and claims	Local check on management plan prescriptions
	(2) Ratios of timber outturn: percent domestic conversion (down-stream): to high quality semi-finished and finished products	Statistics of commerce and trade Labour statistics
	(3) Provision of employment at skilled labour, technical and managerial levels	

sector to adopt long-term strategies of sustainable forest utilization and conservation management
- effective and up-to-date monitoring to control annual cut levels and to protect habitat and environment, supported adequately by scientific and application-oriented research
- maintenance of active communication, co-operation and generally good relations with local populations.

Six appendices give mainly terminological clarifications and some additional detailed explanations. The guidelines certainly point in the right direction, but possibly may not be sufficiently concrete and detailed enough to be easily translated into practical action. But, looking at the present chaotic and disastrous situation in most logging areas, their adoption, even in their admittedly very broad terms, would already bring progress toward improvement, provided that there is adequate commitment at the highest political levels in the producer and consumer countries alike.

2.3. *The guidelines for planted forests*

The case of planted forests is more complex and complicated than that of natural forest. "Natural forest" is well-defined and, despite of its extreme natural heterogeneity and complexity, the management is more uniformly geared towards maintaining the natural pristine forest condition as nearly as possible by cautious selective harvesting and silviculture. Planted forests, by comparison, have much more diverse conditions and objectives. The legal situation of land tenure and customary rights is often complicated. Crop types are simpler but of much more varied types than in the productive part of the natural forest. Correspondingly it is more difficult to define principles of management which would apply generally. Therefore, the guidelines became more complex and elaborate. The ITTO Guidelines (ITTO 1993) include 64 principles and 72 recommended actions compared with 41 principles and 36 possible actions in the guidelines for natural forests. At the core of the guidelines for planted forests is the principle that ecological, economic and social risks are interdependent and must be minimized. To achieve this, holistic concepts and approaches must be adopted. The natural dynamics of the ecosystem (vegetation, soil and climate) must be utilized as much as possible. To work against the natural dynamics of the ecosystem is both costly and risky. Naturalistic silvicultural management is generally preferable to monocultures. Monoculture means the succession of generations of pure crops of the same tree species. A typical example is the industrial eucalypt plantation system. Naturalistic silviculture means mixed crops and/or succession of alternating pure or mixed crops composed of successively different species and species mixtures. Monocultures should be restricted to well-defined and well-considered economic objectives and to exceptionally favourable and suitable sites, such as fertile soils on flat ground.

The guidelines are structured similarly to the natural forest guidelines and cover the following aspects, adopting throughout a very comprehensive perspective, a holistic concept and an appropriate regard to societal needs, aspirations and expectations:

— policy and legislation
— comprehensive assessment of ecological, economic, societal and environmental feasibility
— principal approaches to be adopted for the establishment
— post-establishment management, including protection and monitoring.

The principles and possible actions are phrased, as much as feasible, in a similar manner as those for the natural forest guidelines. Six appendices give concise definitions of terms and more detailed explanations.

3. Conclusions

The ITTO Guidelines will have served their purpose if they succeed in:

— replacing the presently prevailing rapacious, predatory, reckless and wasteful timber mining by proper and orderly sustainable management and utilization of the natural forests as a renewable economic and indispensable environmental and cultural resource
— adjusting forest development to the general land-use and economic development, aiming at full integration between all sectors involved
— reducing the economic, social and environmental risks and damages arising from poor planning of plantations by proper allocation of sites, species and stand types.

Success will depend on the fulfillment of 6 requirements which are:

Firstly and above all a determined national and international political and public will and the forthcoming actual support for improvement without which the guidelines will remain a theory.

Secondly, the integration and coordination of activities within forestry, especially between forest management and forest product trading, and between forestry and the other sectors of land use and the economy which is generally indispensable for success. This includes acceptance by the domestic and international trade of appropriate grading rules and terms of trade consistent with sustainable resource utilization. The presently prevailing unfair and one-sided trading practices dictated by the buyer must be replaced by fair trade between equal partners.

Thirdly, the support and consensus of the local rural and urban populations is essential which can only be secured if their well understood interests are adequately met and sustainably served by forestry.

Fourthly, vigorous regional economic development. Sustainable forestry and land use development is dependent on the successful development of urban-industrial centers providing infrastructural support, regional markets and technical and commercial skills, and providing employment opportunities as an attractive alternative to the resource-wasting traditional rotation-cum-shifting agriculture, otherwise no forest will be left in due course.

Fifthly, in the field of research and development, the notorious gap between research and practice must be bridged (Bruenig 1984) and the common habit of short-term isolated and specialized sectoral research and development projects must be replaced by comprehensive system-oriented and problem-solving programmes of integrated long-term projects. In bilateral and multilateral cooperative projects, the same principles must apply as in national projects and partnership must be on equal terms and conceived with a long-term and comprehensive perspective which accords with the aspirations of the people (Bruenig 1991).

Sixthly, on an efficiently operating integrated and multisectoral data sequestering and information processing system of situation analysis and development monitoring which guides but also controls and binds all actors involved in land use and those responsible for sustainability of natural resources use and economic and social development (Bruenig 1992).

The ITTO Guidelines set principles and recommend possible actions precisely on these terms. The guidelines consistently adopt a holistic, systemic and comprehensive approach. Principles and recommended possible actions are firmly embedded in the network of manifold interdependencies and interactions within and between natural, technical, economic, political and socio-cultural levels at local, regional, national, international and global scales. The ITTO Guidelines accord with the principles that prevention is better and much more cost-effective than rehabilitation or restoration, and that problems must be solved at the root of their systemic linkages and not by repair of isolated symptoms. It has to be seen to what extent the ITTO Guidelines will be accepted and implemented in their true spirit, or whether they will be manipulated and circumvented by governments, forest-based industries, timber traders, and aid agencies and be used as an alibi and smoke-screen behind which short-term narrow self-interests can be comfortable pursued as hitherto.

The ITTO Guidelines for the sustainable management of natural and planted tropical forest will, if implemented, bring back the principles and standards of traditional multiple-purpose forestry on which diverse forest-based industries and trades can safely rest for the long-term material benefit and spiritual well-being of the people of the less-developed tropical producer countries.

References

Barnett, T. 1989. Commission of enquiry into aspects of the timber industry. Report to the Government of Papua-Niugini, 2 volumes, not published (an abstract has been published privately by an Australian NGO).

Bruenig, E. F. 1984. Forest research and planning in South and Southeast Asia. Applied Geography and Development, Vol. 23, pp. 46–54.

Bruenig, E. F. 1991. Sustainable development from developing countries perspective. Lankawi Commemoration Inaugural Lecture, Kuala Lumpur, 27.10.1991. Bonn, MAB-Mitteilungen (in press).

Bruenig, E. F. 1992. Sustainable utilization, management and conservation of tropical rainforests (in press).

Dawkins, H. C. 1958. The management of natural tropical high forest with special reference to Uganda. Imperial Forestry Institute, Oxford University, Oxford.

Dawkins, H. C. 1963. The productivity of tropical high-forest trees and their reaction to controllable environment. Dissertation, Oxford University and Commonwealth Forestry Institute, Oxford.

Evans, J. 1982. Plantation Forestry in the Tropics. Oxford, Oxford University Press, pp. XV + 472.

Figuera Colon, J., Wadsworth, C. & Branham, S. (Eds.). (1978). Management of the Forests of Tropical America: Prospects and Technologies. Institute of Tropical Forestry, Southern Forest Experiment Station, in coop. University of Puerto Rico, pp. V + 469.

Forest Department Sabah. 1972. Manual of silviculture in Sabah for use in the productive forest estate. Kota Kinabalu, Sabah, 60 S.

Francke, A. 1942. Zur Kenntnis bisheriger Verjüngungssysteme im tropischen Regenwald unter besonderer Berücksichtigung ihrer Grundlagen im afrikanischen Äquatorialwald. Manuskript, Universität Hamburg und Bundesforschungsanstalt für Forst- und Holzwirtschaft, Hamburg, 432 + XLIV S. (Habilitation thesis, unsubmitted and unpublished due to the death of the author in World War II).

Graaf de, N. R. 1986. A silvicultural system for natural regeneration of tropical rainforest in Suriname. Agricultural University, Wageningen, The Netherlands. XV + 250 S.

Hamilton, L. S. (Ed.). 1983. Forest and Watershed Development and Conservation in Asia and the Pacific. Boulder, Westview Press, pp. XXIV + 559. ISBN 0-85531-534-5.

ITTO. 1992. ITTO Guidelines for the sustainable management of natural tropical forest. Yokohama, ITTO Policy Development Series 1, pp III + 18.

ITTO. 1993. ITTO Guidelines for the establishment and sustainable management of planted tropical forest. Yokohama, ITTO Policy Development Series 4, pp. IV +38.

Leslie, A. J. 1971. Economic Problems in Tropical Forestry. Rome; FAO.

Leslie, A. J. 1987. A second look at the economics of natural management system in tropical mixed forest. Rome; Unasylva 155: 47–58.

Marshall, G. 1990. The political economy of logging: The Barnett inquiry into corruption in the Papua Nuigini timber industry. The Ecologist, 20 (5): 174–181.

Williams, I. & Hamilton, L. S. 1982. Watershed Forest Influences in the Tropics and Subtropics. A Selected Annotated Bibliography. Honolulu, East West Center, pp. XII + 217.

13. Conversion of exploited natural Dipterocarp forests into semi-natural production forests

JÜRGEN HUSS and MAMAN SUTISNA
Institute of Silviculture, University of Freiburg, Germany

Keywords: Different light conditions, early thinnings, enrichment plantings, fertilizer, natural regeneration, soil preparation

Abstract. According to Indonesian regulations Dipterocarp forests of Kalimantan should be treated by the Indonesian selection system. It assumes, that in a 35-year cycle merchantable trees of >50 cm diameter can be harvested sustainably. *De facto* most stands are heavily exploited by successive cuttings and subsequently either left untreated, leading to succession, or being planted with exotics after clear-cut and burning. The Indonesian selection system obviously does not meet the ecological demands of many light-demanding Dipterocarps. Clear-cutting and burning on the other side lead — besides other undesirable effects — to the loss of indigenous species and forest types. Only few attempts have been made so far to convert heavily exploited stands into production forests with indigenous species.

Three types of experiments have been carried out in East Kalimantan to improve natural regeneration, enrichment plantings, and early thinnings with the following results:

— Soil preparations did not increase germination and survival rates of seeds of 6 Dipterocarp species remarkably.
— Growth and survival rates of young plants of 5 Dipterocarp species after enrichment plantings were extremely poor under closed canopy. Under light shelter planting shock and competition by grasses and lianas was less than in the wide gap giving the best results in the first year. Light shelter, however, has to be opened up soon because of increasing light demand of most of the Dipterocarp species.
— Many stands obviously bear reasonable amounts of naturally regenerated young plants, which should be liberated specifically by selective tendings from competition by ground vegetation and pioneers — earlier than being recommended so far. In our experiments even small trees showed noticeable responses to early thinnings. Thus even small numbers of young plants per area unit may be adequate to form new stands.
— Costs for the lay-out of paths and the tending operations are only roughly 10% as compared with exotic tree plantations, which need clear-cutting, slash-burning, planting and weeding.

Heavy exploitation cuttings therefore must not necessarily always be disadvantageous to establish semi-natural production forests, if tended early and selectively.

1. Problems of conversion

Primary forests of the moist tropics today suffer different fates as illustrated in Figure 1.

The forests are mainly used for one of the five following purposes:

1. There are strong tendencies to save at least a portion of the natural forests as nature preserves, national parks, and reserved forests. The proportion however is small. Mainly only remote areas with specific site conditions are left and even if officially protected, they are, in fact, often exploited and damaged through illegal cuttings by the local population or contractors.
2. Most of the remaining primary forests are however — the other extreme — converted permanently into non-forestry land use systems such as agriculture, settlements, industries and traffic routes.
3. Shifting cultivation is another major reason for spoiling large areas of these forests.
4. Exploitation with clear-cutting, burning and planting of fast-growing species is of increasing importance, since newly established pulp and saw wood industries cause a rapidly increasing demand for timber.
5. Unfortunately only very few examples of primary forests are being properly managed according to selection systems. This approach has been widely discussed and officially decreed in some countries. Nevertheless little

H. Lieth and M. Lohmann (eds.), Restoration of Tropical Forest Ecosystems, 145–153.

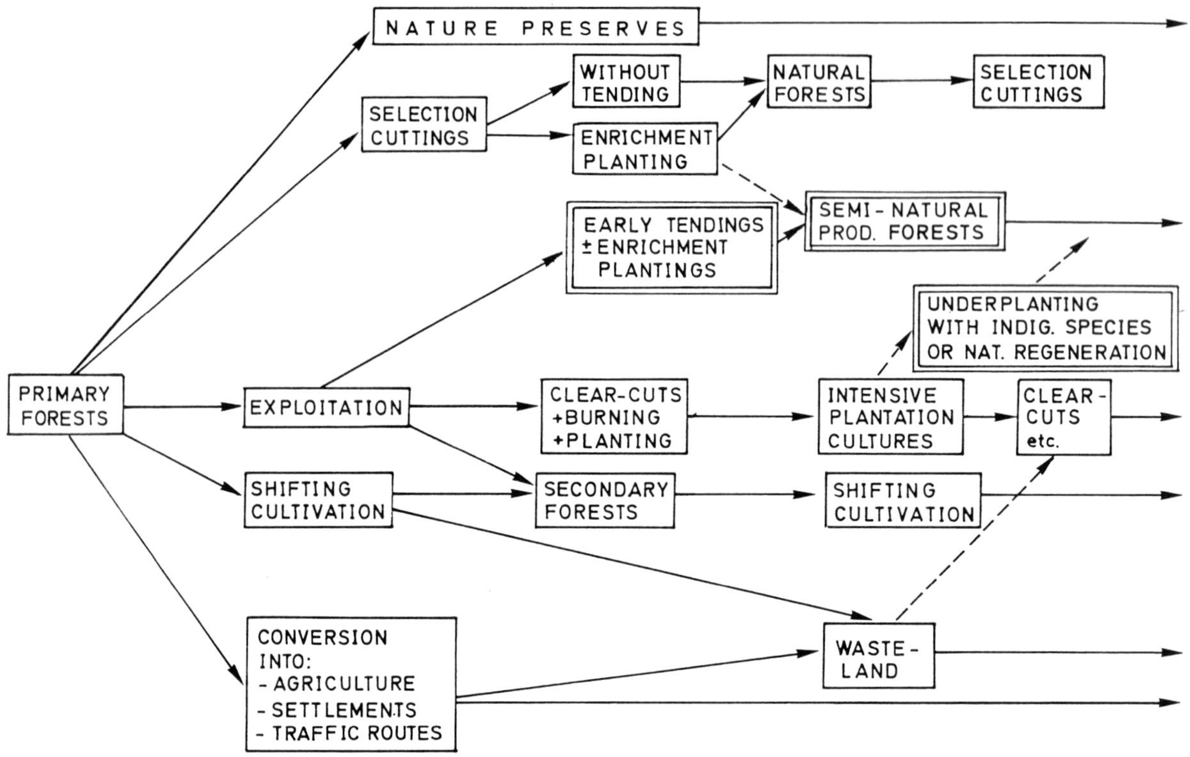

Fig. 1. Conversion of primary forests into different forests and land use systems. (Double bordered boxes show the approach of our experiments.)

success has been gained so far. Details will be discussed later.

There are relatively few attempts to find a way between total preservation and total destruction of the indigenous forests (Weidelt 1990). The formation of semi-natural production forests seems to be such a way in between, and this was our approach. This type of forest has been developed in Central Europe and there will be many analogies with the prospective tropical forests. To build up semi-natural production forests in Central Europe is today's main and generally accepted goal for all forest operations. Here for more than 100 years bare lands have been restocked with pioneer species such as pines. In addition, more or less exploited forests which still contained indigenous species have been converted into productive forests. Now both types offer the chance to develop semi-natural forests of high stability for multifunctional purposes including high value production, either by natural regeneration or by planting indigenous species underneath the shelter of the pioneers. Regeneration, structure and texture of these forests are to be directed as much as possible towards natural dynamics. The proportion of species, as well as stem and crown qualities, however may be altered according to human needs. Forests of this type, therefore, are a compromise between the conservation of natural ecosystems and sustained use with as little disturbance to the ecosystems as possible.

As mentioned before, little has been done in the tropics in this regard. At the Institute of Silviculture in Freiburg we have been trying to follow this direction for several years in primary and more or less exploited natural forests of Chile, Brazil, Turkey and Indonesia.

In the following some of the results of the trials and investigations which were carried out in East-Kalimantan between 1987 and 1989 will be presented. The project was developed within the framework of an Indonesian–German cooperation program of the German Technical Cooperation Service in Samarinda, East-Kalimantan.

It was outlined in the field by the first author during a sojourn in Kalimantan in 1987 together with the Indonesian counterpart M. Sutisna. The latter collected the data and evaluated them for his doctor's thesis, which he submitted to the forestry faculty of Freiburg (Sutisna 1990). Von der Goltz (1988) took part in the investigations collecting data for his diploma thesis.

2. Conditions and management of forests in East-Kalimantan

The experiments were carried out in the "school forest" of the forestry faculty of Samarinda near Balikpapan in the territory of ITCI, one of the biggest Indonesian timber concessioneers.

This area is located just south of the equator with a typical tropical climate: 2500 mm of rainfall evenly distributed over the year, even temperatures, almost no seasons. Its topography is formed by flat or slightly hilly ground at about 100 m above sea level with deeply decomposed soils. The wet evergreen tropical forests are generally composed of 50 to 70% of Dipterocarp species. Since the sixties Dipterocarps with fairly light timber, which could be floated, were increasingly harvested. The heavy Dipterocarps, the sinkers, and other merchantable species, however, became attractive just recently after road building gradually took place.

Officially most primary forests in the lowlands have been or should be managed according to the Indonesian Selection System TPI, which was introduced in 1972 and slightly altered in 1989. This selection system permits — after an inventory — selective cuttings of up to 20 stems per ha with diameters over 50 cm. At least 25 Dipterocarps of 20–50 cm dbh however should be left, assuming that these stems will produce 70 cm stems after a 35-year period. Then, a further selective cutting would be allowed.

In reality this selection system has not yet been proven feasable. This is partly due to the fact that the first cycle will not be finished before the year 2007, 35 years after the system was decreed in 1972. More important, however, is the fact that most contractors will go through the forests two, three or more times shortly after the first cutting in order to collect all those stems, that were not regarded profitable in the first run. The forests are often then left abandoned (see Figure 2, outer left side). Most of these over-exploited forests are, however, used in a tremendous proportion for shifting cultivation and change of land use, leading to losses of 500,000 ha annually in the late eighties.

The establishment of new industries created a permanently rising hunger for veneer/plywood, for pulp and saw wood of even smaller dimensions. Since some years, therefore, this has led to immense clear-cuttings in the heavily exploited forests. After clear-cutting, usually the slash is burned and intensive plantation cultures are established with fast-growing species such as *Acacia mangium, Eucalyptus deglupta* and others. Two million hectares are aimed at.

These operations, however, are cost- and time-consuming: Burning, planting, and especially weeding which has to be done almost every two months up to six times altogether, sum up to about 100 man days/ha (see Figure 2, inner left side).

In addition, from an ecological viewpoint, it is doubtful whether selection systems of this type can work in the long run. At least light Dipterocarps are opportunists and therefore light-demanding species. They need gaps or bigger openings in the forests in order to establish themselves. This may not be possible under the fairly closed canopies of forests where only a few trees have been harvested. Heavy exploitation, therefore, might be more advantageous for those species in order to favour their early growth and establishment. The same is true for all those forests, which contain species with high economic value that have the same autecological characteristics (mahogony and teak for instance).

3. Approach for own investigations

In a three times cut-over forest in the region mentioned, only 40% of the standing volume was left. The remaining trees either consisted of bad-shaped and decayed specimen or of species

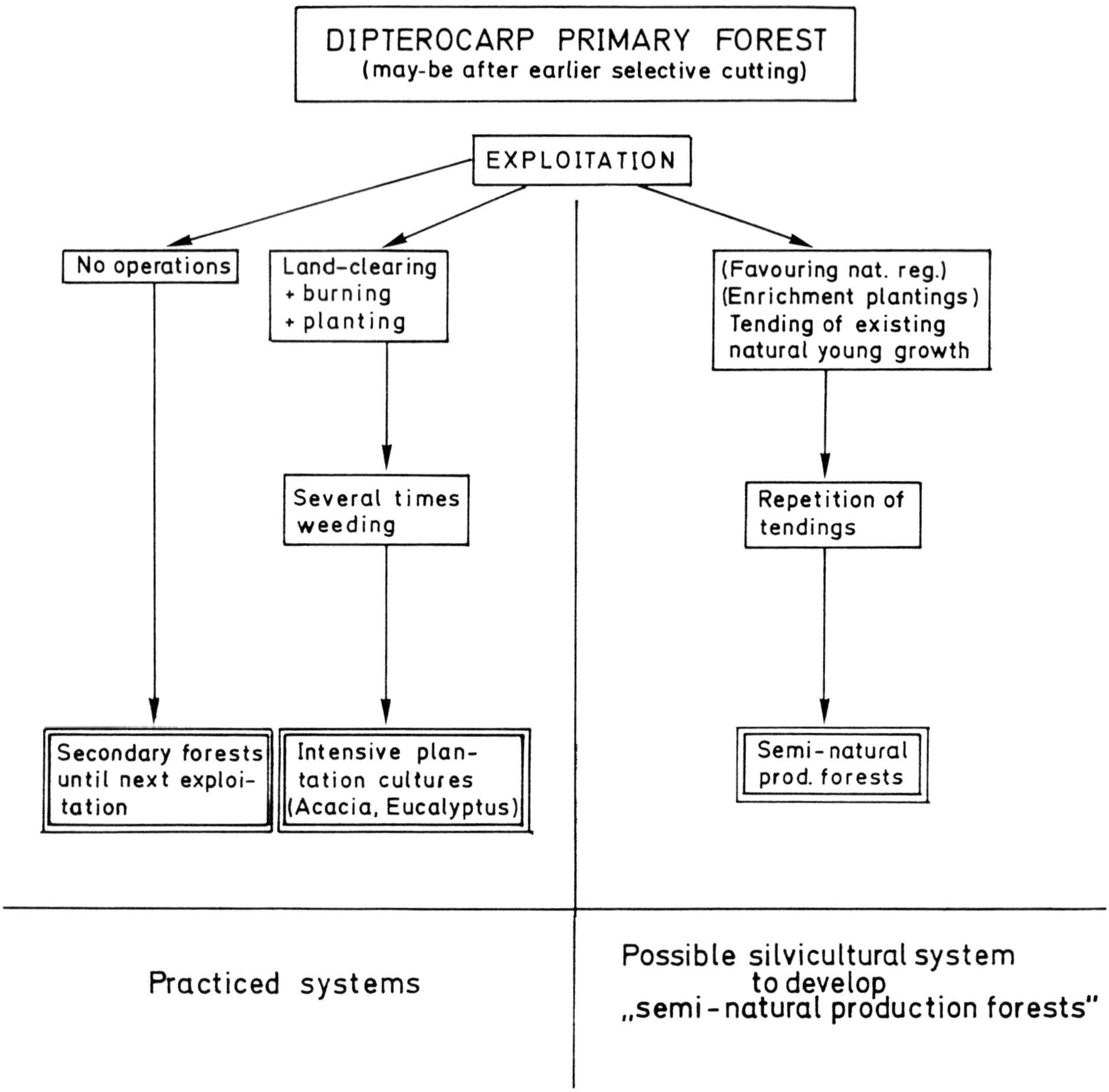

Fig. 2. Ways of treatment of Dipterocarp primary forests (right column shows our approach).

of no use. Slash was distributed over most parts, as the last felling operation took place only one year previous to our experiments. Here we tried to find answers to the following questions:

1. Is there any natural regeneration, and how do the young trees develop?
2. Can ground preparation favour germination and early establishment?
3. Is it necessary to carry out enrichment plantings and how can this be done?
4. How should the light conditions be regulated to help the young growth get started?
5. Is the nutrient supply sufficient in the deeply weathered soils, or might it be necessary to help the young trees overcome competition of ground vegetation by fertilizing?
6. What effect will early liberation have on the proportion of harvestable species, as well as on the proportion of valuable stems?

Three types of experiments were carried out to

favour and tend natural regeneration (see Figure 2, right side):

1. Support of natural Dipterocarp regeneration by soil preparation (see Section 4).
2. Planting Dipterocarps under different light conditions and with different fertilizer applications (see Section 5).
3. Early liberation (see Section 6).

4. Support of natural Dipterocarp regeneration by soil preparation

In Central Europe on most subsoil types germination rates of forest tree seeds can be improved by ameliorating the seed bed. In general this can be achieved by opening up the mineral soil. This ensures a better water supply for the seeds and young plants.

Material and methods

Seeds from 6 Dipterocarp species were available, since they could be collected in neighbouring stands (see Table 1).

In our experiment we introduced 4 different variates of site preparation such as (1) control, (2) clearance of ground vegetation, (3) clearance of ground vegetation + removal of litter, (4) like (3) + cultivation of the upper soil layer.

Results

The seeds of most of the species showed a reasonable germination capacity. Obviously the time of seed storage influenced the germination rate. Direct contact with the mineral soil improved the germination rate slightly. There were, however, differences between species: Small seeds had more difficulties to penetrate the layers of litter, and thus profited more from intensive soil preparations than the larger ones. It may however be kept in mind that the smaller ones had been stored longer. Losses of seedlings were severe within the first three months and were caused by attacks of insects and fungi. They equalized all other influences, diminishing the survival rates of all germinated seedlings down to 5–25%.

Discussion

Dipterocarp seeds generally show an extremely short storage capacity and therefore should be sown immediately after collection. Unfortunately only few seeding experiments have been carried out so far in the forests nor was it tried

Table 1. Germination and survival rates of Dipterocarp species (in order of seed size) included in the germination experiment (between March and June 1987). Different letters behind mean values symbolize statistically significant differences at $p \leq 0.05$

Species	Germination capacity after collection	Duration of storage before sowing	Maximal germination rate within a 3-month period	Time until 90% of seedlings have germinated	Survival rate related to	
					Seeds	Germinated seedlings
	(%)	(days)	(%)	(weeks)	(%)	(%)
Hopea mengarawan	90	4	46 a	9	9b	19
Shorea leprosula	90	3	60 b	8	8 b	14
Shorea ovalis	88	2	73 c	8	11 b c	15
Shorea johorensis	48	2	47 a	8	2 a	5
Shorea smithiana	80	2	72 c	3	9 b	12
Dipterocarpus cornutus	74	2	72 c	8	18 c	25

to improve conditions for natural regeneration. Taguchi and Dalmacio (1987) found on the Philippines almost no response of 5 Dipterocarp species to soil preparation, whereas the removal of litter and undergrowth significantly improved the survival rates. There are, thus, obviously differences according to soil types (or just by chance). For the time being it may only be stated, that there might well be chances of favouring natural regeneration. Many ecological factors, however, which affect survival rates, are obviously unknown so far.

5. Enrichment plantings under different light conditions and with fertilizer applications

Dipterocarp seeds tend to spread not further than 50 m away from the crowns. Therefore numerous regeneration plants may develop under the crowns of the seed-bearers, whereas none will come up at farther distances.

Wildlings can be collected and may be planted into bare spaces some weeks after having been transplanted into containers. Unfortunately wildlings cannot be dug out with adjoining earth balls because the subsoil layers are usually excessively streaked with roots. Bare-rooted wildlings on the other hand are very sensitive to heat and drought and cannot be transported even over short distances.

Material and methods

In our experiment we planted containerized stock, raised from seedlings of 5 Dipterocarp species (*Shorea leprosula, S. laevis, S. johorensis, S. xanthophylla, Dryobalanops lanceolata*). They were kept under three different shading conditions (gap, light shelter, closed canopy).

Different intensities of a phosphor fertilizer were applied into the planting holes, and NPK-fertilizers were added some weeks after planting (variates see Table 2).

Results

Regulation of light conditions had the main effect (see Table 2). *Survival rates* did not differ remarkably shortly after planting. They were highest under light shelter, where the young plants were protected against light-shock and desiccation. Dense shelter on the other side led to increasing losses presumably because of interception due to the closed canopy. In the large gap, high transpiration may have led to reduced survival.

Table 2. Influence of light regime on young planted Dipterocarps. Different letters behind mean values symbolize statistically significant differences at $p \leq 0.05$

Variates	Light intensity (in % of full light)	Survival after 1 month (%)	Survival after 1 year (%)	Height increment after 1 year (cm)
Opening	70	88 b	84 b	73 c
Light shelter	12	94 a	94 a	52 b
Dense shelter	1	84 c	59 c	3 a
Significance		+	+	+++

$+ = p \leq 5\%$; $+++ = p \leq 0.1\%$.

One year later, the survival rates differed even more: In the gap grasses developed vigorously and caused competition, whereas light deficiency may have been the reason for serious losses under the closed canopy.

Growth under the closed canopy was extremly poor, although the shelter had been opened slightly by the earlier exploitations. *Fertilizer application* in general gave little response. Phosphate given into the planting holes caused slightly raised losses according to higher fertilizer amounts and decreasing height increment (see Table 3). NPK only led to a slight increase in height, not significant over all variates. Some Dipterocarp species showed reactions only in sufficient light as is also illustrated in Table 3 by *Shorea leprosula.*

Discussion

Our observations confirm that obviously most Dipterocarps are light-demanding or at most slightly shade-tolerant. They therefore support the above-mentioned statement, that natural re-

Table 3. Height increment of 5 Dipterocarp species after 12 months according to P- and NPK-fertilizer (means of all species and shading variates) and of *Shorea leprosula* as influenced by light conditions and NPK-fertilizing

Gradation variates (g/plant)	P-Fertilizer*	NPK-Fertilizer**			
			Shorea leprosula in		
	Mean of all variates (cm)	Mean of all variates (cm)	Gap (cm)	Light shelter (cm)	Close canopy (cm)
0	47 b	40 a	72 a	75 a	4 a
10	42 a	41 a	96 b	84 a	3 a
20	42 a	42 a	95 a b	88 a	5 a
40	41 a	43 a	105 b c	90 a	3 a
80	40 a	45 a	122 c	84 a	5 a
Significance	++	–	+++	–	–

* 46%.
** Mixture in relation 2:1:1 of 46% CO $(NH_2)_2$: 46% P_2O_5: 50% K_2O.

generation of most Dipterocarp species is promoted by intensive cuttings and careless loggings.

In the case of enrichment plantings the stock should, therefore, be removed once the young plants have reached the thicket stage (i.e. ca. 2 m of height).

Obviously the nutrient supply is adequate for the young plants. However almost no experiments have been carried out in this regard, so that a generalization is not possible so far.

6. Early thinnings

Timber stand improvements have been very successful, according to Weidelt (1986) and Wöll (1988), when carried out in Dipterocarps in the pole-stage 10 years or more after the exploitations. Their operations took place in Philippine stands, which are generally very rich in stem number and proportion of Dipterocarps. In Kalimantan conditions are less favourable, but there are some million hectares, where stand improvements are urgent.

Being liberated shortly after the exploitation it might be possible to rescue many young trees that are in danger from competition by ground vegetation and lianas.

Materials and methods

For our experiments we used a stand that had been exploited several times in relatively short intervals. Indonesian foresters mentioned, that it could be used for nothing but conversion into fast-growing exotic tree plantations.

The experiments included the following steps:

1. *Establishment of footpaths* of 1 m width and 15 m distances from one to the next. This system proved to be easy to survey the stands and to control the operations.
2. *Choice of potential crop trees.* From stems of 1 m height until to some with more than 30 cm in dbh we selected about 200 potential crop trees/ha that were vigorous, straight, without stem injuries and belonged to the Dipterocarp or other merchantable tree species.
3. *Cut of the competitors.* Smaller competitors were cut with machetes; thicker ones however had to be killed by *use of arboricides.* Those who did not die after 6 months had to be felled with power saws.

Results

After a two-year period of observation the following results were gained:

Table 4. Annual mean increment of diameter at breast height (dbh) in mm of crop trees of different groups of species without and after liberation. – = n.s.; + ≤ 0.05; ++ ≤ 0.01; +++ ≤ 0.001

Range of DBH cm	Light Dipterocarps		Heavy Dipterocarps		Non-Dipterocarps	
	Control	Liberated	Control	Liberated	Control	Liberated
0–2	3	+9 +++	2	+7 +++	3	+7 +++
2–5	6	+7 +++	3	+8 +++	5	+6 +++
5–10	8	+10 +++	6	+7 ++	5	+10 ++
10–30	10	+8 ++	8	+9 +++	8	+2 –

Table 5. Time and costs expenses for tending young growth after exploitation (according to v.d. Goltz, 1988) (1.750 Rp. = $1 US in 1988)

Operation	Time (man days/ha)	Costs (Rp./ha)
Path establishment	1.0	3,000
Choice of crop trees	1.0	3,500
Cuttings with machete	2.5	8,000
Cuttings with power saw	1.6	40,500
Total	6.1	55,000

— There were sufficient young Dipterocarps to reach a percentage of 80 of the whole number of potential crop trees.
— Without liberation fellings only 36% of the small potential crop trees reached a height of 1.5 m, whereas 82% did this after the tending operation. Height increment of the very young trees consequently increased remarkably.
— Survival rates did not differ significantly.
— Diameter growth was striking (see Table 4). Especially stems with small diameters profited much. Increment partly quadrupled. Obviously a much higher diameter growth is possible than we first assumed. Therefore production periods may be remarkably shorter than 70 years.
— Labour time was about 7 man days/ha. Costs were highly influenced by machinery costs when a power saw had to be used (see Table 5). Later experiments with Roundup (glyphosate) used as an arboricide showed chances to eliminate the shading trees more easily and therefore saving money.

Discussion

According to the results of their investigations on the Philippines and in Malaysia, Weidelt (1986) as well as Fox et al. (1982) recommended thinnings a decade or longer after the fellings, a time when the young stands are penetrable and the candidates can be recognized. However we think the following recommendations are justifiable for the somewhat inferior conditions in Kalimantan:

— Early thinnings obviously offer favourable chances of ensuring a satisfying number of evenly distributed potential crop trees, if the thinning operations are carried out very early. This is a new approach to this problem.
— Moreover, stems of a wide range of diameters and heights can be selected, provided they are standing far enough from each other.
— Apart from rescuing the young plants as later crop trees, they will gain a remarkable diameter increment. Production periods of 50 years therefore do not seem unrealistic.
— One of the most remarkable advantages however is the immense difference in labour time: Less than 10 man days/ha compared to about 100 man days for the establishment of intensive plantation cultures.

Accepted that the thinning operations have to be repeated twice or three times after 5 years respectively they nevertheless will outclass the intensive plantation establishment by far.

There are some disadvantages however:

— Slash on the ground within the tending areas

even one year after the selective logging is a hindrance. Nevertheless the workers by far preferred to work with these obstacles under light shelter instead of weeding in full sunlight.

— To distinguish tree species, training of the workers by a specialist is necessary, but after a short while this proved — in our investigations — to be no longer a problem.

Tending, if done early and competently, will give much cheaper and safer results than enrichment plantings or other silvicultural means.

Here we can see a promising approach towards acceptable forest types, provided many companies and foresters learn to think and work more ecologically instead of schematically.

7. Conclusions

The findings of the above-mentioned three experiments permit to draw the following conclusions:

— Heavily exploited Dipterocarps forests in Kalimantan obviously show a much higher regeneration capacity than is generally assumed. In most cases it therefore seems to be possible to form high productive semi-natural forest stands out of them by means of early tendings instead of conversion after burning and establishing intensive plantation cultures.

— Germination can be slightly improved by removal of litter, and survival rates as well as height increment may be favoured negligibly by fertilization. Much more important however is the right dose of light. Most of the Dipterocarps seem to need much more radiation than generally alleged. Though full light will favour the growth of the young plants best, ground vegetation and lianas will be stimulated also and become more severe competitors.

— In the thicket stage at the latest, sufficient light should be supplied to the young Dipterocarps and release of the candidates from competitors being secured.

— Apart from light regulation early tendings are by far the most promising and effective means, equally time- and cost-saving compared with methods of converting the exploited forests into exotic tree plantations.

— Semi-natural production forests seem to be a good and probably the only approach to ensure the production of high value timber of indigenous species.

— From a more ecological standpoint the natural outfit with plants and animals in the successive stands will not differ much from natural forests, because the proportion of tree species will be altered only slightly in the upper crown canopy. There will be however almost no direct anthropogen influence in the second or third canopy floor or on ground vegetation.

References

Fox, J. E. D. & Chai, D. N. P. 1982. Refinement of regenerating stand of the Parashorea tomentella/Eusideroxylon zwageri type of lowland dipterocarp forest in Sabah — a problem in silvicultural management. Malaysian Forester 45 (2): 133–182.

Goltz, H. v. D. 1988. Arbeitswissenschaftliche Untersuchung eines Umwandlungs- und eines Überführungsssstems für mehrfach exploitierte Dipterocarpaceen-Naturwälder in Kalimantan, Indonesien. Unpublished diploma thesis, Forstwiss. Fakultät, Univ. Freiburg. 147 S.

Sutisna, M. 1990. Ansätze für die Bewirtschaftung exploitierter Dipterocarpazeen-Wälder in Ostkalimantan, Indonesien. Doctor theis, Forstwiss. Fakultät, Univ. Freiburg. 190 S.

Taguchi, Y. & Dalmacio, R. 1987. Aid to natural regeneration experiments. In: Kostermans, A. J. G. H. (Ed.), Proceedings of the Third Round Table Conference on Dipterocarps. Jakarta: UNESCO, pp. 237–240.

Weidelt, H. J. 1986. Die Auswirkungen waldbaulicher Pflegemaßnahmen auf die Entwicklung exploitierter Dipterocarpazzen-Wälder. Habil.-thesis, Inst. f. Waldbau, Univ. Göttingen. 166 S.

Weidelt, H. J. 1990. Die nachhaltige Bewirtschaftung der Dipterocarpazeen-Wälder Südostasiens — Eine Illusion? Holz-Zentralblatt 9: 110–112.

Wöli, H. 1988. Struktur und Wachstum von kommerziell genutzten Dipterocarpazeen-Mischwäldern und Auswirkungen von waldbaulicher Behandlung auf deren Entwicklung, dargestellt am Beispiel von Dauerversuchsflächen auf den Philippinen. Doctor thesis, Fachbereich Biologie, Univ. Hamburg. 265 S.

14. Some effects of selective logging on forest productivity and ecology

HANS-JOACHIM WEIDELT
Institut für Waldbau, Abt. II: Waldbau der Tropen und Naturwaldforschung, Büsgenweg 1, D-3400 Göttingen, Germany

Keywords: Ecological effects, timber production, tropical forests

Abstract. The few scanty data that so far exist on the effects of selective logging on stand structure, residual stands, future yield, soil, climate and hydrological cycle, nutrient budget, species diversity of fauna and flora indicate a wide spectrum of impacts that ranges from almost complete annihilation of ecological forest functions to hardly noticeable interventions. The paper based on field data and hypothesis allows the conclusion that selective logging only under special conditions can be regarded as a sustainable form of land use.

Introduction

Timber cutting in most of the moist tropical forest areas today is supposed to be conducted by numerous modifications of selective logging systems.

The cutting rules of most countries regulate timber harvesting by fixing the minimum cutting diameter, sometimes differentiated by species, the maximum volume allowed to be harvested per unit of area, the minimum number of undamaged residuals required and the length of the cutting cycle. Besides this there are regulations regarding harvesting techniques and silvicultural treatment of logged-over areas. In practice however it is often a problem to implement and enforce these regulations due to inadequate staffing and budgets of most forest services.

In contrast to the so-called uniform systems, forms of which are practiced only in a few tropical countries today and which plan a next cut only after a full rotation of 80–100 years, selective logging intends to conduct a harvest in cutting cycles of 20–40 years. It is hoped, that during this period a sufficient number of smaller trees has grown into harvestable dimensions.

According to estimates of Lanly (1982), 4.4 million ha of primary forest ("undisturbed productive closed forest") are selectively logged annually. The exact area is difficult to assess in cases of very light logging, even by aerial photographs or remote sensing where only the network of logging roads shows up clearly. More intensive exploitation, however, such as that under the Selective Management System (SMS) in Malaysia is clearly visible on LANDSAT TM Band 7.

Any kind of timber harvesting has a more or less serious impact on the forest ecosystem, among others, especially on

— stand structure, residual trees and future yield
— the soil
— climate and water regime
— nutrient status
— species richness of fauna and flora.

These problems will be discussed in the following sections.

The effect of logging on stand structure, residual stand and future yield

There is no timber harvesting in a natural forest under polycyclic management without damage to the remaining trees.

The felling and yarding damages depend, among other factors, on the logging removals. Figure 1 demonstrates the effect of logging removals as independent variable on residual stand and logging damage expressed in terms of basal area per hectare.

Other factors that may influence logging damage are:

— yarding technique, type and power of equipment
— topography
— training standard of staff

H. Lieth and M. Lohmann (eds.), Restoration of Tropical Forest Ecosystems, 155–161.

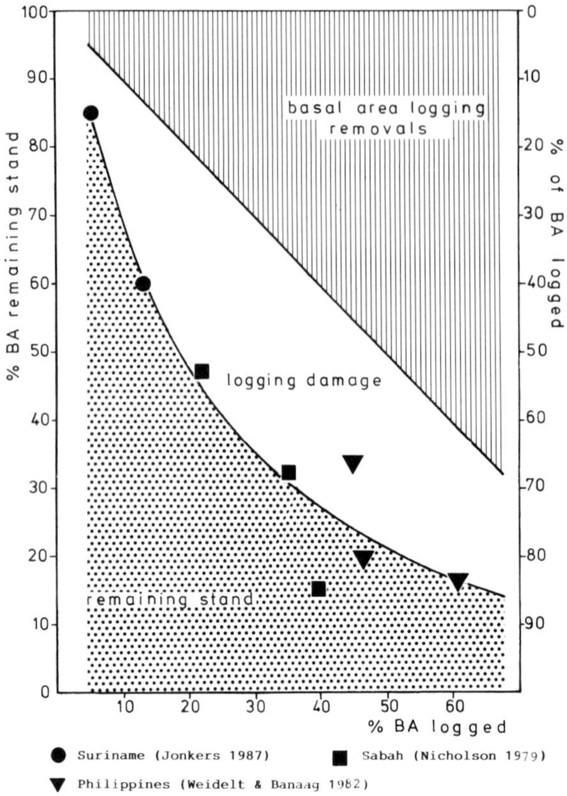

Fig. 1. The approximate relationship between basal area removed, b.a. of the residual stand with or without only minor injuries, and logging damage.

organization and supervision of logging
— operations.

Logging always creates gaps which depending on size provide an ecological niche for light demanding pioneers and less light demanding gap opportunists. The secondary vegetation shows a different character when in addition to increased light the soil has also been disturbed (e.g. mass appearance of *Anthocephalus chinensis* in East Kalimantan on disturbed soil along logging roads).

In contrast to a virgin forest, a logged-over forest temporarily will show increased increment, which is partly due to the invasion of fast growing pioneers after logging and faster growth of non-pioneers due to reduced competition.

Of special interest to the forester is the increment of the trees below the minimum cutting diameter in the period between cutting cycles. Their ability to respond to more light and space depends on two factors: species and age. Light demanding pioneers and gap opportunists are usually less able to respond to more light than shade tolerant species. Many of those trees left behind after logging may be considered retarded rather than advanced growth, and after long periods of suppression are not able to react to more light and space.

Young trees, however, show a strong reaction, but would be mature only after a full rotation of 80–100 years. Therefore, all in all usually only a rather low harvestable volume will be put on by the trees intended for the next cutting cycle.

Even under the favorable conditions of East Mindanao, the Philippines, only a harvestable volume between about 35–80 m^3 ha^{-1} was put on by the trees that passed the minimum cutting diameter within the cutting cycle of 35 years, which means only 1–2 m^3 ha^{-1} y^{-1} (Weidelt and Banaag 1982).

However, if we count all species and the increment of undersized commercial trees, the total stem wood increment may be in the range of 5–10 m^3 ha^{-1} y^{-1} under average conditions in logged-over forests.

On very good sites and under intensive silvicultural treatment the current annual increment for a logged-over forest may reach 15–20 m^3 or more (Weidelt and Banaag 1982).

There are a number of very valuable economic species such as the Meliaceae (*Swietenia* and *Cedrela* in Latin America, *Khaya* and *Entandophragma* in West and Central Africa) which are represented in a natural forest often almost exclusively as mature and overmature trees with only few trees below the minimum cutting diameter. After logging, very few trees of those species may be left. Without a considerable silvicultural input no sustained yield of these species can be expected.

Some of those commercially important Meliaceae are generally considered gap opportunists. The hope that their number increases after logging in some cases did not materialize and practically no regeneration of *Entandophragma utile* and *E. cylindricum* was found in primary as well as logged-over forests in the Republic of Congo in spite of the presence of seed trees (Fickinger, Ph.D. thesis in preparation). The reasons for this are not yet fully understood.

Swietenia macrophylla appears to profit more directly from openings by logging and regenerates more easily (Lamb 1966); however in Venezuela this was not the case (Plonczak 1989).

The effect on soil

The effect on soil is demonstrated in Fig. 2 by comparing data collected from intensively logged-over forest with those from undisturbed forest. The data show the logical sequence from soil compaction, reduced infiltration, increased run-off and erosion. The rate of erosion in the quoted case is 17 times higher in logged-over forest as compared to virgin forest.

SOIL COMPACTION (Hamzah 1978)		
tractor path	depth 0 cm	1,67 g/cm^3
	20 cm	1,45 g/cm^3
undisturbed forest soil	0 cm	1,08 g/cm^3
	20 cm	1,33 g/cm^3

⇩

REDUCED INFILTRATION RATES (Abdulhadi et al. 1981)	
compacted forest soil	0,63 (0,01-2,26) cm/min
undisturbed forest soil	4,62 (2,26-6,66) cm/min

⇩

INCREASED RUN-OFF (Kellman 1969)	
logged-over forest	1,73% of annual RF
primary forest	0,26% of annual RF

⇩

SOIL LOSSES BY EROSION (Kellman 1969)	
logged-over forest	84 g/m^2/y
primary forest	5 g/m^2/y

Fig. 2. Effects of intensive logging on forest soils in the dipterocarp forest region of East Kalimantan (Hamzah 1978, Abdulhadi et al. 1981) and Mindanao (Kellmann 1970).

Soil disturbance usually only covers part of the area. Data from the Philippines with logging volumes of ±100 m^3 ha^{-1} show 35–40% disturbed soil surface under tractor skidding.

However, most foresters — perhaps not all — may agree that a forest ecosystem can tolerate a certain disturbance in small scattered patches as long as the disturbance and destruction is not concentrated on large areas (Nicholson 1979).

The effect on climate and hydrological cycle

When only a few trees are cut and felling and yarding damages remain within limits, the climatical effects on a local, regional and global level will hardly be noticeable.

Where logging results in large-scale forest destruction, the effects on climate are manifold and may even affect the chemical composition of the atmosphere (Crutzen 1990). Here only the effects on temperature and hydrological cycle will be dealt with.

Effects on temperature (Fig. 3)

Forest destruction by logging may accelerate warming of the lower troposphere mainly for two reasons:

- the net release of CO_2 adds to the greenhouse effect, which however is dampened by increased surface albedo of heavily logged areas and reduced water vapor in the atmosphere. Since water vapor is the strongest greenhouse gas, its decrease reduces the greenhouse effect to a certain degree
- reduced evapotranspiration from open logged-over forests permits more radiation energy to be transformed directly into heat.

All this will lead to a net increase of temperatures at the lower troposphere.

At the upper atmosphere the effects are reversed and will result (or have already resulted) in a lowering of temperatures, because forest destruction increases surface albedo, which means more energy is reflected and lost. Furthermore, less heat will be absorbed by atmospheric moisture originating from tropical forests.

The effects of gaps on microclimate were investigated among others by Schulz (1960) and Longman and Jenik (1974).

We must consider that the main cause for climatical changes is burning of fossil fuels; destruction of forests is estimated to account for only about 20% (10–30%) of the increase of greenhouse gases (Deutscher Bundestag 1990). Logging activities are most likely responsible for only a very minute share, since most of the tropical forest destruction is caused by shifting cultivation and pasture making.

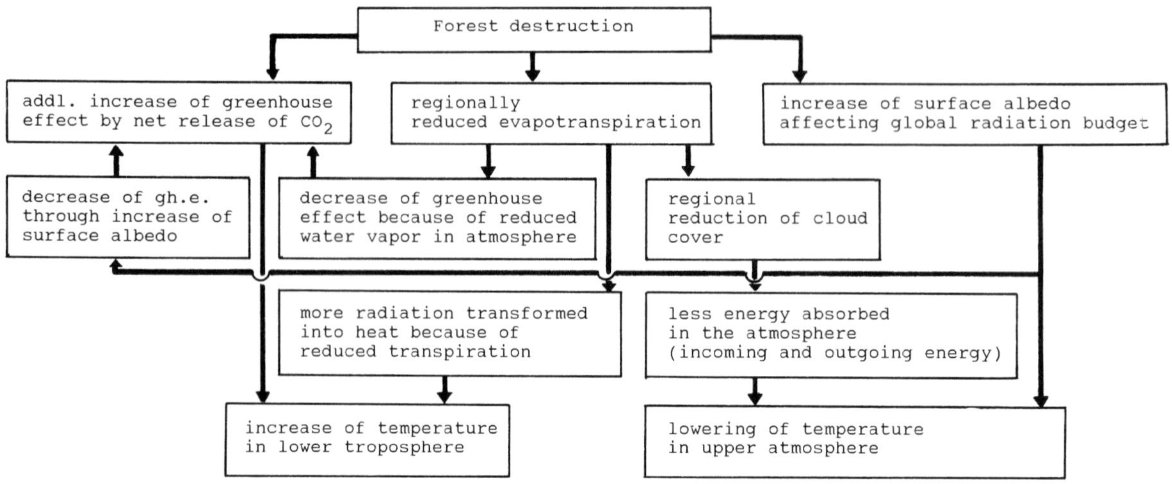

Fig. 3. Effects of large-scale forest destruction in the tropics on global temperature (based on: Deutscher Bundestag 1990).

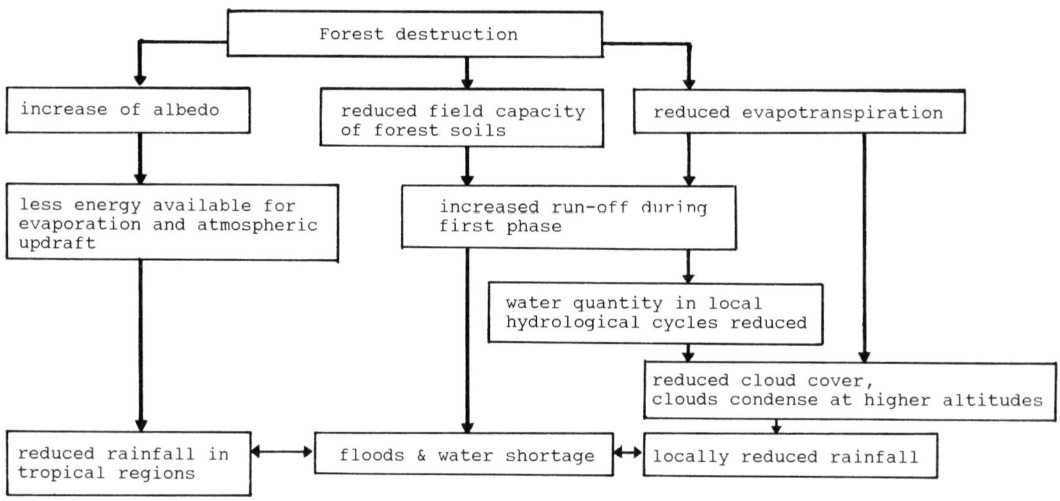

Fig. 4. Effects of large-scale forest destruction on the water cycle (based on: Deutscher Bundestag 1990).

Effects on water cycles (Fig. 4)

Climatologists seem to agree that in spite of temperature rises at the lower troposphere, due to increased albedo, there will be less energy available to fuel air circulation especially in the innertropical convergence (ITC) which may result in an overall decrease of rainfall in the tropical regions. It is still unknown how far this may effect even the climate (even temerature) outside the tropics.

On the other side forest destruction will impair the regulating effect of the forested watersheds causing floods and water shortage down the river.

Reduced evapotranspiration and increased run-off will reduce the quantity of water in the local hydrological cycles and therefore reduce local rainfall. This will be felt in the interior of formerly large forest areas like the Amazon basin which receive about half their precipitation within the local hydrological cycle.

These effects however will be felt only where logging results in large-scale forest destruction, which under orderly management or under very light selective logging will not be the case.

Impact on the nutrient cycle

Numerous studies have demonstrated, that about 80–90% of the essential plant nutrients with exception of N and P in a tropical forest ecosystem are stored in the vegetation (Jordan 1985; Ruhiyat 1989). The question is, if such a delicate system can tolerate a nutrient export by logging. The nutrient budget for the Indonesian Selective Logging System is demonstrated by Fig. 5.

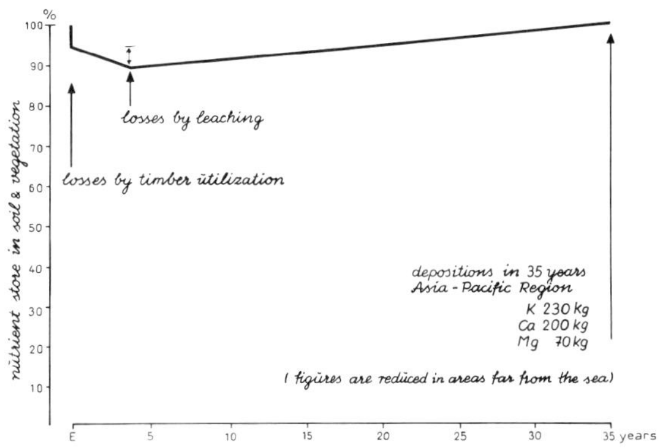

Fig. 5. Regeneration of nutrient store under selective logging by harvesting 80 m³ at a cutting cycle of 35 years, East Kalimantan (after Ruhiyat 1989).

The direct export of nutrients by the removal of 80 m^3 logs ha^{-1} only amounts to 2–5% of the total store. Further losses to the system may occur through rapid decomposition of dead biomass from logging debris, whose nutrients cannot be absorbed by the soil and remaining stand.

These losses however are expected to be minimal under extensive logging (Jonkers 1987) or in the range of about 5% of the total store under intensive logging in the dipterocarp forest (Ruhiyat 1989).

The total losses of about 10% may however easily be compensated during the cutting cycle of 35 years by atmospheric depositions (Ruhiyat 1989). The atmospheric depositions are high in SE Asia because of volcanic activity and island situations. In regions far from the sea with precipitation mainly from the small hydrological cycle the situation may not be so favorable.

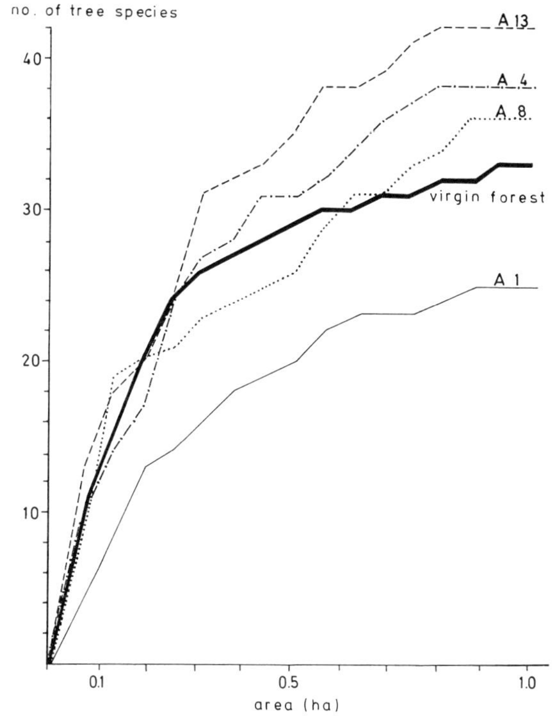

Fig. 6. Species/area curves for logged-over Llanos forests of different ages in Venezuela (after Plonczak 1989).

Impact of logging on species diversity of fauna and flora

The tropical forest is the habitat of hundreds of thousands of animal and plant species, that form an intrinsic network of interactions. Conservationists all over the world raise doubts, if such an ecosystem can tolerate anthropogenic disturbances such as timber utilization.

We are still not in the position to assess the long-term effects of selective logging on species diversity, but we already have a few scanty data on the immediate, short-term effects on species density of limited areas.

A place, that was occupied by one large tree before logging, provides growing space for many small trees of different species after the old giant was cut. The species might have been present before logging as seedlings in waiting position, as dormant seeds buried in the soil, or the seeds were brought in by wind or animals.

If one considers all trees above 10 cm dbh, logged-over forests can have a higher species density than primary forests (Fig. 6). Because of interspecific competition later the species density will certainly adjust to a lower figure.

The impact of intensive logging on the fauna on the island of Borneo was investigated by Harrison (1969) and Wilson and Wilson (1975): small rodents, mostly *Rattus* spp. (after Harrison 1969)

— selectively logged forest	7.1 animals/ha
— primary forest	4.5

hornbills, Bucerotidae (after Wilson and Wilson 1975)

— selectively logged forest	18.0 observ./day
— primary forest	18.3

primates (after Wilson and Wilson 1975)

— selectively logged forest	39 animals/km^2
— primary forest	51

The small rodent population obviously had increased after logging. The number of hornbills, who are considered to be sensitive indicators for human disturbance, was practically not affected. Only the number of primates (orangutans, gibbons, leaf monkeys) decreased by about 20%. No studies are known to the author that show the effect of logging on the species diversity of insects. The eradication of tree or animal species as a direct consequence of selective timber cutting appears to be unlikely or a rare event.

Theoretically it could be possible, that by logging a rare endemic tree species without natural regeneration at the time of cutting, the reproduction chain of trees, pollinators and seed vectors for a number of other species may be interrupted causing some kind of domino effect. This would be the case in ecosystems with highly specialized interacting partners, while in those with less specialized partners there always remain alternatives (Fig. 7).

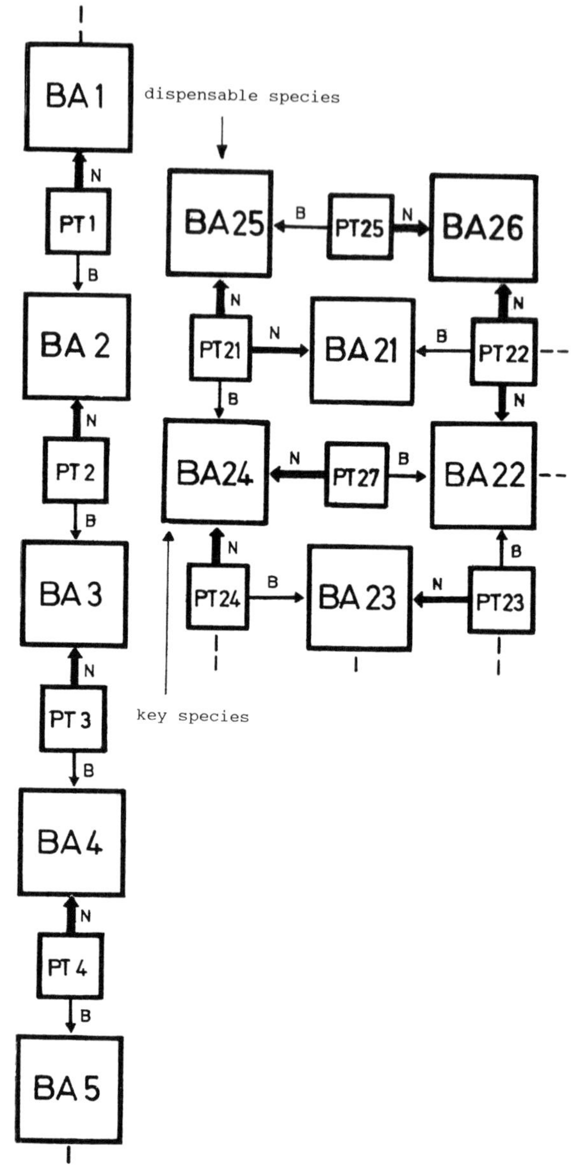

Fig. 7. Models of highly specialized (left) and less specialized reproduction chains (right) of tree species (BA) and pollinators (PT). "N" means dependence of pollinator on a tree species as a food plant; "B" means dependence of a tree species on a certain pollinator.

Unfortunately so far only the reproductive biology of very few tropical tree species is known, and we are by far not in a position to construct even parts of reproduction chains, so that all considerations of that point will still remain theoretical.

We should also remember, that because of the longevity of tropical forest trees it could take centuries until it becomes obvious, that the chain of biological dependences has been broken at some point.

A famous example was the dependence of *Calvaria major* (Sapotaceae), a rare tree from Mauritius Island on the flightless Giant Pigeon or Dodo (*Raphus cucullatus*), the last survivor of which was killed over 300 years ago. The dependence of *Calvaria*, whose population has declined to a few old trees, on the Giant Pigeon for germination was only discovered by accident very recently (Jacobs 1981).

All in all up to now only rather scanty data exist on the impact of logging on the environment.

There is a need for research in most aspects of environmental effects of logging so that it becomes clearer for what logging is responsible and for what not, and management can be improved.

Conclusions

Selective logging covers wide areas of tropical forests; annually about 4.4 million ha of hitherto undisturbed forests are logged-over.

Though selective logging may increase the overall increment of a forest, the increment of the trees intended for the next cutting cycle is often not satisfactory. In many cases also natural regeneration of the desirable species is a major problem, so that the systems of selective logging have to be improved by higher silvicultural inputs in order to be sustainable. The ecological impact of selective logging may be considerable on a local basis, but may contribute only to a small degree to the expected global climatic changes as long as a forest cover continues to exist after logging.

Selective logging systems, even intensive forms as practiced in SE Asia do not seem to pose a threat to the nutrient balance of tropical forest ecosystems, because of compensation of nutrient removals by depositions over the cutting cycle. Although the short-term effects of logging on biodiversity do not indicate annihilation of tree and animal species, the long-term effects are difficult to assess because of the longevity of trees, the dominant constituents of tropical forest ecosystems.

References

Abdulhadi, R., Kartawinata, K. & Sukardjo, S. 1981. Effects of mechanized logging in the lowland dipterocarp forest of Lempake, East Kalimantan. The Malaysian Forester 44: 407–418.

Crutzen, P. J. 1990. The international global atmospheric chemistry project (IGAC). Summary of an established project of the IGBP. In: Corell, R.W. & Anderson, P.A. Global environmental change. NATO ASI series I, vol.1.

Deutscher Bundestag. 1990. Zweiter Bericht der Enquete-Kommission. "Vorsorge zum Schutz der Erdatmosphäre" zum Thema Schutz der tropischen Wälder. Drucksache 11/7220.

Hamzah, Z. 1978. Some observations on the effects of mechanized logging on regeneration, soil and hydrological conditions in East Kalimantan. In: Proceedings Symposium on the Long-Term Effects of Logging in Southeast Asia. Bogor, Biotrop, Special Publication 3: 73–87.

Harrison, J. -L. 1969. The abundance and population density of mammals in Malayan lowland forests. Malayan Nat. J. 22: 174–178.

Jacobs, M. 1981. The Tropical Rain Forest. Springer, Berlin.

Jonkers, W. B. J. 1987. Vegetation Structure Logging Damage and Silviculture in a Tropical Rain Forest in Suriname. Wageningen.

Jordan, C. F. 1985. Nutrient Cycling in Tropical Forest Ecosystems. John Wiley & Sons.

Kellmann, M. C. 1970. Some environmental components of shifting cultivation in upland Mindanao. J. Trop. Geogr. 28: 40–56.

Lamb, F. B. 1966. Mahogany of Tropical America. University of Michigan.

Lanly, J. -P. 1982. Tropical Forest Resources. FAO, Rome.

Longman, K. A. & Jenik, J. 1974. Tropical Forest and its Environment. Longman Group Limited, London.

Nicholson, D. I. 1979. The effects of logging and treatment on the mixed dipterocarp forests of Southeast Asia. FAO Forestry Division, Rome.

Plonczak, M. 1989. Struktur und Entwicklungsdynamik eines Naturwaldes under Konzessionsbewirtschaftung in den westlichen Llanos Venezuelas. Göttinger Beiträge zur Land- und Forstwirtschaft in den Tropen und Subtropen 43.

Ruhiyat, D. 1989. Die Entwicklung der standörtlichen Nährstoffvorräte bei naturnaher Waldbewirtschaftung und im Plantagenbetrieb, Ostkalimantan (Indonesien). Göttinger Beiträge zur Land- und Forstwirtschaft in den Tropen und Subtropen 35.

Schulz, J. P. 1960. Ecological Studies on Rain Forest in Northern Suriname. Amsterdam.

Weidelt, H. -J. & Banaag, V. S. 1982. Aspects of management and silviculture of the Philippine dipterocarp forests. GTZ-SR Bd. 132.

Wilson, C. C. & Wilson, W. L. 1975. The influence of selective logging on primates and some other animals in East Kalimantan. Folia Primatologica 23: 245–274.

15. Utilization and protection of the forest belt of Mount Kenya (East Africa)

J. R. RHEKER
Department of Geography, University of Bonn, Meckenheimer Allee 166, D-5300 Bonn 1, Germany

Keywords: Afforestation, indigenous forest, non-timber products, taungya, timber

Abstract. The forest belt of Mount Kenya is of great economic and ecological importance. Mount Kenya Forest has always experienced great pressure both on its products and for conversion to agriculture and thereby threatening its conservation and catchment protection functions. Though forest management has so far prevented agricultural clearing, Mount Kenya Forest has nevertheless been heavily modified by softwood plantations (mostly cypresses and pines), which today constitute the so-called 'productive forest'. The indigenous forest has been declared 'protective forest' to restrict further exploitation.

Mount Kenya Forest is managed by 16 forest stations implementing specific planting-, tending- and protection programmes. Utilization of forest resources is granted to various licensees, who concentrate on either timber or non-timber products (e.g. fuelwood). The delicate balance between utilization and protection is jeopardized by recent protection ordinances, which may be counterproductive and create even more dangers to the forest in the near future. These ordinances include the eviction of squatters and forest workers and the abolishment of the 'shamba-system' (a taungya system).

1. Introduction

The extended highland plateaus of East Africa are topped with isolated volcanic cones, which rise from semiarid to semihumid savannahs like ecological islands. Although small in their horizontal extension, these volcanoes show a marked altitudinal zonation with characteristic vegetation belts, the lowest of which being commonly referred to as the 'montane forest belt'. The relatively humid forest belt is of great ecological importance as a living space for plants and animals and as a water regulator for the surrounding landscape. However, it also provides an important economic function by supplying timber and other raw materials. Well-balanced long-term strategies aiming at the sustainable use of natural resources are required to conserve the multi-functionality of the forest belt for future generations.

Problems of forest policy and forest management, which are quite urgent considering the varying demands and interests of a rapidly growing population, will be exemplified for Mount Kenya. Owing to its geographical position immediately on the equator, its central situation within the East African mountainous region and its good accessibility, the mountain stands out as an ideal study object. Although Mount Kenya has been the most thoroughly investigated single mountain within Eastern Africa, research has largely concentrated on ecological, morphological and glaciological studies above the upper timberline (Rheker et al. 1989). The recent "Workshop on Ecology and Socio-Economy of Mount Kenya Area" (Winiger et al. 1990) called for more research into the forest belt. The present paper considers the utilization and protection of this highly productive and sensitive ecozone.

2. Delimitation and differentiation of Mount Kenya Forest

The forest belt of Mount Kenya is limited downwards by agriculturally transformed savannahs, upwards by the so-called 'afro-alpine belt' with its characteristic plant communities, consisting of rosette plants (*Senecio brassica, Lobelia keniensis*), ericaceous bush, and tussock-grasses (see Rehder et al. 1988). Together with the adjacent periglacial and nival belts, the afro-alpine belt belongs to Mount Kenya National Park (admin-

H. Lieth and M. Lohmann (eds.), Restoration of Tropical Forest Ecosystems, 163–172.

istered by the Ministry of Tourism and Wildlife), whose official border corresponds to the 10 500-ft contour. Though this border generally follows the timberline, it has been twisted along the two most frequented climbing routes (Naro Moru-, Sirimon Track) in order to safeguard two protected corridors inside the forest (compare Figs. 1 and 2).

While the course of the upper timberline at Mount Kenya is more or less natural, i.e. principally determined by climatic influences, and on the average coincides with the 8°-annual-isotherm (Fröhlich and Willer 1977; Winiger 1986), the lower timberline reflects hygric conditions as well as human interferences, which have led to a considerable reduction of the original stock of trees. The twisted western forest boundary and the complete absence of forests on the northern slopes of Mount Kenya may be due to precipitation patterns and land use history. The large-scale agriculture practiced by British colonists within the 'White Highlands' to the west of Mount Kenya contributed to both the destruction of forests and the conservation of trees along river-beds (Kohler 1986). The forest-gap in the north may originate from the spread of fires set by the Masai to improve the grazing potential of the bordering dry plateaus (Wimbush 1930). The bulge of the forest belt west of Chuka can be explained by the fact that this particularly humid region was a no-man's land during the inter-tribal war between the Meru and Embu and thus escaped clearance (Abraham 1958).

The forest belt of Mount Kenya is administered by the Forest Department in subordination to the Ministry of Environment and Natural Resources. It is divided into Mount Kenya Forest Reserve and the small north-eastern Meru Forest Reserve. In the present paper, the term 'Mount Kenya Forest' will refer to both these reserves. As a 'gazetted forest', i.e. as an area reserved for forestry purposes, Mount Kenya Forest comprises indigenous forests and plantations as well as some bush- and grasslands. With a total area of about 2200 km^2 it is managed by 16 forest stations on the basis of the regulations and recommendations formulated in "Sessional Paper No.1 of 1968" (Republic of Kenya 1968), which centre around the principle of sustainable utilization of forest resources.

The floristic composition of Mount Kenya's indigenous forests results from the prevailing thermo-hygric conditions (compare Figs. 1 and 3). Due to rather cool temperatures, the upper timberline at about 3300 m is formed by low, partly stunted stands of *Hagenia abyssinia* and *Hypericum leucoptychodis*, which are succeeded downslope by pure or mixed bamboo communities. Further down, the 'montane forest' proper extends. On the more humid eastern and southern slopes, between about 1800 and 2400 m, it is composed of *Ocotea usambarensis* (Camphor), *Pygeum africanum* (Red Stinkwood) and *Podocarpus milanjianus* (Yellow Wood, Podo). On the drier western slope, between about 2200 and 2700 m, it is dominated by *Juniperus procera* (Pencil Cedar), *Olea hochstetteri* (Kenya Olive) and *Podocarpus gracilior* (Podo) (Trapnell et al. 1976; Wimbush 1930).

Resulting from differences in exposition to the easterly trade-winds, the considerable variances in the annual totals of precipitation and the marked seasonalities of rainfall also determine the practicability of afforestation programmes as do the properties of relief. Thus, the wetter, steeper, more dissected eastern and southern slopes are largely covered by indigenous forests ('protective forest'), while the drier, plain western footzone of Mount Kenya is lined with plantations ('productive forest').

3. Species-composition and distribution of plantations

Although they only amount to 7.4% of Mount Kenya Forest (Fig. 4), plantations constitute the productive forest proper, on which work within the forest stations is concentrated. In order to meet the demands of the numerous new landowners who invaded the White Highlands with the coming of Independence, the Forest Department decided to transform the already exploited forests on the western slope of Mount Kenya into fast-growing plantations (Wawiye 1989, unpubl.). The favoured species were exotic softwoods, especially pines (*Pinus patula, P. radiata*) and cypresses (*Cupressus lusitanica*), both of which furnish timber for construction and join-

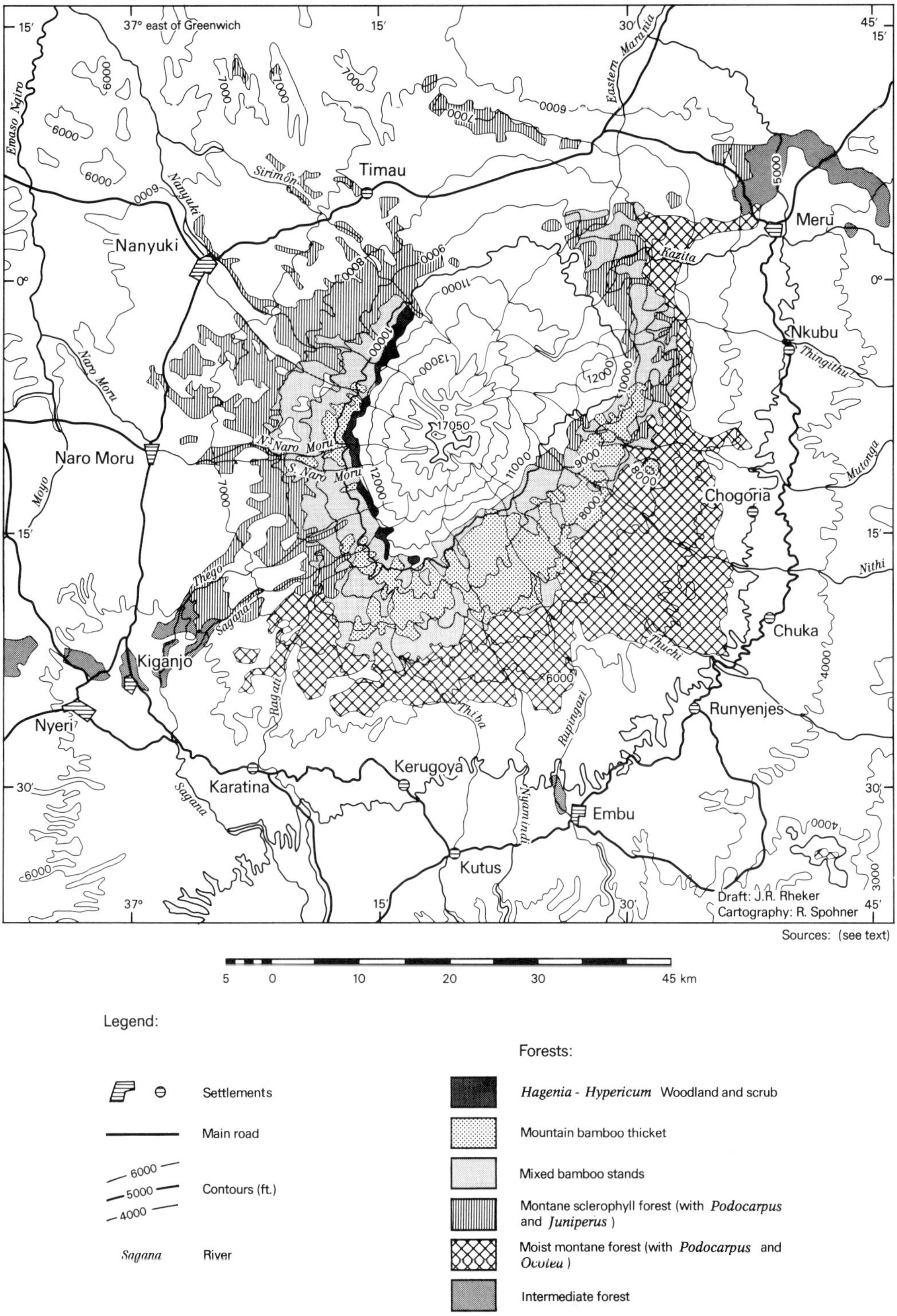

Fig. 1. Mount Kenya Area: Topography and natural forests. [*Source*: Trapnell et al. 1976; War Office 1955; both modified]

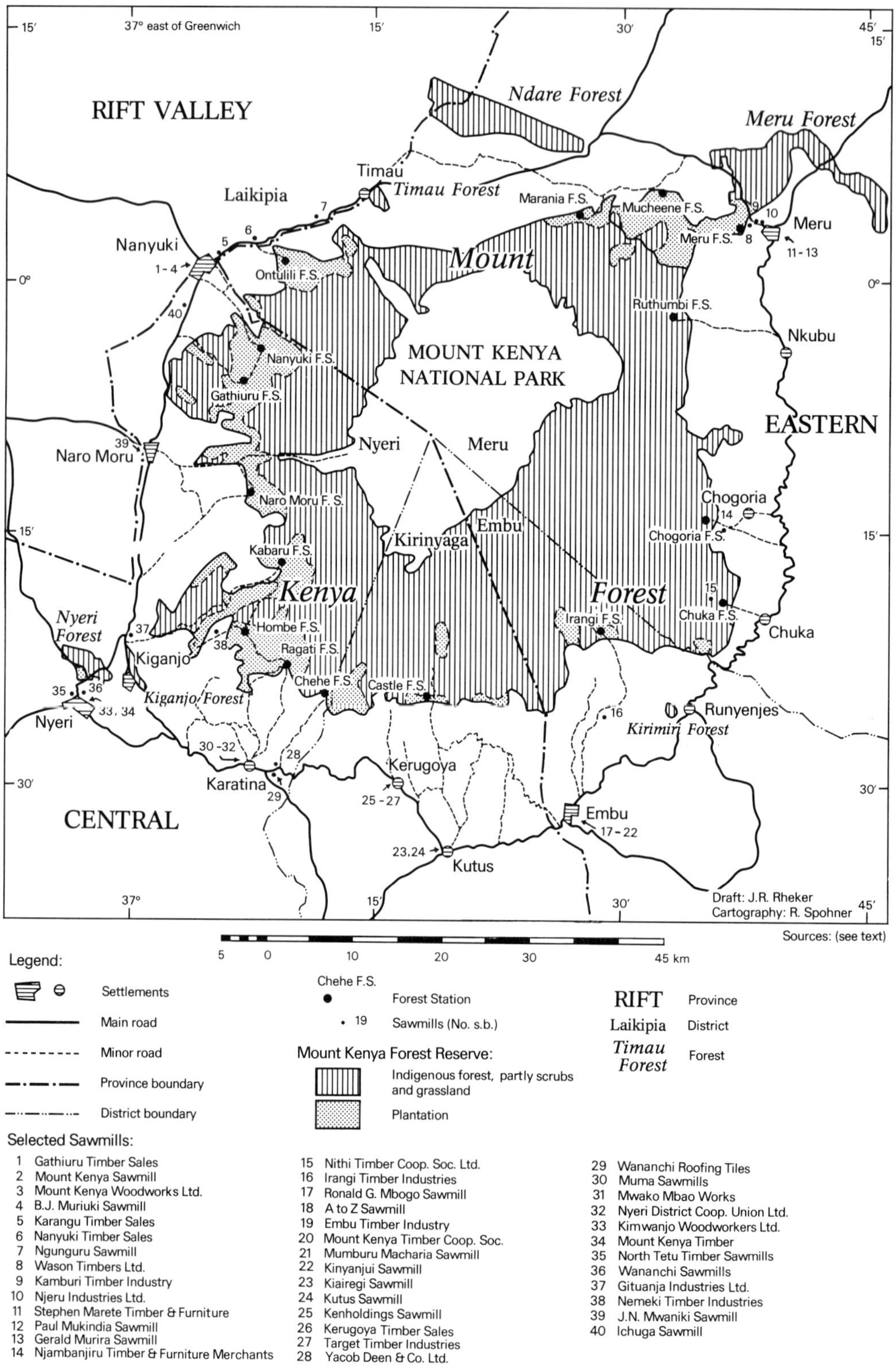

Fig. 2. Mount Kenya Area: Administrative division, road network, location of forest stations and of selected sawmills. [*Source*: Forest Department 1989, unpubl.; War Office 1955; both modified; own investigations]

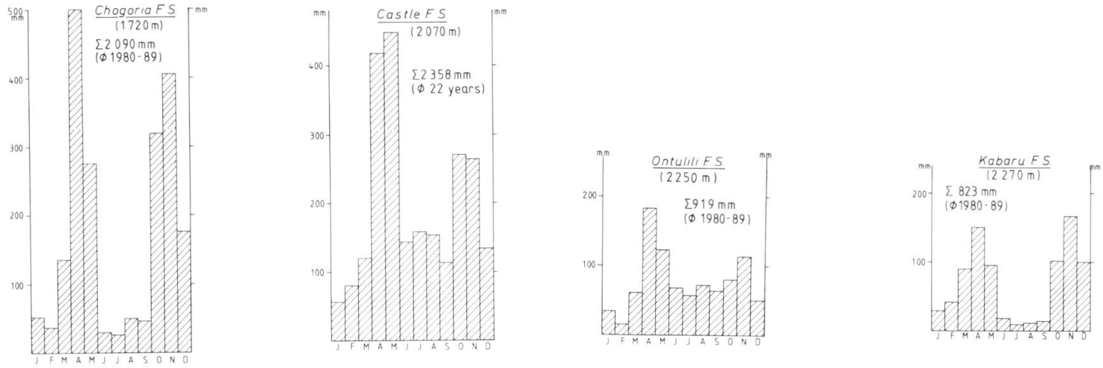

Fig. 3. Precipitation patterns at four selected forest stations [*Source*: Rainfall records of the forest stations]

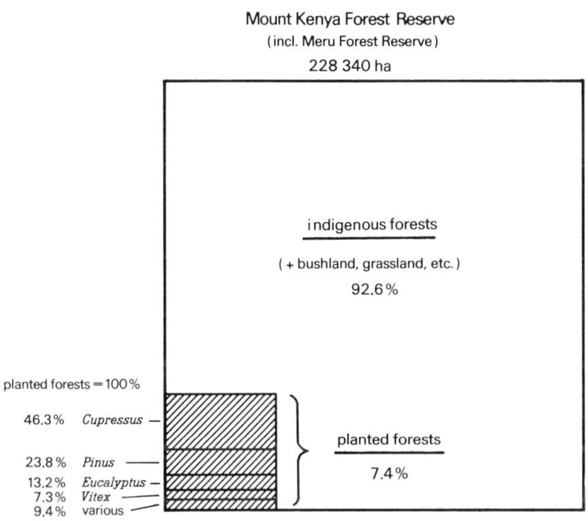

Fig. 4. Composition of Mount Kenya Forest. [*Source*: Plantation maps 1:10 000, prep. by Forest Department Survey Branch, Nairobi; unpubl.]

ery with a rotation length of rarely more than 30 years. The need for powerline and telegraph poles as well as the demand for fuelwood by some neighbouring tea factories is primarily served by eucalyptuses (*Eucalyptus globulus, E. saligna*), which owing to their lofty growth are also used as windbreaks or bordermarks on some forest stations. As the only indigenous species, Meru Oak (*Vitex keniensis*) is planted on a larger scale. It is highly regarded for its valuable wood. Unfortunately, the tree requires at least 60 years to reach maturity.

Pinus was the prevalent plantation species during the late 50s and early 60s, but was slowly surpassed by *Cupressus*, which dominated afforestation during the 80s. However, discontinuities in the age-class structure of plantations are quite frequent. They can be explained by various factors, including station-specific management plans, the availability of planting sites, seedlings and manpower, and the survival after planting.

Nanyuki F.S. provides an example of the spatial organization of the plantation system (Fig. 5). The forest village includes public buildings, houses for the workers and their families as well as the nursery. Nearby, cedar plantings dating from the 20s and 30s constitute the core of the forest station. Further afield, compartments with younger stands of cypresses, pines and eucalyptuses alternate in a dominant SE–NW orientation. Roads, tracks and grassy firebreaks provide access. Along the rivers, spurs of the eastern indigenous forest have been conserved as protectors against soil erosion.

4. Shamba-system and squatter problem

The afforestations at Mount Kenya were traditionally carried out by means of the so-called 'shamba-system', an adoption of the 'taungya-system' developed for the teak plantations in Burma in the middle of the last century (see Hesmer 1966; Lamprecht 1986). The shamba-system, which integrates agriculture and forestry, is characterized by a typical sequence of cultivation practices. First of all, the areas assigned for afforestation are cleared. This can be done either by licensees, especially if there are merchantable trees, or by peasant farmers, who in any case are responsible for the subsequent

Fig. 5. Plantation map of Nanyuki Forest Station, State: 2/1990. [*Source*: Plantation map 1 : 10 000 'Mount Kenya Forest', sheets 4 & 11, prep. by Forest Department Survey Branch, Nairobi; unpubl.]

clearing of the plots alloted to them. After the remains of the vegetation cover have been burned, the farmers, with the onset of the rainy season, start the cultivation of food crops, which have to be interplanted with trees during the following two years. Mixed culture has to be abandoned after another one or two crop harvests, when the rapid growth of trees prevents further raising of food crops. The farmers must then shift to other plots to start a new cultivation cycle.

At Mount Kenya, the shamba-system was practiced with considerable success for some decades to the mutual benefit of both the forest workers and the Forest Department. The forest workers were provided with land for traditional subsistence agriculture (maize, potatoes), and the Forest Department profited by free establishment and initial maintenance of plantations.

In addition to the official employees and their families, the forest stations were, however, settled by other groups of people. Though generally apostrophized 'squatters', these people were allowed to participate in the shamba-system as an additional working force, especially during periods of increased afforestation activities. Among the squatters were 'retired workmen', who would remain on the forest stations, friends and relatives of the official employees, and sporadically also workers of sawmills situated within the forest stations. The continuous influx of squatters led to various trespassings (e.g. illicit felling and grazing), which impede proper forest management.

To dispose of these nuisances, the Kenyan Government concluded the eviction of squatters from the forest stations in 1989. Simultaneously, the shamba-system was abolished. The extent of squatting is indicated by the fact that about 10 000, 5000 and 2500 people had to leave Gathiuru-, Ontulili- and Nanyuki F.S., respectively. Resettlement programmes had not been implemented. Instead, it was assumed that people would return to the place from which they had come.

These decisions of the Government affect not only the former squatters. Management practices on the forest stations will have to change as well, since cheap or free labour will no longer be so readily available. The ecological and technical consequences resulting from the abolition of the shamba-system, which provided near optimal conditions for tree planting, cannot be predicted yet. Moreover, the depopulation of the forest stations at Mount Kenya may not be beneficial, but may have a negative impact in leaving the forest unprotected against the destructive influences of man and animals.

The management problems are compounded by the fact that in connection with the eviction of squatters the Forest Department also required the official employees to leave the forest stations. Although nuclear teams have remained on all stations, the majority of workers had to look for new domiciles outside the forest. For the people concerned this does not only mean tiresome daily marches, but also a breach with the social community previously existing within the forest villages. For forest management it means less scrutiny of clandestine operations, and a slower response to wildfires.

5. The utilization of Mount Kenya Forest

The Kenyan Forest Department is not only responsible for the cultivation and protection of forests, but also for the commercialization of the products either planted or naturally available within the forest stations. Licences for the utilization of forest produce are sold to individuals, institutions and enterprises. A distinction is commonly made between timber and non-timber products. 'Timber' consists of those logs which are cut by sawmillers and pitsawyers for further processing. All other types of exploitation are included under 'non-timber products'.

5.1. Timber harvesting

In order to obtain licences for felling trees, sawmillers – whose plants are located at the margin of the forest stations, along the main roads or within the surrounding villages – have to apply to the respective district forest offices. Selected sub-compartments of the forest blocks are assigned in accordance with the total number of applicants, the working capacities of the sawmills, and the felling cycles prescribed for the individual forest stations. Whilst 5-year 'long-term licences' dominated in the past, sawmillers will henceforth have to contend with 1-year 'short-term licences', thus envisaging increased concurrence and economic insecurity.

These temporary licences are issued for clear-fellings and thinnings of plantations, and for salvages of merchantable species within the indigenous forest. Salvaging is mostly performed by pitsawyers, who cut the selected trees (mainly Camphor) on the spot into transportable pieces. Since the felling of live indigenous species was banned in 1986, pitsawyers have to rely on 'dead-fallen material', which has resulted in a steep decline in their business.

Because of the limited supply of raw material, most sawmillers hold licences for two or three forest stations. Moreover, they do not only produce standardized boards and ledges, but sometimes also manufacture furniture, doors, windows, prefabricated houses, etc. Waste resulting from the processing of timber is commercialized as well. Offcuts are either sold piecemeal as construction wood or, if the quality is bad, as fuelwood by head-, cart- or lorry-loads. Sawdust is generally thrown up to dumps and occasionally used as fuel, litter, and a filler for potholes.

5.2 Non-timber products

Non-timber products may include a range of products from powerline and telegraph poles (eucalyptus plantations) and smaller roundwood (first thinnings) to medicinal plants and honey (indigenous forests). Specific licences are issued for each of these products.

To meet the high demand for fuelwood, foresters sell 'monthly fuelwood licences', which entitle the licensee to collect one headload of fuelwood each day during one month. The gathering of fuelwood is beneficial, because it removes much of the easily inflammable brushwood. However, monthly fuelwood licences have recently been restricted on some forest stations to check the destruction of trees.

Fuelwood licences are also granted to merchants or wholesalers, who harvest fuelwood by cart- or lorry-loads from selected sub-compartments, and who prefer big customers such as public institutions. The licence fees are calculated on the basis of 'stacks' (1 stack is about 3 m^3), prices being lower for the exotic than for the better burning indigenous woods.

While the collection of fuelwood often takes place after clear-fellings or thinnings, twigs ('withies') are generally provided by prunings and locally used as building material. Most forest stations also authorize the cutting of fodder grass, the demand of which has increased, since grazing of livestock within Mount Kenya Forest was banned at the end of the 80s. With the exception of Ontulili F.S., where a large sheep farm still holds respective rights, grazing licences are no longer issued to prevent damage to trees.

The restriction of monthly fuelwood licences and the ban on grazing may contribute to the conservation of forests, but impose hardship on the local population, whose economic prospects are still further limited.

6. Dangers and protective measures

Most of the dangers and destructions encountered within Mount Kenya Forest are affected either by human activities or by attacks by big game. Menaces by parasites or diseases are comparatively unimportant. However, the 'cypress aphid' recently affected hedges and trees all over Kenya and may pose a problem in the near future.

All exploitation within the forest belt requires special permission, and any unlicenced activity is considered an offence. Detection of such activity by forest guards, watchmen or patrolmen results in fines or court action. Illegal activities include the removal of forest produce, grazing of animals, and damage to young trees. The indigenous forests in particular are said to be infested by tree poachers, whose nocturnal activities cannot be stopped because of lack of security staff.

Fires also threaten the forests on the western and nothern slopes of Mount Kenya especially during the dry season. In 1960, the Kenyan Forest Department elaborated a 'Mount Kenya Fire Plan', which was succeeded by a detailed control and action plan for the highly endangered Ontulili F.S. (Mather and Muchiri 1980, unpubl.). According to internal statistics, all of the 19 fires registered during the eighties within the area administered by Ontulili F.S. (excluding Ndare- and Timau Forest) had inflamed between December and March and destroyed a total of 7240 ha. Neighbouring farmers and honey gatherers, who frequently smoke out hollow branches or trunks without proper care, top the list of malefactors.

Many preventive measures have been put into practice. These measures include firebreaks, patrolling and – on Ontulili F.S. only – daily fire danger ratings based on meteorological data. The grass cover of firebreaks should be maintained by regular grazing, but this may be hampered by the abolition of grazing licences.

The younger plantations are constantly endangered by wild animals (especially elephants and buffaloes), which invade from the indigenous forests and frequently uproot or debark the tender trees. Since these animals are protected by law and may not be shot in general, they can either be turned away (which rarely works) or be prevented from entering the tree plantations. For this purpose, a 'game-moat' has been dug between the indigenous- and the plantation forest (comp. Fig. 5). Because it is damaged in many places, the game moat can, however, hardly fulfil its protective function.

In order to shield Mount Kenya Forest, especially its wetter southern and south-eastern sections, from the advance of smallholders, the Kenyan Government has founded the Nyayo Tea Zone Development Corporation, which since 1984 is charged with transforming the lower margin of the forest at a breadth of about 100 m into cooperative tea-farms. Although the buffer-like Nyayo Tea Zone has largely been completed up to now, official information about its effectiveness as a protective and at the same time productive belt are still lacking.

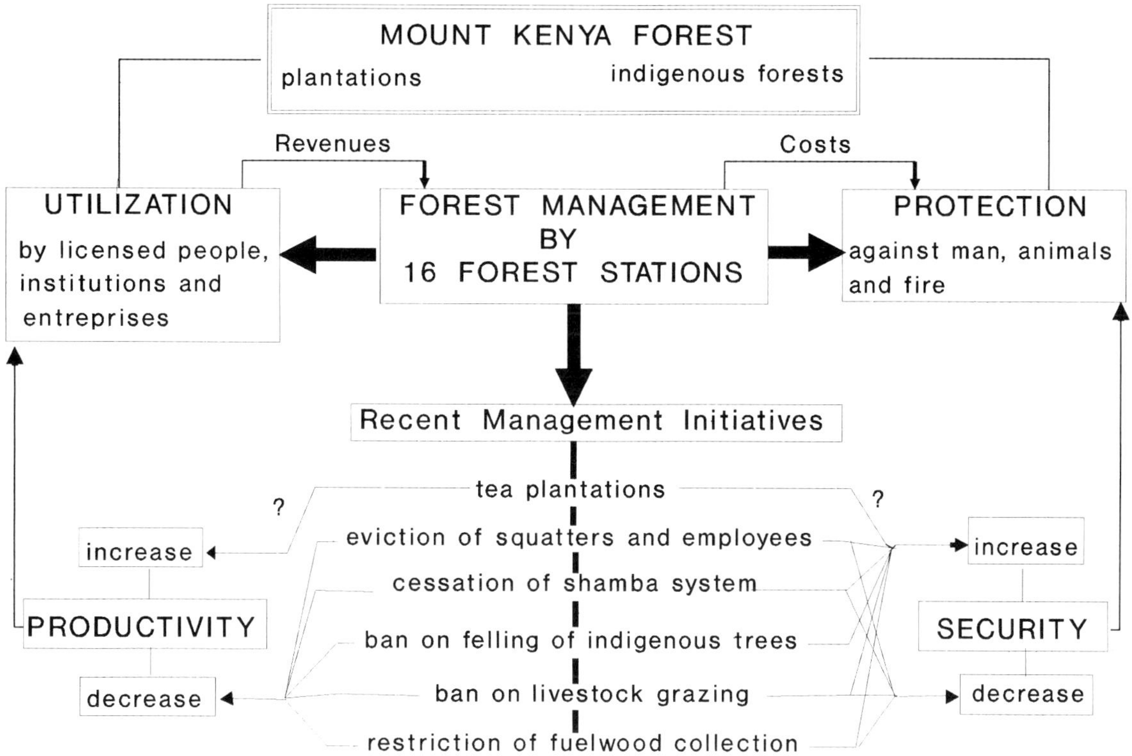

Fig. 6. Variables and measures concerning the utilization and protection of Mount Kenya Forest.

7. Management and research needs

The crucial problems for the management of Mount Kenya Forest have been summarized in a diagram (Fig. 6). It becomes evident that many of the recent initiatives, which were officially enacted for protective reasons, will certainly eliminate some security risks, but at the same time reinforce or even provoke other dangers. On the whole, security within the forest stations may neither increase nor decrease. Unfortunately, most of the initiatives are counterproductive and will provoke detrimental economic consequences for the forest stations as well as for the local people. This creates a conflict, not a balance between production and protection, the latter not sustaining but restricting the former. Instead of the reactive initiatives implemented during the last few years, more innovative measures should have been conceived.

The crucial problem for the management of Mount Kenya Forest is, of course, a financial one. With sufficient money, it would be easy to improve the forest stations' infrastructure, nowadays characterized by bad loose-surface roads, lack of vehicles and telephones as well as by inadequate tools and implements. With the same prerequisite it would also be possible to employ more people to accomplish the plantation and protection works required. However, increased funds will only come from increased production.

In order to optimize the utilization and conservation of Mount Kenya Forest, it will moreover be necessary to (re-)define superior management units, each serving one of the following purposes:

— afforestation with indigenous or exotic species aiming at the establishment of an economically effective plantation system; afforestations should be confined to areas already transformed into plantations or to bush- and grasslands
— enrichment plantings to secure species diversity and long-term possibilities for selective fellings; enrichment plantings, which up to now have rarely been carried out, should be intensified in degraded sections of the indigenous forests
— ecological conservation with unaltered species composition and natural regeneration; additional areas for conservation could be obtained by lowering the borderline of Mount Kenya National Park and/or by establishing protected corridors along the other climbing routes.

The demarcation of planning units and the formulation of their management plans requires more detailed investigation into the ecological and economic differentiation of the forest belt. Interdisciplinary studies, monitoring programmes and international comparisons of scientific results are badly needed. It is to be hoped that a future Mount Kenya Programme to be implemented under the auspices of UNESCO-MAB will produce firm development concepts aiming at a well-balanced integration of production and protection on behalf of the mountain's most crucial eco-zone: the forest belt.

Acknowledgements

The author wants to thank the Deutsche Forschungsgemeinschaft for financing two journeys to Kenya in March/April 1990 and from February to April 1991.

References

Abraham, M. F. H. 1958. The East African camphor forests of Mount Kenya. East Afr. Agric. J. 24: 139–141.

Fröhlich, W. & Willer, M. 1977. Bodentemperaturen und obere Waldgrenze. Vorläufiger Bericht über Untersuchungen an den Hochbergen Ostafrikas. Die Erde 108: 347–355.

Hesmer, H. 1966. Der kombinierte land- und forstwirtschaftliche Anbau. Vol. I: Tropisches Afrika. Wiss. Schriftenr. des Bundesm. f. Wirtsch. Zusammenarbeit, Vol. 8. Stuttgart.

Kohler, T. 1986. Mount Kenya: The forest belt and its utilization. In: Winiger, M. (Ed.), Mount Kenya Area. Contributions to Ecology and Socio-Economy. Geographica Bernensia, African Studies Series, Vol. A 1. Berne.

Lamprecht, H. 1986. Waldbau in den Tropen. Hamburg/Berlin.

Rehder, H., Beck, E. & Kokwaro, J. O. 1988. The afroalpine plant communities of Mt. Kenya (Kenya). Phytocoenologia 16: 433–463.

Republic of Kenya. 1968. A forest policy for Kenya. Sessional Paper No. 1 of 1968. Nairobi: Government Printer.

Rheker, J. R., Taiti, S. W. & Winiger, M. 1989. Bibliography of East African mountains. Laikipia Reports, Vol. 13, Univ. of Berne.

Trapnell, C. G., Birch, W. R., Brunt, M. A. & Lawton, R. M. 1976. Vegetation map of Kenya, Sheet 2. Nairobi: Survey of Kenya.

War Office. 1955. East Africa 1:250 000. Map GSGS 4801. Special Sheet Nyeri, Edition 1. Nairobi.

Wimbush, S. H. 1930. The forests of Mount Kenya. In: Dutton, E. A. T. (Ed.), Kenya Mountain. London.

Winiger, M. 1986. Causes and effects of the thermohygric differentiation of Mt. Kenya. In: Winiger, M. (Ed.), Mount Kenya Area. Contributions to Ecology and Socio-economy. Geographica Bernensia, African Studies Series, Vol. A 1. Berne.

Winiger, M., Wiesmann, U. & Rheker, J. R. (Eds.). 1990. Mount Kenya Area. Differentiation and dynamics of a tropical mountain ecosystem. Geographica Bernensia, African Studies Series, Vol. A 8. Berne.

16. Research and development programme for forestry in Sarawak: A pilot model approach towards sustainable forest management and economic development

E. F. BRUENIG
University of Hamburg, Leuschner Str. 91, D-2050 Hamburg 80, Germany

Keywords: Conservation, integration, marketing, multi-sectoral coordination, trade, utilization

Abstract. The Research and Development Programme for Forestry in Sarawak is the response of the Government to a critical situation which has arisen over the past 15 years. It comprises 23 thematic project areas which are closely integrated. Vertical and lateral integration of the programme is affected by a project "System Analysis and Modelling and Hierarchic Geographic Information System (GIS) for Research, Management and Development in Sarawak". The programme aims to initiate and stimulate progress on a broad front. The long-term focus is on the essential aspects of macro-economic development, environmental protection, species and nature conservation, and social and political aspects of forestry. Its basic philosophy is to integrate the development between all sectors of land use, commerce, small- and large-scale industry and administration. An important aspect is to harmonize methodologies and data sets, and to link forestry research closely to forestry practice. In this unconventionally comprehensive and fully integrated multi-sectoral approach to solve the manifold economic, social and cultural problems of development, Sarawak is unique in the tropical world. The programme has the potential to become a pilot model of sustainable utilization, management, conservation and development of natural and human resources.

1. Background

Sarawak occupies a land surface of 12.3 million hectares. The population of 1.5 million people is concentrated in the coastal lowlands and along rivers. There are 23 major and 10 minor ethnic groups. Apart from the Chinese and Indians, the population is racially of Malay stock who occupied the lands in several waves of migration. About 300 Malay stock people are still practicing the nomadic life of forest-dwelling hunters and gatherers. They live in large forest tracts in northeastern Sarawak, including Mulu National Park, Pulong Tau N.P. and 3 special Penan community forests, totalling 285 815 ha, or almost 1000 ha per person. It is still a matter of contention among experts whether these people are primary forest dwellers or have been pushed into the forest by more aggressive tribes. There are no aboriginal negritoes on Borneo. The processes of migration and land occupation are still in a very active phase and the country in this respect is still in a pioneer phase of settlement and land allocation. At this time, about 0.5 million ha of land are under agricultural tree and non-tree crops. The area under shifting cultivation has grown from 2.3 million ha in the sixties to about 3.5 million ha now, of which in 1987 about 0.12 million ha were within the Permanent Forest Estate (Sarawak Forest Department 1990a).

Table 1. Population structure in Sarawak in 1987 (Sarawak Forest Department 1990)

Ethnic group	Number of people
Bidayuh	130 000
Chinese	453 000
Iban	460 000
Malay*	319 000
Melanau	90 000
Other indigenous ethnic groups	83 000
Others (incl. Indian)	18 000
Total	1 553 000

* Distinguished from the other Malay stock ethnic groups by their Islamic faith.

H. Lieth and M. Lohmann (eds.), Restoration of Tropical Forest Ecosystems, 173–183.

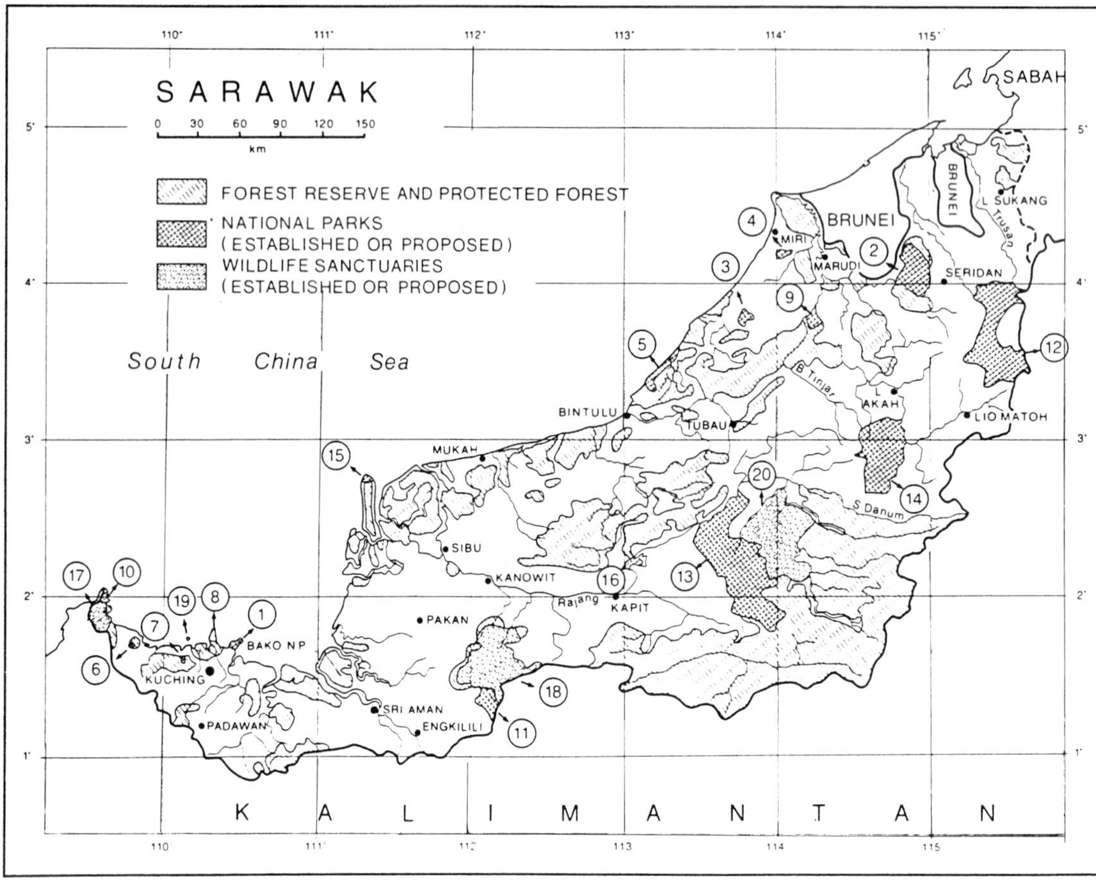

Fig. 1. Permanent forest estate in 1989/90 including production (forest reserve, protected forests) and totally protected areas (national parks and wildlife sanctuaries). No. 18 is the Lanyak Entimau Wildlife Sanctuary, the Sarawak part of a very large transboundary sanctuary. No. 2 is Mulu National Park and No. 3 Lambir N.P., both proposed sites for Long-Term Ecosystem Research (LTER) and potential keystones in the international LTER network. Adjacent production forest areas are suitable as Model Management areas, parts are also designated as Penan Community reserves. (Federal Research Center for Forestry and Forest Products, Hamburg, prepared from data supplied by the Forest Department, Sarawak)

The climate is typical non-seasonal to weakly seasonal wet equatorial rainforest climate but subject to sporadic extreme droughts in connection with the ENSO (el Niño southern oscillation) anomalies. More severe climatic condi- tions are expected to develop as a result of "Global Change".

There are more than 8000 recorded species of flowering plants and over 20 000 species of animals. Sarawak is one of the world centres of species richness and diversity with Mulu National Park (Fig. 1) probably being the most diverse rainforest ecosystem and landscape in the world.

2. History of forestry

Forest utilization for home consumption and export of such products as hornbill and rhinoceros horn, bird nests, resins, medicines, and speciality timbers, has a long history which goes back over two thousand years. Documented forestry history covers less than 200 years and six periods can be distinguished in the history of forestry and of forest development:

I pre-1841 Brunei suzerainty, traditional native forest exploitation, partly for export to Europe and China.

II 1841–1918 Sovereign Raj, pre-Forest Department forest utilization begins to be regulated.
III 1919–1940 Forest Department under the Third Rajah, Consolidation.
IV 1941–1945 Japanese occupation, exploitation.
V 1945–1963 Forest Department under the British Colonial Government, implementation of conservation and sustainability concepts and first steps towards integrated multi-sectoral surveys and management planning.
VI since 1963 Forest Department Sarawak in Malaysia.

The history of forestry in Sarawak during periods II–V has been described by Smythies (1963) and details on forest inventories and forest management by Bruenig (1963 and 1965). Forest research and organized forestry development began early in period III and continued after the Japanese occupation in periods V and VI. During the later parts of phase II and early parts of phase III, a socially and economically very important development was the improvement of traditional native methods of forest utilization. Destructive and wasteful practices of exploitation were replaced under government regulation by more efficient and sustainable utilization. Examples are the prohibition to fell Guttapercha trees for collecting the latex and introduction of tapping methods (Dunn 1975) and the prohibition of the destructive, wasteful and dangerous native form of traditional Kuda-Kuda logging by proper harvesting and logging methods in the peatswamp forests (Bruenig 1965).

Regular forest research, sustained-yield management and conservation in Sarawak have their origin in the twenties. Regulated management of mangrove and upland forest for the production of chicle (*Dyera lowii*), cutch (*Rhizophora spp.*), fuelwood and timber for local use (Belian — *Eusideroxylon zwageri*), and export (Sempilor — *Dacrydium pectinatum*) was then introduced. Experimental plantings of native species such as, among others, Merbau (*Intsia palembanica*), Rhu (*Casuarina nobile*), Kapor (*Dryobalanops beccari, D. fusca, D. oblongifolia*), Engkabang (*Shorea spp.*) and Durian (*Durio zibethius*) already began in the twenties. In the fifties, the Declaration of Forest Policy, the institution of a Commission on Customary Native Rights and the introduction of formalized procedures of sustainable management and silviculture in the Peatswamp forests were milestones of development towards environmentally, economically and socially sustainable forestry in Sarawak. In addition customary rights and cultural needs were taken care of by the provision that any development programme had to be based on ethnological studies which were initiated for all racial groups in Sarawak in the fifties, and continue to be vigorously pursued under the helm of the Sarawak Museum.

Formalized procedures for sustainable forest management by working or management plans were developed and introduced by the Forest Department in accordance with the British Empire Forest Policy (Schlich 1922) as follows:

- Mangrove forests: working plan for the Sarawak Mangrove Forest Reserve to produce poles, fuelwood, charcoal and cutch since the twenties
- Peatswamp forest: Sempilor (*Dacrydium pectinatum*) Swamp Forest, management of the conifer Sempilor for export and domestic markets, simple felling plan and natural regeneration, also since the twenties
- Riparian and Alluvial forest: belian (*Eusideroxylon*) production under a coppice system by small local licencees, simple felling prescriptions, since the twenties; later export of belian was banned to assure sustained supply of domestic demands
- Peatswamp forest: Working Plans and Regional Development Plans for Mixed Ramin (*Gonystylus*) Peatswamp forest in the permanent forest estate supplemented by plans for phased conversion of adjacent forest areas on state land, since the fifties, including strict control of harvesting and based on integrated, multi-purpose site and forest growing stock surveys
- Extractives: since twenties, but commercial interest faded during the last few decades when the markets for chicle (*Dyera lowii*), cutch and the various types of dammar collapsed. But even before, the native people were no longer attracted by the hard work of

collecting in the forest for very small financial reward.

In the seventies, the FAO undertook a conventional, commercial growing stock inventory in the mixed Dipterocarp forest (upland, lowland and hill forest). This was mainly used as source of information where timber was available for exploitation and less as tool of management planning and control. As a result, the eighties saw a cascading rush for timber and extensive timber exploitation in Mixed Dipterocarp forest. Even the fragile Kerangas and Alan Peatswamp forests were engulfed. The rush was fuelled by the interaction between information supplied by the FAO inventory and external market temptations and pressures mainly from East Asian consumer countries.

The rapid expansion of exploitation of Mixed Dipterocarp, Agathis-bearing Kerangas and Alan Peatswamp forests exceeded the technological and administrative capacities of the Forest Department by far. The result has been heavy revenue losses, but much more serious was a strong decline of the economic and non-material values of the forest resource due to:

— heavy and lasting damage to soil, water bodies and forest stands
— consequently, danger of long-term instability and reduction of biological productivity
— excessive waste of merchantable timber due to poor technology of harvesting and unfair grading and trading practices causing great socio-economic losses to the country
— unsatisfactory regeneration and growth of presently commercial and potentially commercial tree species which makes the proposed 25-year felling cycle untenable with present harvesting practice
— general overcutting beyond probable sustainable yield levels.

Ecological, economic and social sustainability of forestry were therefore rightly questioned and current practices suspected to be incompatible with the particular societal setting and the traditions of Sarawak. The International Tropical Timber Organization (ITTO) was consequently invited by the Sarawak State Government in 1989 to commission an expert group to review and report on the state of forestry with respect to the promotion of sustainable forest management. The Mission reported in 1990 that in comparison to other tropical and temperate countries, Sarawak has achieved and maintains high standards of forestry generally. The Mission also noted that "the quality of management in the Peatswamp forests is effective in respect to sustained timber yield for all species except ramin; management planning for the Hill Dipterocarp Forests is good but falls down in execution." The latter was not due to lack of ecological knowledge or of management or silvicultural know-how. Reasonably adequate basic scientific knowledge was available as shown by the extensive literature on the soils and vegetation of Sarawak. The adequacy of silvicultural know-how before the timber rush is demonstrated by among others Lee and Lai (1977). The cause for failure in the hill forests was chiefly lack of political support to the forest service, lack of sense of responsibility among managers and other decision-makers and lack of data, facilities, expertise and authority for analysing and convincingly demonstrating the unavoidable outcome of the chosen course.

Since the formation of the Federation of Malaysia in 1963, the accelerating economic development in Sarawak and the fast growth of population combinedly lead to an increase of forest clearing, mainly for traditional shifting (shifting-cum-rotation) agriculture. Valuable forest growing stock was destroyed for meagre short-term gains on soils and sites which were absolutely unsuitable for any kind of agriculture. Conflicts arose between different land uses for allocation of land. Consequently, the need for effective coordination and integration became increasingly more urgent:

— between government agencies and departments
— between forest management and conservation planning and the actual development of land use, especially in the areas where traditional agriculture was practiced.

3. Problems

Effective coordination and integration of forestry planning, implementation and control essentially requires

— adequate and reliable information based on sufficient and accurate data with efficient access and free flow of information within and between all sectors involved
— efficient instruments for situation analysis, land use and forestry planning and implementation, and monitoring and evaluating actual achievements.

Serious deficits existed in both areas within forestry and generally in all sectors of land use. Particularly complex and difficult problems have arisen in recent years as a result of expanding timber logging, increasing timber prices and the development of infrastructure. The value of having access to the usufruct of timber and the value of speculative land ownership have caused local unrest and friction. Questionable customary rights were claimed even for areas within the Permanent Forest Estate and the Totally Protected Forest Area which had already been legally settled long ago during the statutory legal constitution of the areas. A review of existing knowledge and data, and a well designed spatial information system is therefore urgently required not only in view of the international rainforest discussion, but particularly in the interest of the local communities and socially and environmentally compatible economic development.

4. Overall objectives of research and development policy

The principles and objectives of the long-term integrated research and development programme are based on the Statement of Forestry Policy in Sarawak (1954) which has been deliberately reconfirmed by the present government. The programme emphasizes the need to:
— conduct fundamental and application-oriented research to provide the scientific data base which is essential for continuous improvement of integrated forest management
— enable the government to achieve sustainability of timber production even before the ITTO "Target 2000"
— provide for sustainable production of all forest products from natural forests, supplemented where necessary and feasible by plantations
— provide for the preservation of the richness and biodiversity of the forest flora and fauna (including water bodies) and of the landscape diversity for species and nature conservation, preservation of genetic resources and for environmental protection
— achieve a sustainble balance between various kinds of rural land use and urban-industrial development compatible with national interests; maintaining a sustained source of state revenue and private income from natural resources and preserving the environment to keep it inhabitable in the interest of all inhabitants now and in the future
— increase timber utilization by improved harvesting and processing techniques and practices
— develop and implement high standards of timber logging (felling to haulage) to preserve economic, social and environmental values
— improve marketing of the timber and timber-based products in Malaysia and abroad by providing intelligence on market opportunities, trends and malpractices
— provide through improved planning and monitoring the information needed at state level for policy-making for short-term and long-range programmes of social and economic development
— study and develop low-cost, low-risk systems of silviculture which promise high productivity and profitability in natural forests (first priority), agroforestry (second priority), native-tree planting (third priority) and industrial plantations (low priority in view of the serious ecological and economic risks and uncertainties connected with tree monocultures in the humid tropics)
— develop criteria and indicators for the objective assessment of compliance with the principles of sustainable yield and management, particularly with respect to maintaining the multiple functionals of the forest ecosytem and its value for the social and natural environment
— facilitate the immediate application of research results in forestry practice, especially in natural forest management, wasteland afforestation and reforestation, timber utilization, by effective data access and dissemination of information.

- to adjust and improve efficiency of round-timber export and downstream timber conversion by improving log grading to suit industry best and reduce waste in harvesting and conversion; reducing the presently excessive waste in the whole utilization and conversion process; utilizing unavoidable waste; improving the value of the production by competitive quality at competitive costs; improving conditions of trade which are presently strongly biased in favour of the buyer
- establishment of a comprehensive multi-purpose data bank and an integrated hierarchic Geographic Information Sytem (GIS) in the form of an Integrated Multisectoral Spatial Information System (IMSIS) to secure comprehensiveness and transparency of planning and implementation of forest management, to integrate and monitor forest research and development at local and regional levels, and to serve as an efficient tool for preparing and checking high-level political decisions.

5. The integrated research and development progamme for forestry in Sarawak

The mission of the International Tropical Timber Organization (ITTO) which reviewed the state of forest management in Sarawak in 1989/90 suggested in their report to the Sarawak Government that forestry in the hill forests of Sarawak fell short of targets, was not sustainable and that action was needed. In response to the recommendations given in the report by the ITTO Mission to Sarawak, a comprehensive research and development programme was prepared by the Forest Department in 1990. This programme extends and integrates the then existing research programme and is fully consistent with the Forest Policy of Sarawak, 1954. It sets clear priorities in relation to the major current problems of forestry and land use and economic development in Sarawak. The programme aims at improving the efficiency and effectiveness of research and development among other things by:

- harmonizing data collection, data storage and data processing
- providing direct data linkages between projects of research, development, management and monitoring, and between different government departments and institutions.

This design of the programme should ensure that full use is made of all data obtained by research and forestry practice. It also will ensure rapid information flow from research to practice and vice-versa and provide means of efficient information retrieval for policy decisions at high level.

The programme applies to all categories of forests in the state, which includes non-forested land within the Permanent Forest Estate (PFE) and Totally Protected Area (TPA), as well as state land as long as it is still forested or after forest has been established by afforestation or natural succession. The total natural forest area at present is 8.7 million ha (70% of land area), declining by 60 000 ha annually as a result of shifting cultivation. An area of 250 000 ha natural forest is designated for regular conversion to settled agriculture, of this 140 000 ha to the year 1995 and 110 000 ha from then onward. The PFE area is being extended to 6 million ha and the TPA to 1 million ha (Statement of the State Government of Sarawak, Malaysia, on the 11th Session of the International Tropical Timber Council, 28.11.-04.12.1991, Yokohama; Table 2).

6. Structure and content of the R&D programme

The Research and Development Programme covers the whole range from basic research to monitoring regional development. The programme is integrated by data and information linkages between the different levels of

- specialized basic research (sectoral process research)
- natural ecosystem research (holistic, basic and systemic research)
- socio-economic system research (holistic but applied research)
- sociological and ethnological research (Sarawak Museum)
- research in regional development (applied technological)
- application in management and land use practice (planning, implementation, monitoring)

Table 2. Total area of permanent forest estate (PFE) and totally protected area (TPA) in 1990 before the decision to increase the area of both categories

Classification	Forest type and area (ha)		
	Mangrove forest	Peat swamp forest	Hill MDF
Natural forest permanent forest estate:	36 992	761 704	3 698 916
(1) Forest reserves	25 057	357 169	471 553
(2) Protected forests	11 909	401 283	322 981
(3) Communal forests	26	3 252	2 382
Totally protected area:	—	—	279 953
(1) National parks	—	—	104 362
(2) Wildlife sanctuaries	—	—	174 851
(3) Wildlife rehabilitation centre	—	—	740

— monitoring of regional development (continuous analysis of ground and remote-sensing data).

The Research and Development Programme covers the major relevant disciplines or subjects with a total of 23 project areas (Fig. 2). The subjects or disciplines and the projects are integrated by means of the core project Nr. 15 "System Analysis and Modelling, Data Structuring and Storing, Integrated Multisectoral Spatial Information System (GIS — IMSIS, Fig. 3). The interdependence between the 23 projects is rated in a 23×23 matrix with 253 interactions from 0 to 5 as follows:

Rating	Inter-dependence	Interaction	Percent of cases
0	negligible	none	1
1	weak	indirect information link	19
2	moderate	information exchange	29
3	strong	direct data link	28
4	very strong	unified methodological and common operational approach	13
5	complete	spatially and operationally fully integrated	9

The purpose of the matrix is to draw attention to the need for cooperation and to counteract the common tendency to keep projects isolated. The research programme also integrates the ongoing research and development and focusses on the most immediate pressing problems and critical questions which face sustainable forestry not only in Sarawak:

— the long-term ecological effects of various forms of interferences and management of natural forest, esp. the effects on stability, biomass, growth and multiple production, esp. of high value timber (projects 1–4)

— the level of sustainable annual yield of timber for the whole state and for management units (projects 5 and 7).

In the practice of forestry the most urgent need is to implement the best practices which are broadly known and described in the ITTO Guidelines and to enforce forestry and other laws and regulations and the stated governmental policy.

This requires

— enforcement of state-of-the-art best practices for immediate improvement of timber harvesting, utilization and marketing by drastically reducing waste of timber and damage to site and forest stand (projects 9, 13, 14) and by improving labour performance

— further long-term improvements of techniques in a pilot model management area (projects 6, 8, 9, 16)

— strengthening of the forest department in size and efficiency for effective enforcement (project 22)

— completion of the long-term conservation programme (biodiversity project 3).

The utilization and management of non-timber products has a long tradition in Sarawak. Adequate experience exists in the rural communities and in the department. The most crucial problems are related to externalities (markets, alternative employment) which cannot be influenced. Similarly, establishment and management of man-made forests for the production of timber and non-timber products are mainly problems of application and implementation of known principles (see ITTO Guidelines) and available local experience; the uncertainties are again in the externalities.

Table 3. Research and development programme for forestry in Sarawak: problems, priorities and project serial number of those projects which have higher rank of priority among the 23 projects

Problem	Priority	Project (No.)
Sustainability threatened by — overcutting — under-utilization — encroachment — land rights and claims	Yield assessment for the Permanent Forest Estate (PFE) production forest component in terms of present conditions and of biological potential	5. Data evaluation, model simulation, increment and yield assessment 7. Biological yield level vs. societal constraints
Degradation of natural resource (Socially, economically, ecologically) by inappropriate practices (technical and economic)	Improvement of harvesting and utilization, best practice of management and silviculture (ITTO Guidelines)	9. Model Management Area 11. Trade practices 13. Marketing and 21. Training labour and management staff 22. Strengthening Forest Department
Resource degradation as a result of lack of coordination in land use	Establishment of: — Data base — IMSIS (Integrated Multisectoral Spatial Information System)	15. System analysis and modelling, IMSIS/GIS 16. Integrated monitoring
Lack of implementation of best practice, rules and regulations	Strengthening of Forest Department's capacity, number and efficiency of staff, organization	22. Training and restructuring of Forest Department, Enforcement
Continual Encroachment of PFE, incl. Totally Protected Areas (TPA), by indigenous local farmers	Development of sustainable permanent agricultural systems, agroforestry	10. Social Forestry Agroforestry, integration Forestry: Agriculture, Penan Reservation
Deficiencies in basic ecological knowledge (especially of interactions!)	— Basic Long-Term Ecological Research (LTER) and multidisciplinary ecosystem research — continued establishment of National Parks, Nature Reserves, Wildlife Sanctuaries	1. Ecosystem research (Lambir, Mulu, Sabal) 2. Physio-ecology of commercial key species 3. Biodiversity research and protection, species screening, demography and dynamics of forests
Public misconception and disinformation	Raising awareness and understanding among the public and politicians	23. Public education and information

7. The core concept: integration

Efficient implementation and eventual success of the research and development programme hinge on effective integration. The essential feature of the R&D programme is the combination of comprehensive "top-down" with specific "bottom-up" approaches within a coherent system (Fig. 2). Operational and data and information linkages are provided:

a) *vertically between:*
— fundamental ecosystem research
— applied growth, production and yield research, silvicultural and utilization research and experimentation

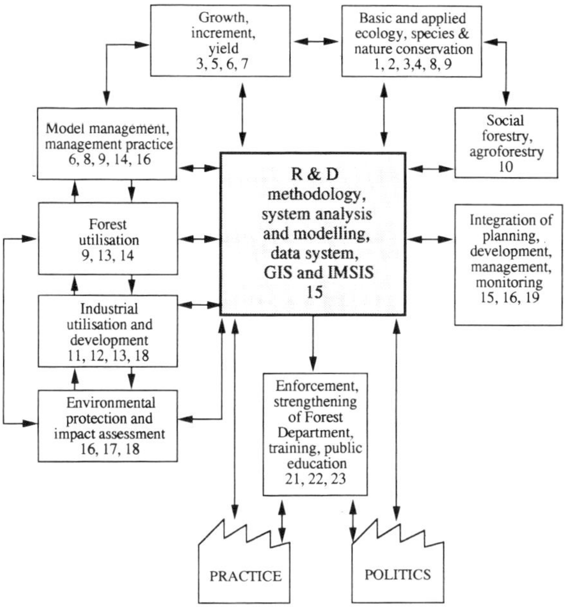

Fig. 2. Structure of an Integrated Research and Development Programme for forestry at the levels "forest ecosystem or forest management unit" and "regional forest and land management". In brackets: number of major projects of the Sarawak R&D programme for the period 1991 to 1996. Rugged top edge indicates external systems. [*Source:* Research and Development Programme for Forestry in Sarawak, 1990. Forest Department (Kuching, unpublished department report)]

— management of productive forests, national parks, wildlife sanctuaries and nature reserves
— forestry and other sectors of land use planning, implementation and monitoring
— forestry and the socio-cultural environment, esp. local communities
— political decision-making.

b) *horizontally between:*
— researchers and technicians of the various disciplines employed within complex projects
— research and development projects on the same or different subjects
— different public and private sectors and government departments involved in the development processes.

An Integrated Multisectoral Spatial Information System (IMSIS) is being developed for this purpose by the Sarawak Forest Department in cooperation with other governmental departments and some German scientists. The hierarchic

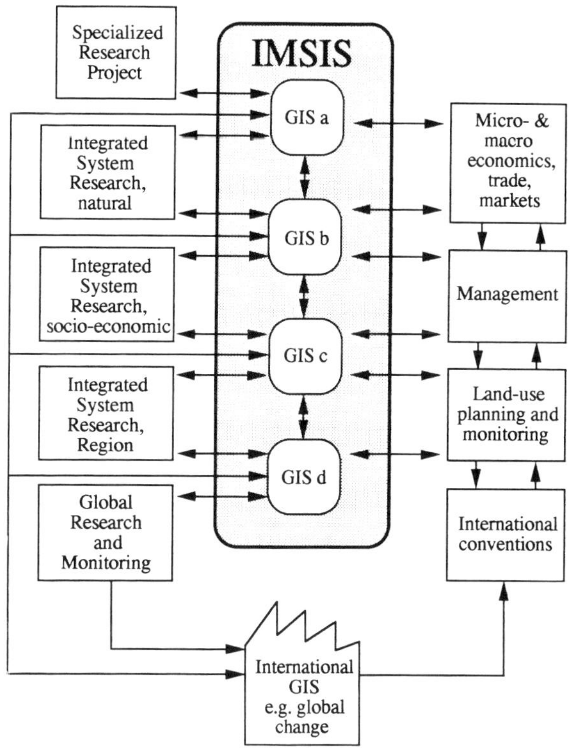

Fig. 3. Hierarchic structure of the R&D programme for Sarawak with the comprehensive 'Integrating Multisectoral Spatial Information System' (IMSIS) compounding the specialized GIS (a to d), which serve individual research or development projects and land use or economic sectors (unpublished departmental report).

structure and comprehensive nature of the information system integrates specific Geographic Information Systems (GIS) which are already or will be installed at "lower" level for special purposes (Fig. 3). The operational linkages with system analytical simulation models permit immediate testing of different options and policies for different scenarios at management-unit, regional and state levels, based on continuously updated spatial information.

8. Implementation

The R&D programme is the conceptional and operational framework within which research and development projects will be designed and implemented. The limited capacities of the forest department and of potential collaborating foreign institutions restrict the speed and scope of implementation. Implementation must be step-by-step and strictly adhere to the priorities (Table 3). As a first step, 6 projects were submitted to ITTO and were approved by the council in 1990. Four projects on biodiversity conservation, social forestry, watershed management and management model area received funding from several donor countries including Japan, Switzerland, Sweden and Norway. The two particularly urgently needed priority projects for achieving sustainability are "Forest increment and yield studies for sustainable Management" (see Primack 1991) and "System Analysis and GIS". These were earmarked and designed for cooperation with Germany which in both fields offers high standards of expertise and technology. Unfortunately funding has not yet been forthcoming. While the funding is generally not a chief and primary issue, without foreign funding it is difficult to procure required foreign expertise and technology.

9. Expected results

Eventually, the execution of the vertically and horizontally fully integrated R&D programme will achieve

- sustainable forest utilization and conservation and environmental preservation
- assured supplies to export markets and to domestic forest-based industries of timber and other produce
- fulfilment of the Sarawak Government's reconfirmed commitment to "Target 2000" of ITTO
- greater effectiveness and quality of inter-agency land use planning, development and monitoring which improves prospects for effective conservation of whole ecosystem types, special habitats, plant and animal species and the environment, while at the same time providing the basis for sustainable social and economic development
- providing permanent skilled employment to local communities as an alternative to destructive shifting cultivation
- improvement of the information base for high-level policy and medium-level management decisions and for checking of achievements and performance.

Thus the Sarawak R&D programme can be seen as a pilot model and an example for eventual implementation throughout the ASEAN region. It is expected that the consistently applied comprehensive approach by means of IMSIS will also overcome deficiencies of operational and administrative cooperation and will improve adaptation to specific cultural and natural site conditions. It will eliminate the unrealistic and irrational perspectives which lie behind the notorious failures of many private and public research and development projects in all parts of the world.

10. Outlook

The societal and political scene in Sarawak is favourable for the R&D programme to achieve success, probably more favourable than in any other tropical country in the world. The native people want economic development and improved standards of living and better quality of life. The Sarawak Government was in full accord with the people when, making the historically unique commitment in the "Statement of the State Government of Sarawak, Malaysia" to the 11th session of the ITTC in Yokohama, 2nd Dec. 1991, it insisted that the timber industry has a major role to play in the economic and social advance of the country, but that in spite of this the annual cut will be reduced to a tentatively estimated sustainable yield level, accepting the unavoidably adverse social and economic consequences. Better estimates are needed, but require further growth and yield assessments. Optimization of forest management and conservation within the whole diverse complex of land use and economic, social and cultural patterns of Sarawak requires new and sophisticated, comprehensive, integrated and multi-sectoral approaches. The Government has decided to develop the necessary technology for which it requires collaborative partnership. State Secretary Hamid Bugo in concluding the statement said: "Sarawak accepts the challenge to achieve sustainable forest management. Will others also seize the opportunity and match words with deeds? I must continue to believe that the international community, supported by all responsible governmental and non-governmental organizations, will also rise to the occasion." If they do, Sarawak can become a pilot model demonstrating that forestry and forest-based industry in tropical rainforest countries must not be declining "sunset industries" but can be emerging "sunrise industries".

Acknowledgements

I am grateful for assistance from the Sarawak Forest Department (information), Dr. Thomas Smaltschinski (computer graphic), Anja Köhler (tables), Thea Schweizer and Frauke Ruhnke (manuscript processing) and Mr. Wohltorf (drafting).

References

Bruenig, E. F. 1963. The history of forest inventories in Sarawak. The Malayan Forester 26 (3): 141–159.

Bruenig, E. F. 1965. The management of forest estates through working plans. The Malayan Forester 28 (1): 46–55.

Dunn, F. L. 1975. Rainforest collectors and traders, a study of resource utilization in modern and ancient Malaya. Monogr. 5, Mal. Branch, Royal Asiatic Soc., Kuala Lumpur, 151 pp.

Lee, H. S. & Lai, K. K. 1977. A manual of silviculture for the permanent forest estate of Sarawak. Forest Department, Sarawak, Silv. Pamphlet S. R. 1/77, 57 pp.

Primack, R. B. 1991. Logging, conservation and native rights in Sarawak forests from different viewpoints. Borneo Research Bulletin 23: 3–13.

Sarawak Forest Department. 1990. Forestry in Sarawak, Malaysia. Kuching, Forest Department Sarawak, 46 pp., 5 maps.

Schlich, Sir W. 1922. Forest Policy in the British Empire. Manual of Forestry, Vol. I, Bradbury, Agnew & Co., London, XI + 3425 pp.

Smythies, B. E. 1963. History of forestry in Sarawak. The Malayan Forester 26: 4.

17. Environmentally sound timber harvesting

Logging guidelines, conservation reserves and rehabilitation studies

J. K. VANCLAY
Professor of Tropical Forestry, Department of Economics and Natural Resources, Royal Veterinary and Agricultural University, Thorvaldsensvej 57, DK-1871 Frederiksberg C, Denmark

Keywords: Sustainable harvesting, management zoning, regeneration

Abstract. Timber harvesting guidelines formulated for north Queensland rainforests allowed economically viable harvests with minimal ecological impact. Harvesting procedures sought to minimize soil erosion, silting and turbidity in streams, destruction of trees in the residual stand, and disruption of natural habitats and processes. Forest officers selected the trees to be harvested, indicated the direction of felling, and supervised the design, construction and drainage of roads.

The long-term impacts of harvesting are not certain, so it is desirable to reserve some areas for conservation and monitoring. Existing data may help to indicate priority conservation areas, but errors and anomalies may be misleading. Some strengths and weaknesses of such data and techniques are discussed.

Degradation by soil erosion and weed infestation can be difficult and expensive to rectify. Without intervention, such areas may revert to high forest only after several decades, resulting in lost timber production. Reduced floristic diversity and inferior habitat for wildlife also imply substantial non-tangible losses. Rehabilitation may not be financially viable unless industrial plantations are feasible. Thus degradation should be avoided by adopting harvesting practices suited to the silvicultural characteristics of the forest, and by minimizing soil loss and invasion by weeds. In short, prevention is better than cure.

Introduction

Tropical forests are being exploited at an unprecedented rate, and deforestation now approaches 17 million hectares annually (Lanley et al. 1991). Often this exploitation is associated with conversion to other land uses, but sometimes the land remains within the forest estate. Such lands may be degraded by inappropriate harvesting operations which may cause soil erosion and weed infestation. This study looks at ways to minimize the impact of harvesting, and at some consequences of degradation caused by negligent logging and other disturbance. The prospects for rehabilitation are also considered.

Any logging, even environmentally sound logging, may cause changes in the forest, and it is desirable to set aside some areas for conservation, recreation and scientific study. Although guidelines for the design of such reserves are well established, procedures to resolve their placement are rather subjective, and this paper considers a number of objective ways to plan the location of such reserves.

The present study does not examine the merits of conserving tropical forests, or the consequences of harvesting them, as these aspects are covered adequately elsewhere (e.g. Poore and Sayer 1991). Rather, it examines ways to reduce the impact of timber harvesting and considers some of the resulting implications.

Harvesting guidelines

Natural forest management should be ecologically sustainable, economically viable and socially acceptable. Here I shall consider only the ecological aspects of timber harvesting. Any harvest (timber, fibre, fruit or exudates) from any forest (tropical or temperate) should be gathered in a way which minimizes environmental impacts. Four basic principles are necessary, but not always sufficient, to achieve this. These are to minimize soil loss, minimize silting and pollution of streams, minimize destruction of trees in the residual stand, and minimize disruption of natural habitats and processes. Of course, if the forest is to be converted to some other land use, only the first two of these principles are relevant. Thus an important prerequisite is a carefully formulated land use plan indicating the tenure,

H. Lieth and M. Lohmann (eds.), Restoration of Tropical Forest Ecosystems, 185–192.

capability and proposed use for all land in the region.

Soil erosion can be reduced relatively easily, by not logging steep slopes, by locating and constructing extraction tracks with easy grades and adequate drainage, by minimizing soil disturbance, and by ceasing operations during periods of heavy rain. Tracks of exposed mineral soil should not be made to each log. Rather, the organic material of the soil surface (and any understorey plants) should be disturbed as little as possible. Winches, grapples and logging arches to lift the leading end of the log clear of the ground may help to minimize soil disturbance during extraction. Whilst these provisions are easy to define in principle, they are more complex to implement in practice. Blanket rules fail to take into account differences in soils and equipment, and a worse-case provision may be impractical. In the seasonal tropics, a wet season cessation of logging operations may be beneficial, but only if drains and culverts are maintained before and during the wet season. Soil erosion occurs not only during logging operations, but also following the cessation of logging. To minimize this erosion, it is essential to check drains and culverts on completion of logging, and to install cross-drains to intercept water running down roads and tracks. Silting of streams will also be reduced by these provisions, and can be further reduced by maintaining buffer strips along streams, and by using bridges and culverts (designed to cope with expected peak flow) rather than fords. Trees should not be felled into streams, and any obstruction to stream flow caused by logging should be cleared on completion of logging.

Productivity may be lost, not only through the physical loss of soil by erosion, but also through changes in soil structure caused by compaction, impeded drainage and destruction of the soil profile (e.g. bringing subsoil and rocks to the surface). Obviously, soil disturbance should be minimized, but it is not always clear if tracked or rubber-tyred machinery is preferable. Smaller specialist machines are preferable to general purpose heavy earth-moving machinery, and it is important that the blade be no bigger than necessary. However, in practice, it is usually not the machine but the operator who determines the extent and nature of disturbance, and training and incentives may do much to reduce impacts. Blanket rules are not sufficient. Rather, it is necessary for operators to understand the principles and intent of the guidelines, and to have the motivation to do a good job. This implies effective supervision, penalties for noncompliance, and performance clauses for contractors and purchasers. The harvesting guidelines used in north Queensland (available from the author) were based on these principles, were effective in these seasonally wet forests, and provided the basis for the ITTO Guidelines of Best Practice (ITTO 1990a; Bruenig 1991).

Tree felling and extraction also requires the skill and the will of operators to minimize destruction of trees in the residual stand. However, some blanket guidelines can be given. In Queensland, trained Forest Service staff marked and sequentially numbered all trees to be removed, and indicated the direction in which these trees were to be felled. The direction of felling was chosen so as to concentrate the crowns of felled trees into groups, and to avoid damaging trees required for the residual stand. The system of numbering trees and logs helps ensure that no merchantable logs are overlooked, and the sequential numbering makes it easier to find these logs. The CELOS system (Jonkers and Schmidt 1984) also required the preparation of maps showing the location of logs, to assist relocation and to minimize unnecessary travel by the skidder. In Malaysia, climber cutting a year or more prior to felling has also been shown to reduce damage to trees required for the residual stand. Care in extracting logs is necessary to avoid damaging the bark of trees in the residual stand, since such damage assists entry of disease and decay. Extracting short log lengths rather than tree-length sections helps reduce soil disturbance and damage to the residual stand. It is false to assume that shoddy logging operations are more profitable. On the contrary, good logging practices may reduce costs by US $8 per cubic metre harvested and reduced damage to the residual stand may yield a 30% increase in value of the next harvest (ITTO 1990b).

Whilst it may be a desirable aim to "minimize disruption to natural habitats and processes", this is perhaps the most difficult provision to satisfy. It is inevitable that any logging will alter the stand structure and likely that the relative species composition may be affected transiently if not permanently. The art of silviculture is to minimize the impact of such changes, and to promote the rapid development of a new stand with desirable characteristics similar to the original stand. Many "weeds" (I refer to exotic and indigenous plants capable of multiplying rapidly and dominating a site where previously absent or present only in small numbers; not to economic importance) including bamboo, some palms and many vines, are light-demanding, and too much disturbance may favour invasion and infestation by such weeds. These may form a stable subclimax and impede regeneration of tree species for many decades. Thus in many cases, minimal disturbance is the safest approach.

In parts of Africa, a "do nothing" approach has worked well. This method allowed logging of a single species with minimal disturbance and then protected the forest from all human disturbance until the canopy had recovered and the market sought a new species. However, this system was only successful where protection of the forest was assured, and adequate time elapsed between harvests.

Queensland foresters were fortunate to have valuable tree species which were relatively light-demanding and regenerated readily, and to have few problems with bamboo, vines and other weeds. Even in these forests, the minimal disturbance approach seemed to be the most reliable, and logging guidelines stipulated that not more than 50% of the canopy was to be removed in harvesting. In practice, canopy disturbance was often much less than this permitted maximum (e.g. Crome et al. 1992). The timber harvesting guidelines formulated for Queensland rainforests provided for (Vanclay 1990):

a) logging guidelines sympathetic to the silvicultural characteristics of the forest, providing for adequate regeneration of commercial tree species and discouraging invasion by weeds (principally climbing vines);
b) treemarking by trained staff who specified trees to be retained, trees to be removed and the direction of felling so as to retain vigorous advance growth, to harvest mature and defective trees, and to minimize destruction of the residual stand;
c) incentives for logging contractors to be trained and for appropriate logging equipment to be used, so as to minimize soil compaction, disturbance and erosion;
d) prescriptions to protect adequate stream buffers and steep slopes from logging;
e) sufficient areas for scientific reference, feature protection and recreation to be identified and excluded from logging;
f) for deficiencies in an evolving system to be recognized and remedied, leading to an improved system.

Many studies of the effects of logging in these forests have been published and collectively provide a unique demonstration of one possible approach to sustainable timber harvesting. These studies have investigated the effects on fauna (e.g. Crome and Moore 1989), flora (Crome et al. 1992; Nicholson et al. 1988, 1990), hydrology (Gilmour 1971) and soils (Gillman et al. 1985), and indicate that timber harvesting in accordance with the guidelines is probably benign and that any environmental effects are transient and localized. Simulation studies suggest that timber harvesting based on these guidelines would probably be sustainable in the north Queensland rainforests (Vanclay and Preston 1989; Vanclay 1990).

This silvicultural system is one approach which may be successful, but other alternatives also exist. The Malaysian Uniform System involved complete removal of the overstorey canopy in lowland dipterocarp forests, and where adequate regeneration existed, it provided good recovery and canopy closure by the commercial dipterocarp species (Wyatt-Smith 1963). Secondary forest may also regenerate satisfactorily after shifting cultivation, and this suggests another possibility. The success of regeneration following such cultivation seems to depend upon the same two factors: preventing soil loss and overcoming weed problems. Thus successful forest management will depend more on an understanding of stand and weed dynamics than on blanket prescriptions.

Conservation reserves

It is inevitable that timber harvesting will change the stand structure of the forest, and likely that the relative species composition may also be affected. These changes may be transient or permanent, and it is desirable to provide a monitoring capability to gauge the nature and effect of these changes. Reserves protected from harvesting provide an important benchmark for assessing the nature and extent of such changes. Such reserves are also important in providing a habitat for plants and animals detrimentally affected by logging, and may also provide a wilderness area for recreation. Whilst stream buffers and steep slopes may constitute a considerable area of forest estate from which logging is excluded, such a haphazard allocation of reserves is inadequate. Many reserves have been allocated on the basis of scenic beauty (i.e. waterfalls, viewpoints, etc.). This may be good from a recreational perspective, but is inadequate for conservation and monitoring. It is preferable to explicitly provide for the reservation of a representative area of each habitat.

This raises several questions: how do you define a habitat; what constitutes a representative area, and how big should it be? There are several established guidelines for the design of conservation reserves: for example they should be large, contiguous, with the perimeter small relative to the area (e.g. Diamond 1975). However, it is not easy to resolve where to put reserves, or how much of a forest should be reserved in this way. In practice these issues may be of little consequence, as practicalities such as effective management and protection may be the deciding factors. Prescribed percentages of each vegetation type may fail to consider the total area involved. For example, a tract of 50 000 ha of type Y forest has a different conservation need than the only (or last remaining) 50 hectares of type Z forest. Similarly, the area required is larger when the reserve is surrounded by intensive agriculture, than when it is the core area within a biosphere reserve surrounded by selectively logged forest. The only effective way to preserve species and habitats may be to create carefully located protected reserves within a large matrix of managed forest.

Shafer (1990) gave a comprehensive review of conservation design principles. Petocz (1989) considered the practical application of these principles in the preparation of an integrated conservation strategy for Irian Jaya. He summarized several key principles for the placement of conservation reserves:

- The entire altitudinal spectrum should be covered;
- All known centers of endemism should be protected;
- Representative cross-sections of each habitat should be included;
- Substantial tracts of lowland rainforest may be required to accommodate their rich floristic diversity and to provide for species occurring infrequently;
- Species requiring large areas should be accommodated by creating protected areas within a larger matrix of managed forest;
- Special consideration should be given to the protection of mangroves, vegetational transition zones, and commercial forest types;
- Habitats for migratory species require explicit attention;
- Breeding and feeding sites (including offshore islands and beaches) should also be considered.

It is important that attention is not confined to undisturbed or large tracts. Many habitats have been extensively disturbed, and some of these habitats now exist only as small, disturbed and degraded areas. Past disturbance may not reduce the conservation importance of a habitat, but rather may increase the importancc of protecting the habitat to prevent further degradation, and to protect these remnants within the reserve system.

What information can be used to define habitats, representative areas and conservation reserves? Mackey et al. (1988) argued that three components were relevant: the biota (plants and animals), the habitat (environmental factors), and the niche (responses of the biota to these factors). Nix (1982) argued that complete niche specification was not necessary, and that it was sufficient to consider primary responses to the dominant environmental regimes; namely radiation, thermal, moisture, mineral nutrient and biotic. Thus Mackey et al. (1988) employed a digital elevation model, 23 bioclimatic attributes inferred from weather records, and geological survey maps to define bioenvironments in north Queensland. Their approach allowed objective identification of habitats without additional field survey work, but the design of reserves remains subjective. This approach does not utilize large amounts of prior information which commonly exists. Such data can be efficiently manipulated by Geographic Information Systems (GIS), and in theory, expert system principles could be used to objectively identify both the size and location of conservation reserves. Unfortunately, computer technology does not solve all the problems, and creates some new ones. Prior information is usually patchy, giving good coverage in some parts, and poor coverage in others, and this limits the inferences that can be drawn from such data.

One such source of prior information commonly available includes herbarium records (of flora, and equivalently, museum records of fauna). These may indicate locations of high species richness and of rare and/or endangered species. Unfortunately, most herbarium records are obtained by serendipity rather than from systematic surveys, and this negates the value of this data source. Reliable herbarium records could help to identify "hot spots", either specially rich in the number of species collected or with many records of a single rare or endangered species, but in practice may indicate more spurious locations than true hot spots. False hot spots of the first kind may occur when the recorded location indicates a place name (e.g. the nearest post office) rather than the actual site of collection. Other anomalies of this kind occur when collectors record latitude and longitude to the nearest degree rather than to degrees, minutes and seconds or precise grid coordinates. False hot spots of the second kind (i.e. many specimens of a rare plant) may occur at picnic areas and at easily accessible and well-known locations of "special" plants (e.g. monotypic endemic genera). These anomalies may occur when field trips are arranged for visiting botanists and taxonomists to show them places and plants of interest, and specimens are collected for both the visiting and host institutions, leading to (over time) multiple specimens in the local herbaria. If these anomalies are not rectified, the herbarium records may provide misleading inferences (e.g. may indicate hot spots outside the forested area). Forestry department and nature society records provide one possible avenue to verify official herbarium records, but pose another problem. State Forests may appear to exhibit richer flora than National Parks, and this may be attributed to the relative ease of access and of obtaining permits to collect specimens. But how much subjective editing of the herbarium database can be done before the database itself is no longer an objective record, but rather a subjective artifact of the editing process? What do you do with areas that remain uncollected? Clearly, if herbarium labels are to provide a useful database for inferring conservation significance, the reliability of each record must be encoded into the database. Thus the database must indicate if the specimen was collected during a systematic survey or otherwise, and must indicate the precision and reliability of the recorded location.

The Delphi survey technique provides a formal procedure to draw on, and collate the knowledge of experts in many disciplines (e.g. Schuster et al. 1985), and this approach may be used to identify areas of particular conservation importance. However, this too is not a panacea, but is subject to many limitations. Typically, respondents will vary from those with an intimate knowledge of a few specific areas to those who have an overview of the whole area, and even "experts" may find it difficult to reconcile other interests as diverse as their favourite study site or holiday location, to fears that publicizing sites would lead to exploitation (in particular, illegal collecting of orchids). This leads to the same problem: that of incomplete and incompatible data sets. Typically, the union of all key conservation areas so identified may encompass the entire region under consideration, whilst the intersection may be null. A workable compromise may require considerable subjective input.

Published maps of vegetation or forest types, of geology or soils, and of land tenure or disturbance may be available and should be considered in conservation planning. However, these different sources normally involve different scales, projections and editions of base maps, and introduce new problems in the form of "slivers" in the GIS. These slivers may indicate unusual transitional habitats important for conservation and research, or they may be anomalous and indicate the accumulation of small errors in the GIS. Anomalous slivers most commonly arise with vector-based GIS such as the popular ESRI Arc/Info system. Layers of information captured from independent sources may contain small errors in mapping or digitizing, and when these are overlaid or intersected, lines which should in reality form common boundaries do not line up in the GIS, but form these slivers. There is no way of knowing which of these slivers are real, and which are artifacts of the GIS. Some sensible decisions can be based on the width and area of these slivers, but these decisions are subjective and at best, repeatable but not objective. So do not get bogged down in technology, but use whatever medium is most readily available. If digital data and GIS are available, they will certainly be very helpful. However, if digital data are unavailable, it may be better and quicker to use transparent overlays than to try to digitize all the necessary data. Collate as much information as possible, but be critical of possible errors, especially with locations.

It seems that there is no objective and automated procedure to identify an optimal biosphere zoning for the conservation reserves, and it may be impossible to avoid subjective decisions, political expediency and gut feelings. However, this is not necessarily bad. Our ability to quantify ecology and habitat characteristics remains rather limited, and the subjective opinion of well-informed local experts, after careful consideration of the best available information, may provide a quicker, cheaper and better basis for conservation zoning than the most sophisticated computer technology which gives no guarantee of a good solution. It is also important to have a good understanding of what is possible; there is little point in proposing grand but impractical schemes which cannot be adopted, and it may be better to stick with the possible and strive for quick and effective implementation. There is little doubt that any proposal which satisfies Petocz's principles (1989), includes any remaining large areas still undisturbed, includes those areas with maximum topographic diversity, and encompasses as many habitats as possible in a single contiguous core area, will provide a good basis for a conservation system.

Rehabilitation

For much of the tropical forest, it is too late to pontificate on how to identify areas to set aside, and on how to harvest the forest in an ecologically acceptable way. For many forests, the damage has already been done. What can be done to help these forests? Some studies on rehabilitation have been made, and provide some guidance for rehabilitation of degraded forests. However, it is necessary to consider the scale of such rehabilitation projects. In Australia, volunteers have rehabilitated several small remnants (i.e. less than 100 ha) of rainforest by replanting and controlling weeds (especially vines, repeatedly until canopy closure). Restoration of a dipterocarp forest in Sarawak involved considerable inputs over only 50 ha (Miyawaki 1991; Mori 1991). At the FRIM Campus in Kepong (Malaysia), dipterocarps and other rainforest species have been re-established on mining spoil and abandoned vegetable gardens (Barnard 1954; Appanah 1991). Whilst these examples have been successful, these procedures cannot readily be applied to large tracts of forests in more remote areas. In many tropical countries, taungya systems (e.g. Evans 1982) which combine agricultural production with tree crops may be more feasible, but may not recreate a natural forest.

So what are the objectives of rehabilitation? Objectives may include enhancing timber production, preventing further soil erosion, restoring scenic quality, or restoring a natural ecosystem, and these may be conflicting. The most common forms of rainforest degradation include soil erosion on old extraction tracks, and the invasion and domination by exotic or indigenous weeds such as bamboo, palms and climbers. Procedures for ameliorating soil erosion are well documented and relatively easy to effect. Earth works to slow the speed of the water and to retain particulate matter can do much to alleviate soil erosion. However, weed control may require major intervention and the use of chemicals or heavy machinery, both of which may contribute other detrimental effects. In areas with good access, one financially viable option for bringing degraded forest back into timber production is to convert it into industrial plantations, but this introduces two problems: (1) substantial areas of bare soil may be exposed and liable to erosion during plantation establishment; and (2) most industrial plantations are monocultures, often exotic monocultures, which provide less diversity, and create a habitat for fewer plant and animal species than does the degraded forest. Mixed species plantations and agroforestry schemes may offer greater environmental and economic benefits, but may be financially less attractive for investors. Thus before embarking on a major programme of rehabilitation, the objectives and implications should be carefully examined.

In some instances, rehabilitation may be effected relatively easily if commenced sufficiently soon after logging. In Vanuatu and other areas in the Pacific, the vine *Merrimia peltata* is a serious problem following canopy disturbance, forming climber towers and impeding tree growth. However, it provides good cattle fodder, and grazing following logging can control the vine, provided that grazing pressure and duration are monitored to avoid damage to regenerating trees. Another means of control effective in trial plots of industrial tree plantations was to plant sweet potato, which occupies the site more quickly than *Merrimia*, displacing the latter, but which does not climb trees and is itself eliminated following canopy closure. Other short-lived pioneer shrub species may offer similar potential for controlling weeds through rapid canopy closure without excessive impairment of tree growth. Other species may also offer potential suppressing weeds, provided that they exhibit fast growth, small stature and do not climb the regenerating trees. Some legumes may be suitable, and offer the added advantage of nitrogen fixation.

A study of extensive areas of degraded hill dipterocarp forest in Peninsular Malaysia indicated that the opportunities for intervention were rather limited once woody or aggressive weeds were established (Silviconsult 1990). The most prevalent weeds were bamboo and bertam palm which are rather difficult to eliminate or control. Poisoning may not be effective as excessive canopy opening following death of the original weed crop could provide new opportunities for invasion by the same and other weeds. Enrichment planting may help trees to occupy the site more quickly, but the growth of these trees is slow relative to the growth rates of the weed species. Intervention may not be financially viable (Silviconsult 1990), except for industrial plantations on sites with good access and topography, and these may introduce other environmental problems. Where local people can be involved, agroforestry and community forestry may offer greater environmental and economic benefits, but this may not recreate a natural forest.

The moral of this study is that prevention is better than cure. Harvesting operations in forests prone to infestation by bamboo, palms, vines and other weeds should be carefully formulated to be silviculturally appropriate and to cause minimal canopy disturbance. On completion of harvesting in an area, sufficient drainage should be effected on all extraction tracks to minimize the danger of erosion. Careful harvesting is cheaper, more effective and more certain than rehabilitation.

Conclusion

Rehabilitation of degraded forest may be expensive, slow and uncertain. Once degraded, tropical forests may eventually rehabilitate themselves, but this may take many decades during which timber production may be negligible and the habitat value for other flora and fauna may be reduced. Intervention to hasten natural recovery may not be financially viable, and industrial schemes such as plantations may incur environmental costs. Agroforestry and community forestry schemes may offer greater potential where the local community is willing to participate, but should not be seen as a panacea as significant environmental costs may still be incurred. Clearly, prevention is better than cure. Thus harvesting operations should be customized to the silvicultural characteristics of the forest and, unless intensive silviculture has been proven, should strive to minimize soil and canopy disturbance. Drainage work should be undertaken on roads and tracks at completion of harvesting to prevent soil erosion.

It is inevitable that harvesting will change the structure of the forest, and possible that the species composition may also be altered, so it is desirable that some should be set aside as conservation and reference areas. These conservation areas should provide a representative sample of all forest types and habitats.

References

Appanah, S. 1991. Planting Quality Timbers in Peninsular Malaysia. Ministry of Primary Industries, in press.

Barnard, R. C. 1954. A manual of Malayan silviculture for inland lowland forests. Part IV. Artificial regeneration. Research Pamphlet No. 14, pp. 109–199.

Bruenig, E. F. 1991. The ITTO guidelines for the sustainable management of natural and planted tropical forests. This volume.

Crome, F. H. J. & Moore, L. A. 1989. Display site constancy of bowerbirds and the effects of logging on Mt. Windsor Tableland, North Queensland. Emu 89: 47–52.

Crome, F. H. J., Moore, L. A. & Richards, G. C. 1992. A study of logging damage in upland rainforest in north Queensland. Forest Ecology and Management, 49: 1–29.

Evans, J. 1982. Plantation Forestry in the Tropics. Oxford University Press.

Diamond, J. M. 1975. The island dilemma: lessons of modern biogeographic studies for the design of natural reserves. Biological Conservation 7: 129–146.

Gillman, G. P., Sinclair, D. F., Knowlton, R. & Keys, M. 1985. The effect of some soil chemical properties of the selective logging of a north Queensland rainforest. Forest Ecology and Management 12: 195–214.

Gilmour, D. A. 1971. The effects of logging on streamflow and sedimentation in a north Queensland rainforest catchment. Commonwealth Forestry Review 50: 38–48.

International Tropical Timber Organization. 1990a. ITTO guidelines for the sustainable management of natural tropical forests. ITTO Technical Series 5. 18 pp.

International Tropical Timber Organization. 1990b. ITTO action plan: criteria and priority areas for programme development and project work. ITTO, November 1990. 22 pp.

Jonkers, W. B. J. & Schmidt, P. 1984. Ecology and timber production in tropical rainforest in Suriname. Interciencia 9 (5): 290–297.

Lanley, J. P., Singh, K. D. & Janz, K. 1991. FAO's 1990 reassessment of tropical forest cover. Nature & Resources 27 (2): 21–26.

Mackey, B. G., Nix, H. A., Hutchinson, M. F., Macmahon, J. P. & Fleming, P. M. 1988. Assessing representativeness of places for conservation reservation and heritage listing. Environmental Management 12 (4): 501–514.

Miyawaki, A. 1991. Restoration of native forests from Japan to Malaysia. This volume.

Mori, K. 1991. Global environmental action: restoration of a natural rainforest ecosystem in Malaysia. This volume.

Nicholson, D. I., Henry, N. B. & Rudder, J. 1988. Stand changes in north Queensland rainforests. Proceedings of the Ecological Society of Australia 15: 61–80.

Nicholson, D. I., Henry, N. B. & Rudder, J. 1990. Reply: Disturbance regimes in north Queensland rainforests: A reevaluation of their relationship to species richness and diversity. Australian Journal of Ecology 15 (2): 245–246.

Nix, H. A. 1982. Environmental determinants and evolution in Terra Australis, pp. 47–66. In: Barker, W. R. & Greenslade, P. J. M. (Eds.), Evolution in the Flora and Fauna of Arid Australia. Peacock, South Australia.

Petocz, R. G. 1989. Conservation and Development in Irian Jaya: A Strategy for Rational Resource Utilization. E. J. Brill, Leiden, Netherlands. 218 pp.

Poore, D. & Sayer, J. 1991. The Management of Tropical Moist Forest Lands: Ecological Guidelines. 2nd ed. IUCN, Gland, Switzerland.

Schuster, E. G., Frissell, S. S., Baker, E. E. & Loveless, R. S. 1985. The Delphi method: application to elk habitat quality. USDA Forest Service Research Paper INT-353.

Shafer, C. L. 1990. Nature Reserves: Island Theory and Conservation Practice. Smithsonian Institution Press, Washington D.C. 189 pp.

Silviconsult. 1990. Natural forest rehabilitation study, Malaysia. Asian Development Bank, Manilla.

Vanclay, J. K. 1990. Effects of selection logging on rainforest productivity. Australian Forestry 53 (3): 200–214.

Vanclay, J. K. & Preston, R. A. 1989. Sustainable timber harvesting in the rainforests of northern Queensland, pp. 181–191. In: Forest Planning for People, Proceedings of 13th Biennial Conference of the Institute of Foresters of Australia, Leura NSW, 18–22 September 1989. Institute of Foresters of Australia, Sydney.

Wyatt-Smith, J. 1963. Manual of Malaysian silviculture for inland forests. 2 vols. Malaysia Forest Record 23. Forestry Dept. Malaysia.

18. Conservation and restoration of montane forest communities in Thailand

W.L. WERNER[1] and T. SANTISUK[2]
[1]*Professor Dr. Wolfgang L. Werner, Professur Physische Geographie, Katholische Universität Eichstätt, Ostenstr. 18, W-8078 Eichstätt, Germany*
[2]*Thawatchai Santisuk, Ph.D., Director of Forest Botany Div., Royal Forest Department, Bangkok 10900, Thailand*

Keywords: Conifers, fire, non-timber forest products

Abstract. Montane rain forests form the natural vegetation of Thailand, usually at the elevation above 900–1000 m a.s.l. But these evergreen montane rain forests are heavily subjected to various types of land-use. A frequent and somewhat reversible form of degradation is the lower montane oak forest with an impoverished flora and a dominance of Fagaceae, Lauraceae, Magnoliaceae and Theaceae, as well as other pioneer species following shifting cultivation. Frequent surface fires and poor sandy soil render the invasion of *Pinus kesiya* to form lower montane pine-oak forest. Deciduous dipterocarp forest and pine-deciduous dipterocarp forest with *Pinus merkusii* occurring sometimes up to 1350 m a.s.l. are also maintained by annual surface fire and thereby creating a more or less open forest in nature.

In Northeastern Thailand, the montane forests, luxuriantly developed on the isolated sandstone table-mountains, are effectively protected as national parks or wildlife sanctuaries.

Due to illegal settlements of the non-Thai hill people in the extensive mountain chains of the North since the last century, the natural montane vegetation has been much altered. As a consequence, open secondary vegetation as grasslands are a common landscape in the mountains. To prevent surface run-off and soil erosion, the abandoned degraded fields are planted with *Pinus kesiya*.

Besides the protective role as watershed area and productive role in terms of timber, non-timber forest products in the montane forests of Thailand are of economic significance.

Planting of lesser-known but useful indigenous conifers in the disturbed montane vegetation of Thailand, such as *Cephalotaxus griffithii* or *Calocedrus macrolepis*, should be encouraged and promoted through ecological and silvicultural research.

1. The forest types of the mountains of Thailand

The montane forests of Thailand are situated in an area where taxa from holartic, palaeotropical and Australian origin meet. While the lower slopes of the mountains in Northern and Northeastern Thailand are exposed to a long drought period, and thus are covered by deciduous forest, the regions above 900–1400 m benefit from frequent mists and reduced evaporation, and are thus covered by evergreen forests. Holarctic and Chinese elements are prominent in the top elevation of the ridges, whereas the tropical element reaches up to mid-elevation.

a. (Lower/upper) montane rain forest

Montane rain forests form the natural vegetation of Thailand, usually at the elevation above 900–1000 m a.s.l. In this vegetation type, dipterocarps, of both deciduous and evergreen types are no longer present, and tropical lowland elements are partly replaced by holarctic plant genera. Typical examples are Magnoliaceae, Theaceae, Lauraceae and Fagaceae. **Lower montane rain forests** still survive in some national parks, i.e. Doi Inthanon and Doi Suthep in Chiang Mai, and wildlife sanctuaries, i.e. Phu Luang in Loei, of Thailand. **Upper montane rain forest** with its umbrella-shaped crowns is luxuriantly developed around the top of Doi Inthanon only (Robbins and Smitinand 1966; Santisuk 1988).

Even in regions such as North and Northeast, where deciduous forests extensively cover the lower slopes of the hills, evergreen montane rain forests can be expected above c. 900 m in the moist valleys, as reduced evaporation and more frequent mists reduce water-stress. But these evergreen montane rain forests are much degraded in various ways by the hill tribes.

H. Lieth and M. Lohmann (eds.), Restoration of Tropical Forest Ecosystems, 193–202.

b. Lower montane oak forest

A frequent and somewhat reversible form of degraded lower montane rain forest is the **lower montane oak forest** with an impoverished flora and a dominance of Fagaceae (*Castanopsis, Quercus, Lithocarpus*) and other pioneer species following shifting cultivation, i.e. *Schima wallichii, Anneslea fragrans* (Theaceae); *Lyonia ovalifolia, Vaccinium sprengelii* (Ericaceae); *Helicia nilagirica* (Proteaceae); *Betula alnoides* (Betulaceae); *Engelhardtia spicata* (Juglandaceae).

Frequent fires and sandy soil allow the invasion of *Pinus kesiya*, to form **lower montane pine-oak forest** (cf. Werner 1990) (Figs. 1, 2).

c. Mixed deciduous forest, dry deciduous dipterocarp forest and pine-deciduous dipterocarp forest

Mixed deciduous forest occurs in regions where the seasonal climate does not allow evergreen rain forests, but conditions are more favourable for mixed deciduous species of both teak and non-teak species. Teak (*Tectona grandis*) is the commercially most important timber species of this forest type. The exclusion of dipterocarps is one of the main criteria to distinguish mixed deciduous forests besides the presence of tall bamboos.

But in the more xeric sites governed by moisture stress in the lateritic or gravelly soils, deciduous dipterocarp forest is widespread. The dominant species, belonging to the deciduous dipterocarps and comprising more than 60% of the canopy components, are *Dipterocarpus tuberculatus, Dipterocarpus obtusifolius, Shorea siamensis* and *Shorea obtusa*. Other dominant species are *Pterocarpus macrocarpus* (Leguminosae), *Gluta* (*Melanorrhoea*) *usitata* (Anacardiaceae) and *Aporosa villosa* (Euphorbiaceae) (Fig. 3).

Deciduous dipterocarp forest and pine-deciduous dipterocarp forest with *Pinus merkusii* occurring sometimes up to 1350 m high, are also maintained by annual surface fire; hence a more or less open forest in nature (Stott 1984, 1986; Schweinfurth 1988; Santisuk 1988).

Deciduous dipterocarp forest in an undisturbed site and on deeper and moister soil develops trees of more than 30 m in height and more than 70 cm DBH, e.g. *Dipterocarpus obtusifolius, D. tuberculatus*. But such sites are rare nowadays, hence open, savanna-like forests with small trees become the normal feature (cf. Werner 1993).

2. Forms of degradation

a. Felling

In the plains of Thailand, most lowland forests have been converted into agricultural lands. However, the mountain slopes in the vicinity of Thai villages and towns are traditionally covered by forest, as it can still be observed in the remote areas of Khun Yuam or Mae Hongson in Northern Thailand. These forests are influenced by fire, felling for timber etc., but instead of collecting firewood in the nearby forest, as it is usually practised in India for instance, charcoal is produced in remote forest areas.

In the mountains of Northern Thailand, shifting cultivation is widespread and has increased a lot during this century, after the influx of hill-people from neighbouring countries in the North. The system of land-rotation, which has been practised by people like Karen and Lua; short cultivation-long forest fallow allows the remarkable regeneration of lower montane oak forest. But the system of the Hmong (Meo), Yao, Musser, Ikaw and Lisu, with a consecutive use of the land for about 10 years, has resulted in an abandoned, degraded savanna-land only (cf. Santisuk 1988).

To prevent surface run-off and soil erosion in the watershed areas, the abandoned degraded fields are planted with *Pinus kesiya* by the Royal Forest Department. Once fire is effectively prevented for at least 5–10 years, these pine plantations or other patches of the natural pine stands are sooner or later replaced by the broad-leaved species of lower montane oak forest and hence forming lower montane pine-oak forest.

b. Fire

In the deciduous forests of Thailand fires are common during the dry season (January–

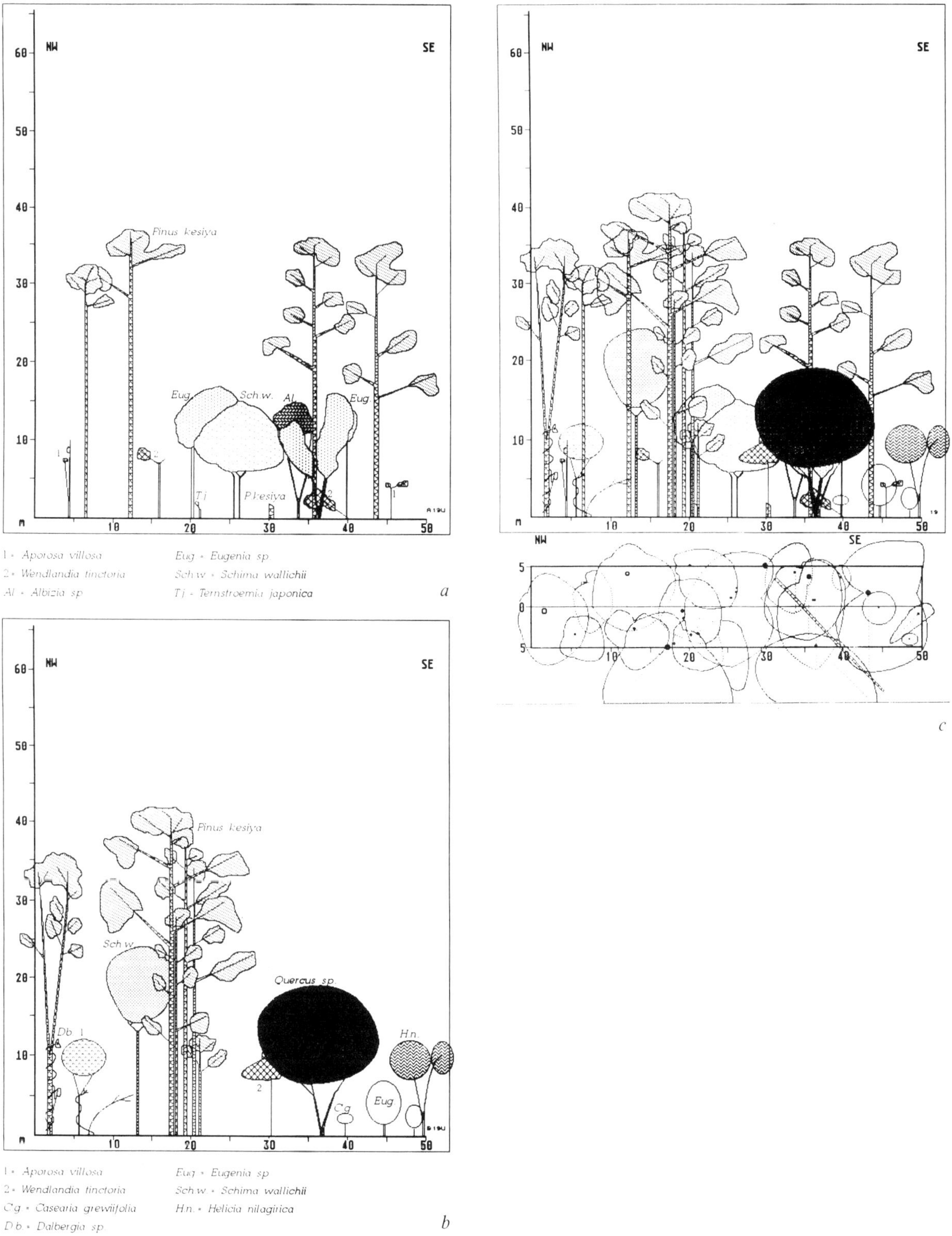

Fig. 1. Profile in lower montane pine-oak forest of Doi Inthanon National Park, 970 m a.s.l. (a) Part A. (b) Part B. (c) Parts A + B (cf. Fig. 2).

Fig. 2. Lower montane pine–oak forest in Doi Inthanon National Park, 970 m a.s.l., site of profile Fig. 1.

March), but the lower montane oak and pine–oak forests are influenced as well. Normally, the first restrict to the wilted undergrowth and the debris on the forest floor, while the standing trees and saplings are protected against the flames because of their thick bark and deep rootstocks. The positive effect of surface fire is the clearing of the undergrowth, so that seeds of trees can germinate on the bare ground after the fire and the seedlings can further develop. Grasses and woody lianas such as certain species of *Derris*, *Milletia*, *Pueraria* or *Spatholobus* have a suffocating effect on young seedlings. Surface fire is one of the decisive factors for the establishment of pines, which normally cannot compete with the broad-leaved species. But frequent fires may as well kill young seedlings and saplings and allow the survival of fire-adapted trees only. Pine plantation plots in Doi Inthanon National Park have shown, that the number of species is increasing, once fire has been prevented for several years. For the rehabilitation of lower montane rain forest, fire protection is essential (cf. Santisuk 1988; Stott et al. 1990).

3. Nature reserves in Thailand (Fig. 4)

Thailand is the only country in the region with a satisfactory national system of protected areas, as well as effective legislation and management. The extensive system of national parks and wildlife sanctuaries is an excellent base for *in situ* conservation of plants.

a. National Parks

National Parks have been established in Thailand since 1962 with Khao Yai and Phu Kradeung being the first reserves. Meanwhile more than 50 National parks with a total area of about 30 000 km^2 have been created in the past three decades. The largest one is Kaeng Krachan National Park, located at the northern end of the Malay Peninsula and bordering Burma with an area of 291 500 ha. Six other parks have an area of more than 100 000 ha. The parks are under the administration of the "National Parks Division" of Royal Forest Department (cf. Arbhabhirama et al. 1987).

The most important park in Northern Thai-

Fig. 3. Dipterocarpus tuberculatus with epiphytic fern and orchids. Doi Inthanon NP, 1000 m a.s.l., western slope above Mae Chaem

land is Doi Dinthanon, formerly called Doi Angka, around the highest peak of the country (2576 m). This reserve, which has been declared in 1978, has an area of 48 200 ha, including all major forest types of Northern Thailand. Doi Inthanon has perhaps encountered with the most serious problems among the national parks of Northern Thailand: conflict of land use by illegal shifting cultivators and various development programmes organised by both government and non-government bodies, regular fire and a great number of visitors. As a consequence, the fragile habitats of the lower and upper montane rain forests have been much threatened. Parts of the primary montane rain forests have given way recently to road expansion and buildings.

In Northeastern Thailand, the montane forests luxuriantly developed on the isolated sandstone table-mountains, are effectively protected as national parks or wildlife sanctuaries. Occasional fire from the lower slopes is the only serious threat, prohibiting natural forest regeneration. Because of no local or tribal inhabitants in the montane zones of those flat-topped sandstone mountains, the natural montane forests namely lower montane rain forest, lower montane oak forest, lower montane pine-oak forest and lower montane scrub are still intact, harbouring a great number of warm temperate and montane species.

Phu Kradeung, with an area of 34 816 ha has been declared a national park in 1962. During the dry season (October–May) the table-mountain is climbed by thousands of visitors, who create the environmental problems of trampling the natural sandstone-rock vegetation of orchids and lichen, and occasionally setting up campfires. The park is further threatened by a future cable-car project, which would increase the number of undisciplined visitors above a tolerable level (Fig. 5).

Khao Luang National Park stands for the montane vegetation of the South, as Khao Luang is the highest peak of the Thai part of the Malay Peninsula, reaching 1864 m a.s.l. The park was established in 1975, covering 57 000 ha. Above 500–800 m, the dense tropical evergreen rain forest is luxuriantly preserved. The lower slopes outside the park-boundary are covered by private agroforestry-plantations, which may likely by an ideal buffer-zone. The economic trees are betelnut-palms (*Areca catechu*), durian (*Durio zibethinbus*), a local variety (champedak) of jack-fruit (*Artocarpus integer*), mangosteen (*Garcinia mangostana*), langsat (*Lansium domesticum*), *Parkia speciosa*, *Archidendron jiringa* etc. On the ground floor coffee, taro, pineapple, banana etc. are cultivated between those fruit trees. As the plantations are devoid of trees of the original forest, which have hitherto been felled, it would be beneficial to promote the forest trees into a multi-cropping system.

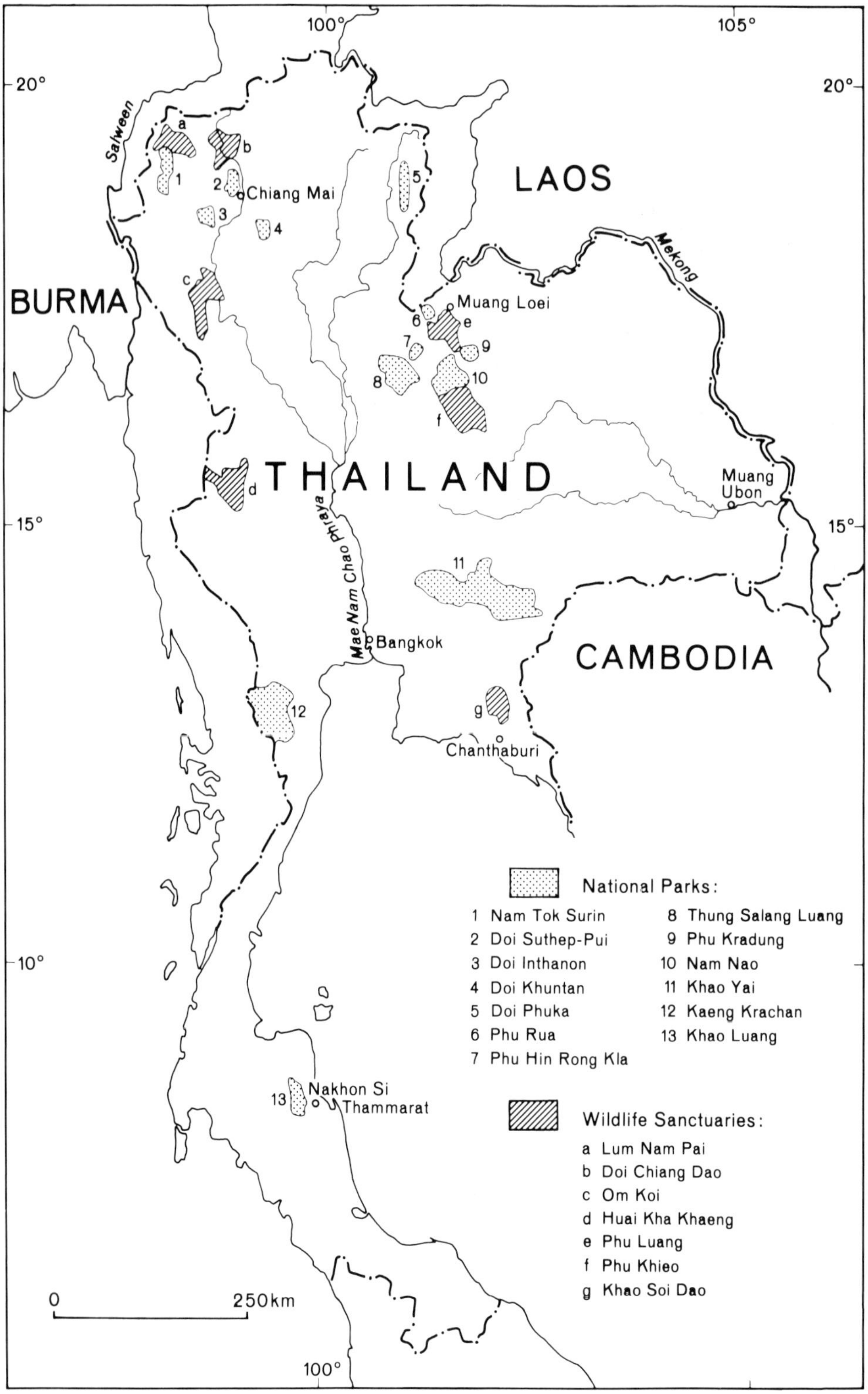

Fig. 4. Map of National Parks and Wildlife Sanctuaries in Thailand containing montane forests.

Fig. 5. Phy Kradeung National Park, Daeng Kliff, 1250 m a.s.l. Lower montane rain forest on the slope, pine-savanna on the plateau.

It is important to note, that this was part of the area where, in November 1988, within 3 days, rainfall of 1000 mm created catastrophic floods and landslides, killing more than 80 people and wiping out several villages. Even on certain steep slopes covered by natural rainforests, landslides also occurred. This disastrous event was the main reason for the Thai government to change its forest-policy and declare the indefinite "logging ban".

b. Wildlife Sanctuaries

Wildlife Sanctuaries are under the administration of "Wildlife Conservation Division" of the Royal Forest Department. While the sanctuaries in Northern Thailand face the same problems as the national parks, those in the mountains of the Northeast with diversified flora and fauna have little human influence.

Phu Luang Wildlife Sanctuary is perhaps the most magnificent flat-topped sandstone mountain (1571 m) in Thailand, with an area of 84 800 ha densely covered by lower montane rain forest, lower montane scrub, and pine-savanna. Besides the two native species of *Pinus* in the savanna, six native species of conifers grow in the lower montane rain forest: *Calocedrus macrolepis* (Cupressaceae), *Cephalotaxus griffithii* (Cephalotaxaceae), *Dacrydium elatum*, *Dacrycarpus imbricatus*, *Podocarpus neriifolius* and *Podocarpus polystachyus* (Podocarpaceae). Phu Luang has a remarkable landscape and unique lower montane scrub with diversified rock-flora of various epiphytic orchids (e.g. *Dendrobium infundibulum*, *Bulbophyllum dixonii*), *Agapetes saxicola* (Ericaceae), *Sorbus granulosus* (Rosaceae), *Schefferopsis subintegra* (Araliaceae), *Wightia speciosissima* (Scrophulariaceae) and lichen (*Cladonia* spp., *Parmelia* spp.) on bare sandstones.

Phu Kieo Wildlife Sanctuary, established in 1972 with an area of 156 000 ha, is covered by lower montane rain forest, pine-deciduous dipterocarp forest with *Pinus kesiya* and treeless savanna. This sanctuary is roamed by deer, elephant and tiger. There is also a trace of the two-horned rhinoceros (*Dicerorhinus sumatrensis*).

Fig. 6. Phyllanthus emblica with fruits. Omkoi plateau, 1200 m a.s.l.

4. Use of non-timber forest products for conservation of the forest ecosystem

The montane forests of Thailand are also important sources of non-timber forest products: fruits and nuts, turpentine, lacquer, benzoin and wild orchids are some examples. The economic significance of non-timber forest products varies locally, but in general they are most important to low income households in rural areas.

The sap of *Gluta* (*Melanorrhoea*) *usitata* is used for lacquer, which is applied on the famous Northern Thai lacquerware, which is manufactured in Chiang Mai. Although the tree is very common in the forests of Northern Thailand, it is not commercially tapped nowadays. The lacquer is imported mainly from Burma instead. The resin of *Styrax benzoides* is sometimes exploited as a substitute for "Siam-benzoin", which is a source of fragrant material.

Tapping of *Pinus merkusii* for resin is commonly practised in the pine–deciduous dipterocarp forests of Northern Thailand, especially in Mae Hongson and Chiang Mai, where a few distillation factories are available. Turpentine is one product of the distillation of the pine resin. Much more important is the solid rosin, which is used for various chemical products like glue and varnish. The Royal Forest Department produces at an experimental scale around 300 tons of resin per year. The pine forests of Thailand would allow an annual production of 3000 to 5000 tons of resin (cf. Werner 1993).

The fruits of *Phyllanthus emblica* (Euphorbiaceae) are rich in vitamin C. The tree is widespread in tropical Asia and used as folk medicine for various symptoms. Many other medical plants exist in the montane forests of Northern Thailand. *Gomphostemma strobilinum* (Labiatae) in the undergrowth of lower montane rain forest is not only a potential ornamental plant, but also claimed by Thai herbalists for the treatment against cancer. *Dichroa febrifuga* (Hydrangeaceae) and *Rhus chinensis* (Anacardiaceae), two common components of the montane forests in Northern Thailand, are occasionally used in folk medicines by both local and tribal people (Fig. 6).

Many ornamental plants exist in the montane forests of Thailand, for example, wild orchids of the genera *Dendrobium*, *Coelogyne* and *Vanda*, *Lilium primulinum*, *Begonia*, *Impatiens*, *Indigofera*, *Rhododendron*, *Prunus cerasoides* and *Acer calcaratum*. The survival of these plants depends on the protection of forests (cf. Werner 1989). Thailand's nature reserves are thus excellent bases for *in situ* conservation of these montane components.

Seeds of forest-trees are often rich in oil and fat or protein. The nuts of some species of *Castanopsis* (e.g. *C. acuminatissima*, *C. diversifolia*, *C. indica*) are locally collected and sold in the market like chestnut (*Castanea sativa*) in Europe. Some other potential fruits of the montane forests in Thailand are: *Docynia indica*, *Rubus* spp. (Rosaceae), *Diospyros glandulosa* (Ebenaceae), and *Myrica rubra* (Myricaceae).

The valuable barks of *Cinnamomum tamala* and *C. subavenium* (Lauraceae), highly esteemed by Thai people as spice, are exclusively collected from the trees growing wild in lower montane rain forest. The indiscriminate exploitation of these cinnamon barks has severely hampered their natural regeneration. Besides, the barks of certain laurels (e.g. *Cryptocarya* spp., *Litsea*, spp., *Persea* spp.) are also collected ruthlessly for the production of quality joss-sticks traditionally used by Thai people in religious ceremonies.

The native tea plants (*Camellia sinensis* var. *assamica*) growing wild in the montane zone of Northern Thailand are selected and planted in the lower montane oak forest for fermented or pickled tea (*miang*). The villagers clear parts of the undergrowth and of the canopy around the wild and planted *miang* trees to promote their growth. In these permanent tea gardens, the trees are harvested with ladders, a contrast to the neatly cut low bushes of tea estates, as they exist in Sri Lanka and other tropical countries. Hence, the tea (*miang*) gardens in Northern Thailand, in general appearance, look like an open lower montane oak forest.

5. Improvement by enrichment-planting?

The tea gardens of Northern Thailand should be a model to promote other potential economic trees by enrichment-planting or multi-cropping, as the wild populations are often not frequent enough, to meet the sustainable production. *Cinnamomum tamala* (Lauraceae), *Rhus succedanea* (Anacardiaceae) and *Styrax benzoides* (Styraceae) are a few examples. Species of *Castanopsis* with edible fruits and of *Cinnamomum* and some other laurels for barks could be additionally planted in a similar way. Besides these species, the density of valuable timber-trees could be increased inside semi-natural forests, thus avoiding monoculture forest-plantations. The montane species of *Calophyllum polyanthum* (Clusiaceae), *Paramichelia baillonii* (Magnoliaceae), *Fraxinus floribundus* (Oleaceae) and some others are suitable for enrichment-planting.

6. Restoration of degraded montane forests — the need of indigenous species for plantations

To restore the degraded watershed areas in Northern Thailand, a further aspect is the planting of indigenous conifers of the montane rain forests of Thailand, like *Cephalotaxus griffithii* or *Calocedrus macrolepsis*, which are useful but extremely rare today. *Cephalotaxus griffithii* is represented by a few isolated trees in the lower montane rain forests in Doi Inthanon and Doi Suthep-Pui National Parks and in Doi Chiang Dao and Phu Luang Wildlife Reserves. The rare and threatend conifer *Calocedrus macrolepis* surprisingly survived in extensive, magnificent stands in Phu Luang Wildlife Sanctuary. The next stands of *Calocedrus macrolepis* outside Thailand are recorded in Southern China and Laos. It seems to be a promising species for plantations in the montane zone. It is interesting to note that seedlings and saplings of *Calocedrus macrolepis* are luxuriantly developed along the edges and trails of the lower montane rain forest of Phu Luang. Silvicultural research on various aspects of this species should be urgently promoted by Royal Forest Department, Division of Silviculture. *Keteleeria* from the nearby mountains of Laos could be introduced for a trial as well.

Besides conifers the various colonizing species of montane forests such as *Betula alnoides*, *Carpinus viminea* (Betulaceae); *Prunus cerasoides*,

Docynia indica (Rosaceae); *Engelhardtia spicata* (Juglandaceae), *Myrica rubra* (Myricaceae), *Litsea cubeba* (Lauraceae) etc. should be used for plantations to rehabiltiate degraded savannaland. For that purpose, the problems of mycorrhiza associated with the roots of seedlings and the short span of germination must be solved by appropriate silvicultural research.

7. Conclusion

The montane forest communities of Thailand are rich in species of various biogeographical regions. They should not only be conserved as a valuable gene-pool, but also for a proper watershed-management. In the mountain areas, where these forests are degraded or replaced by grassland, rehabilitation-measures are essential. Besides the successful monoculture plantations of *Pinus kesiya* in the watershed area of Northern Thailand, polyculture or enrichment-planting should be as well taken into account in order to restore the structure and biodiversity of the natural montane forests.

References

Arbhabhirama, A. et al. (Eds.). 1987. Thailand — Natural resources profile. 300 pp. Bangkok.

Chapman, E. C. & Sabhasri, S. (Eds.). 1983. Natural resource development and environmental stability in the high lands of northern Thailand. Mountain Res. & Development 3 (4): 309–428.

Robbins, R. G. & Smitinand, T. 1966. A botanical ascent of Doi Inthanond. Nat. Hist. Bull. Siam. Soc. 21: 205–227.

Santisuk, T. 1988. An account of the vegetation of Northern Thailand. Geoecological Research 5. 101 pp. Stuttgart.

Schweinfurth, U. 1988. Pinus in Southeast Asia. Beitr. Biol. Pflanzen 63: 253–269.

Stott, P. A. 1984. The savanna forests of mainland South East Asia: An ecological survey. Progr. Phys. Geogr. 8 (3): 315–335.

Stott, P. A. 1986. The spatial pattern of dry season fires in the savanna forests of Thailand. J. Biogeogr. 13: 345–358.

Stott, P. A., Goldammer, J. G. & Werner, W. L. 1990. The role of fire in the tropical deciduous forests of Asia. Ecological Studies 84: 32–44. Heidelberg.

Werner, W. L. 1989. Aspects of conservation of endangered plants in Sri Lanka. Proceed. Int. Symp. Conservation Phytoecology, Beijing. pp. 35–40.

Werner, W. L. 1990. Verbreitung und Ökologie der Kiefern Pinus kesiya und Pinus merkusii in Thailand. 47. Dt. Geographentag, 1989 (Semmel, Ed.). pp. 152–157. Stuttgart.

Werner, W. L. 1993. Pinus in Thailand. Geoecological Research 7 286 pp. Stuttgart.

19. Conservation of some disturbed Indian tropical forest ecosystems

R. S. AMBASHT
Botany Dept., Banaras Hindu University, Varanasi, India

Keywords: Nutrient dynamics, primary production

Abstract. Tropical forests, particularly in India, are fast disappearing due to indiscriminate overexploitation in the absence of adequate restoration steps. Natural recovery of tree vegetation, even in rainy to very rainy places, is poor due to high frequency of repeated biotic and anthropogenic onslaughts, soil impoverishment and accelerated loss of soil, water and nutrients. We have made case studies of certain wet outer Himalayan forest in Darjeeling and dry deciduous forests around Varanasi for the import of energy, water and the nutrients through precipitation, the rates of primary production, litter fall, decomposition, nutrient release and export in form of harvested timber and fodder material, nutrient and water runoff and soil erosion. We feel that there is an urgent and immediate need for the conservation of habitats for adequate restoration and recovery. The tropical forests, despite of the abundance of rainfall and bright sun, high standing crop biomass and rapid turnover of material, are quite fragile due to poor edaphic conditions and are easily undergoing savannization and desertification.

Introduction

The Indian subcontinent is in large parts tropical with wide variation in climatic regimes from very warm to cold and from very moist to dry conditions. Nearly all possible combinations of temperature and moisture are present. Wide ranging soil types, from rich alluvium with thick profile to thin residual poor lateritic soil, are found. Highly degraded saline *usars* (saline-alkaline soils) and eroded ravinous wastelands are extensively distributed. Therefore, all the different types of tropical forest vegetation are found in India, but degradation and depletion of forest cover are rampant everywhere. The rate of desertification and encroachment of Indian forests is phenomenal and extinction of precious gene pools is also rapid. Urgent, drastic, bold and effective steps towards them and conservation are required. The main forest types are:

1. *Coastal Mangroves* particularly in Bengal, Orissa, Andhra Pradesh and Tamil Nadu are dominated by *Rhizophora mucronata, R. conjugata, Kandelia rheedii, Bruguiera gymnorhiza, Sonneratia acida* and *Nipa fruticans.*

2. *The Moist Tropical Forests* are broadly divisible into three main forest types: (a) the *Tropical Wet Evergreen Forests* receiving more than 2500 mm annual rainfall, located in the islands of Andamans. South-Western coast of Kerala, parts of Tamil Nadu and Karnataka in the South and Bengal, Assam, Meghalaya in the North. These climax formations have the richest diversity of flora and fauna. The tree canopy, is adjusted in 4–5 layers, the top layer reaching up to 50 m high with occasional trees of certain species piercing to still higher heights (flag trees). Shrubs, epiphytes, lianas and ground flora make the forest floor almost impenetrable. Giant trees with large flanking buttressed base are common. Important and dominant trees are *Dipterocarpus grandiflorus, D. pilosa, D. indicus, Hopea odorata, H. parviflora* and *Mesua ferrea.* Their boles and branches are covered with the epiphytic growth of orchids and the canopy is often covered with climbers, e.g. *Ventilago, Pothos* and *Caesalpinia,* (b) the *Tropical Moist Semi-Evergreen* forests are better developed in the northern part of the country in Assam, Bengal, Meghalaya and where annual rainfall is about 2000 mm. While quite a few of the moist evergreen trees are found also here, *Artocarpus chaplasha, Michelia champaca, Eugenia jambolana, Terminalia* sp., *Tetrameles* sp. and *Shorea robusta* are predominant. *Odina wodier, Dillenia pentagyna, Sterospermum* sp. are also common.

H. Lieth and M. Lohmann (eds.), Restoration of Tropical Forest Ecosystems, 203–208.

(c) *Tropical Moist Deciduous Forests* are extensively found in the high rainfall areas in the inner parts of the subcontinent in Karnataka, Madhya Pradesh, Uttar Pradesh, Bihar, Bengal and Orissa. These forests have rather closed, multistoreyed canopy reaching 30–40 m high. The most important and first quality timber trees *Tectona grandis* (teak) and *Shorea robusta* (sal) are produced in natural and plantation stands. *Terminalia paniculata, T. cernulata, Salmalia malabaricum, Pterocarpus marsupium, Lannea grandis* and *Adina cordifolia* are other naturally occurring constituent trees. The high revenue-yielding sweet-scented, sandalwood tree *Santalum album* grows in Karnatak. These forests become impenetrable in the rainy season and assume an open appearance in the summer when the tree and shrub leaves fall and annuals die and disappear.

3. *The Tropical Dry Deciduous Forests* occurring on annual rainfall below 1200 mm are extensively found in Punjab, Uttar Pradesh, Bihar, Madhya Pradesh. They provide many of the economically important forest produce such as timber (sal, teak and *Dalbergia sissoo*), bamboos (*Dendrocalamus strictus*) catachu (*Acacia catachu*) used in tannary and eaten with betel leaves (chewing), medicinal fruits of Aonla (*Emblica officinalis*) and Bel (*Aegle marmelos*), matchstick and packing case wood of *Boswellia serrata* and biri leaves (*Diospyros melanoxylon*) used in wrapping tobacco in the manufacture of biri, a cigarette like-smoke most commonly used throughout India. These forests have extensive grass cover and assume savanna appearance particularly in dry seasons. Because of high density of villages around such forests leading to extensive grazing, lopping and collection of fuel wood, these forests are degenerating into open savannas, shrub-savannas or deserts. In drier belt the scrub jungles dominated by spiny thickets and spiny shrubs of *Zizyphus jujuba, Z. xylopyra, Acacia arabica, Prosopis juliflora, Salvadora oleoides, Tecomella* sp. *Capparis aphylla, Calotropis procera* are quite common. Grasses are common and grazing is most extensive in the states of Rajasthan, Haryana and Gujarat.

The tropical and subtropical vegetation of the outer Eastern and Western Himalayas are constituted by broad-leaved evergreen as well as deciduous trees while with the rise in altitude, from about 1500 m, the conifers predominate. There is a distinct difference in the species composition and tree density in the Eastern and Western zones, as the East is more moist and warm. Temperate and alpine forests are extensive throughout on the high Himalayas, the timberline being around 4000 m and snowline around 5500 m altitudes.

Overexploitation and felling of forest trees on massive scale began about 130 years by railway companies as the first railway in India started in 1853. Since then the onslaughts on Indian forests have never declined, but some good afforestation and plantation activities have always existed. In recent decades social forestry involving tree plantation in rural and urban areas has come into existence as the exclusive forest lands have considerably reduced. As low as only 11% of the land in India is under effective forest cover. Much of the extensive forests have been converted into villages, croplands, industrial areas, flooded dams, mining and mine-spoiled lands, highly eroded ravines, rocky outcrops on hills and seasonally grazed pastures. The degradations at many places have resulted into impoverishment and erosion of soil to such an extent that tree regeneration and vegetation recovery are not possible even to provide protection against fresh biotic pressures. Such desertifications are not only confined to densely populated alluvium and hills of central India but also to the extremely wet, hitherto densely covered rain forests of Meghalaya which is among the highest rainfall belts of the world. The main factor of desertification there is the ever shortening of cycle of slash-and-burn practice for short-term use of forest land to raise cereal crops. The land abandoned after harvesting a few years of crops fail to regenerate fully before they are again slashed and burnt. Erosion has rendered such forests into mere rocky wastelands with the topsoil already washed away or in the process of being washed.

In this paper the attention is confined to our studies (1) around Varanasi in the Chakia forests and (2) plantation forests of Darjeeling Hills in the outer Himalayas. We have also taken up the results of our studies made on a selected species for plantation programmes and of some case studies made by others on rain forests and anticipated forest damages due to construction of proposed dams.

Case studies

1. Chakia forests on Vindhyan Hills

We carried out MAB research project on the structure and functioning of natural and modified silvicultural ecosystems in the Chandraprabha sanctuary of Chakia near Varanasi during 1975–1978 (Singh and Misra 1978). Studies were made on near natural forest stand of mixed dry deciduous trees such as *Anogeissus latifolia, Diospyros melanoxylon, Buchanania lanzan, Pterocarpus marsupium, Phyllanthus emblica* and shrubs of *Holarrhena antidysentrica* in paired plots, one protected and the other grazed. Singh and Ambasht (1980) have analysed the biomass, chlorophyll content and litter production of two adjacent teak (*Tectona grandis*) plantation forest (13 years old), one on a grazed and another on a protected plot. Singh, Ambasht and Misra (1979a, b, 1980) and Singh, Misra and Ambasht (1980a, b) also studied a nearby shrub-savanna stands dominated by shrubs of *Zizyphus jujuba* under protection and under grazing. They measured the rate of primary production, litter fall, rate of litter decomposition, energy dynamics and soil conservation efficiency of the vegetation. It is interesting to find that the leaf area index in all the above three vegetation categories, i.e. natural forests, plantation forest and the savanna ranged between 3.1 to 5.0 only but it was always higher in the protected than the grazed plots. The highest biomass of 94 (grazed) to 103 (protected) t ha^{-1} was in natural mixed forest and 10 t ha^{-1} (grazed) to 21.8 t ha^{-1} (protected) in 13-year-old teak plantation. In the shrub-savanna the biomass was only 6.7 (grazed) to 8.9 t ha^{-1} (protected). However the difference in the annual productivity was less sharp, i.e. 14 and 12.5 t in mixed stands, 4.5 to 10.6 t ha^{-1} in the teak stands and 8.3 to 13.1 t ha^{-1} in the savanna stands. Thus there is a progressive increase in the production/biomass ratio from the natural to plantation stands and the shrub savanna stands (Table 1). Singh, Misra and Ambasht (1980a) found that the dominant grasses in the savanna under protection from the grazing effectively conserve the soil from erosion. Ambasht (1985) and Misra (1980) have found that during May 1977 to April 1978, the soil loss from protected savanna site was only 1.3. t to 1.9 t ha^{-1} per year but from bare plots with grasses scraped, the soil loss rose by 30 times ranging between 61.6 to 66.9 t ha^{-1} yr^{-1}. The grass cover had 97.8% conservation value. Besides such heavy soil loss, the water runoff was also considerably higher from the bare plots. They also studied the erosion of soil nutrients against the simulated heavy grazing leading to desertification. The total amount of nitrogen loss through soil and water was 4.27 to 6.18 kg ha^{-1} yr^{-1} from differently dominated vegetation under protection as against 110 to 120 kg N ha yr^{-1} from the scraped plots. The corresponding phosphorus losses were 0.28 to 0.4 kg ha^{-1} yr^{-1} from vegetated and 7.7 to 9.3 kg ha^{-1} yr^{-1} from the bare plots. The process of savannization of the existing forests of Vindhyas of Northern and Central India has already taken place on a massive scale and with continued and sustained onslaughts of biotic pressures even the savannas are being converted into rocky wastelands due to massive erosion of life-sustaining topsoil.

Table 1. Biomass (t ha^{-1}) and net primary production (t ha^{-1} yr^{-1}) of a natural forest, teak plantation forest and a shrub savanna of Chandraprabha sanctuary in Chakia forests in protected (Pr) and grazed (Gr) conditions*

	Natural forest		Teak forest		Shrub savanna	
	Pr	Gr	Pr	Gr	Pr	Gr
1. Biomass (t ha^{-1})	103	94	21.8	10	9	6.7
2. Net primary production (t ha^{-1} yr^{-1})	14	12.5	10.6	4.5	13	8.3
3. Production/biomass ratio	0.13	0.13	0.48	0.45	1.4	1.2

* Data from Singh and Ambasht (1980) and Singh and Misra (Eds.) (1978).

In order to conserve the heavily disturbed forest ecosystems, it is necessary to create protected zones with effective buffers on the one hand and plantation of fast growing trees of multiple use (fuel, fodder and timber) on the other. It is impossible to prevent villagers from

Table 2. Biomass (t ha^{-1}) and net primary production (t ha^{-1} yr^{-1}) of young *Leucaena* plantation stand in the botanical garden, Banaras Hindu University

	Leucaena leucocephala plantation at the end of		
	1 year	2 year	3 year
1. Biomass (t ha^{-1})	26	80	157
2. Net primary production (t ha^{-1} yr^{-1})	26	54	77
3. Production/Biomass ratio	1.0	0.67	0.49
4. Nitrogen fixation (kg ha^{-1})	—	104	—

taking head-loads of fuel wood. Teak and bamboo plantation in this region have repeatedly failed under the existing poor level of protection and poor status soil. Performance of a fast growing nitrogen fixing tropical tree, *Leucaena leucocephala* has been tested. It is found to grow rapidly, produce profuse foliage rich in nitrogen, abundant fuel wood and it coppices very well. It can be repeatedly harvested at a 3- to 5-year cycle. Sandhu et al. (1990) have reported that leaf litter contributed most to the soil nitrogen pool. This species thus quickly covers the ground, binds the soil and enriches the soil nutrient through nitrogen fixation by roots and by gradual nitrogen release through decomposition of litter. *L. leucocephala* plantation at Varanasi attained biomass values of 26, 80 and 157 t ha^{-1} at 1, 2, 3 years of age and the annual net productivity was very high being 26, 64 and 93 t ha^{-1} yr^{-1} at three ages respectively (Table 2). Srivastava is studying the efficiency of actinorhizal nitrogen fixer *Casuarina equisetifolia* on reclamation and restoration of degraded forests in the industrial and mining areas near Varanasi.

2. *Alnus forests on Darjeeling Hills*

Sharma and Ambasht (1984, 1986, 1987, 1988 and 1991) have studied the biomass, productivity, nitrogen fixation by root nodules, nodule and litter decomposition and nitrogen accretion in the five different age groups of plantation forest of *Alnus nepalensis* in the moist (about 2250 mm rainfall) subtropical climate. *A. nepalensis* is an actinorhizal (*Frankia* symbiont) fast growing natural colonizer of degraded and eroded hills of central and eastern outer Himalayas (1500–2000 m alt.). The biomass ranged from 106 t ha^{-1} (7-yr-old stands) to 606 t ha^{-1} (56-yr-old stands) showing a high degree of biomass accumulations potential. But the rate of primary production was highest in the 7-yr stand of 25 t ha^{-1} yr^{-1} and it progressively decreased with the increase of age of the stand. Only 13 t ha^{-1} yr^{-1} production was recorded in thc 56-yr-old stand (Table 3). The production efficiency, energy conserving efficiency and energy efficiency in nitrogen fixation have an inverse relationship with the plantation age. The bole accounted for a little more than 50% of the total biomass.

Table 3. Biomass (t ha^{-1}) and net primary production (t ha^{-1} yr^{-1}) of *Alnus nepalensis* plantation forests of different ages (-years old). Data from Sharma and Ambasht (1991)

	Alnus nepalensis plantation (-years old)				
	7	17	30	46	56
1. Biomass (t ha^{-1})	106	237	349	472	606
2. Net primary production (t ha^{-1} yr^{-1})	25.0	19.7	18.7	16.4	12.9

Darjeeling Hills, because of heavy rainfall distributed round the year and rolling topography, is ideally suited for raising commercial crops such as tea, *Cinchona* and cardamum. *A. nepalensis* is found to be an excellent nurse trcc to provide protection and necessary shade and nutrient-rich leaf litter to cardamum plants. Natives use twig and leaf litter as manure for their crop fields and harvest fuel wood in large quantities. The tree withstands prevailing biotic pressure and survives successfully. Even on landslide affected slopes, *A. napalensis* seedlings are the first colonizers. They effectively bind the soil and reduce the physical impact of torrential rains which are frequent during 8–9 months per year. The biomass accumulation ratio (biomass/net production) ranged between 18.1 and 45.6 in the intermediate to mature age plantations showing that the system supports reasonably high total biomass for a given amount of primary production. Therefore, *A. nepalensis* plantation is found to be very useful in managing the degraded forest land in the moist central and eastern Himalayan tropics and subtropics. In Assam and Meghalaya extensive desertification of the rain forests have taken place mostly because of the overexploitation as well as quicker cycle of slash-and-burn method for raising agricultural crops for short periods. It is quite possible that the process of desertification can be arrested through the wide distribution of *A. nepalensis* seeds one or two years before abandoning agricultural operations following the slash and burn. Combining agriculture with quick growing nitrogen fixing trees are being experimentally studied to assess the total impact of agrosilvic practices on rural lands.

Possible threats from construction of proposed dams

Singh et al. (1984) have studied the famous rain forest of the Silent Valley in the state of Kerala in South India for the purpose of assessing the possible impact of construction of a dam over River Kunthipuzha for generating hydroelectric power and providing water for irrigation. This is a precious virgin rain forest in the high rainfall belt of Nilgiri Hills and is free from human settlement effects. The riparian forests have 5–6 layered canopy distribution. They have studied the likely direct and indirect effect of dam construction and feared that there would be a high rate of erosion and a rapid silting of the proposed dam. A sizeable amount of primary produce is likely to be consumed or destroyed by the population of three thousand workers that would be involved in the construction of the dam. They have strongly recommended the preservation of this natural rain forest in the form of a biosphere reserve to safeguared the ecosystem. Dam construction programme has been given up.

A big controversy has arisen over the fate of Tehri Dam, now under construction on River Bhagirathi in Garhwal Himalaya. On completion it is likely to engulf very large tract of forest as it would be one of the largest dams in Asia. The real controversy now is not over the loss of the huge forest area but the possibility of a large earthquake in the magnitude of eight on the Richter scale which seismologists believe would not be withstood by the dam. In such an event the disaster will be colossal as it may submerge several villages, towns and cities involving millions of human beings.

Conclusion

There is therefore an urgent need for defining national policies on the percentage of land area to be effectively maintained under protected forests, the harvest and afforestation programmes, creation of biosphere reserves, introduction of fast growing trees particularly nitrgoen fixers on fragile and village ecosystems to continuously provide timber, fuel wood, fodder, edible fruits and habitat for diverse fauna. These would reduce pressure on protected or managed forests to a considerable extent. Forest protection in turn will ensure sustained supply of forest products, conservation of soil, balance of hydrological cycle and fertility of soil.

Acknowledgement

I am grateful to Professor Dr. H. Lieth and the organizers of the symposium for invitation and financial support to attend the ISTE/INTECOL symposium on 'Restoration of Tropical Forest Ecosystems' in Bonn, Germany.

References

Ambasht, R. S. 1985. Primary productivity and soil and nutrient stability of an Indian hilly savanna lands. In: Tothill, J. C. & Mott, J. J. (Eds.), Ecology and Management of World's Savannas. Austr. Acad. Sci., pp. 217–219.

Misra, K. N. 1980. Primary production and soil conservation studies of an open forest land. Ph.D. thesis, Banaras Hindu University.

Sandhu, J., Sinha, M. & Ambasht, R. S. 1990. Nitrogen release from decomposing liter of *Leucaena leucocephala* in dry tropics. Soil Biology & Biochemistry 22 (6): 859–863.

Sharma, E. & Ambasht, R. S. 1984. Seasonal variation in nitrogen fixation by different ages of root noduls of *Alnus nepalensis* plantations in the Eastern Himalayas. J. Applied Ecology 21: 265–270.

Sharma, E. & Ambasht, R. S. 1986. Root nodule age-class transition, production and decomposition in an age sequence of *Alnus nepalensis* plantation stands in the Eastern Himalayas. J. Applied Ecology 23: 689–701.

Sharma, E. & Ambasht, R. S. 1987. Litterfall, decomposition and nutrient release in an age sequence of *Alnus nepalensis plantation* stands in the Eastern Himalayas. J. Ecology 75: 997–1010.

Sharma, E. & Ambasht, R. S. 1988. Nitrogen accretion and its energetics in the Himalayan alder. Functional Ecology 2: 229–235.

Sharma, E. & Ambasht, R. S. 1991. Biomass productivity and energetics in the Himalayan alder plantations. Annals of Botany 67: 285–293.

Singh, A. K. & Ambasht, R. S. 1980. Production and decomposition rate of litter in teak (*Tectona grandis*) plantation of Varanasi (India). Review d'Ècologie et de Biologie du Sol. 17 (1): 13–22.

Singh, A. K., Ambasht, R. S. & Misra, K. N. 1979a. Weight loss and energy release of decomposing grass litter in a savanna ecosystem. Japanese Journal of Ecology 29: 369–374.

Singh, A. K., Ambasht, R. S. & Misra, K. N. 1979b. Photosynthetic structure in relation to organic matter production of grassland community, Varanasi. Proceedings of National Academy of Sciences, India 49 (III): 149–151.

Singh, A. K., Ambasht, R. S. & Misra, K. N. 1980. Litter production and turnover of organic matter in *Zizyphus jujuba* shrub community. Japanese Journal of Ecology 30: 204–209.

Singh, A. K., Misra, K. N. & Ambasht, R. S. 1980a. Soil conservation efficiency of three dominant grasses of Vindhyan Hills, India. Journal of Japanese Forestry Society 62 (7): 273–275.

Singh, A. K., Misra, K. N. & Ambasht, R. S. 1980b. Energy dynamics in a savanna ecosystem of India. Japanese Journal of Ecology 30 (4): 295–305.

Singh, J. S., Singh, S. P., Saxena, A. K. & Rawat, Y. S. 1984. India's Silent Valley and its threatened rain-forest ecosystems. Environmental Conservation 11: 223–233.

Singh, K. P. and Misra, R. (Ed.) 1978. Structure and functioning of natural, modified and silvicultural ecosystems of Eastern Uttar Pradesh. Indian MAB Project Technical Report, 161 pp.

PART THREE

Political aspects of timber harvest in the tropics and evaluation of the discussions at the symposium

Introduction

The sponsors of the symposium have identified as one of the major tasks for this meeting the initiation of a discussion between the major user of the tropical timber in Southeast Asia, the local population of the areas to be logged, the governments responsible for the legal logging licenses and the environmentalist groups in industrialized countries.

During the time when we organized the meeting it seemed to be difficult to reach an unemotional discussion. Though, the co-sponsor, Mitsubishi Corporation specifically asked the organizers to invite non-governmental organizations (NGOs), they did fear that their demonstrations would disrupt the scientific part of the symposium. This was not the case, however. We invited several NGOs. Several came and two of them, "Friends of the Earth" and "Robin Wood", presented their issue at the meeting. Several others presented flyers at the symposium, which altogether brought the issues to light and reached the stage for a fruitful discussion. Several governmental representatives in Bonn area sent representatives as observers. After all, this was the first time that a major corporation involved in tropical timber logging and export dared to involve itself in a discussion with environmentalists. This part of the symposium relieved the participants from the stigma they were supporting the unecological behaviour of a corporation interested only in whitewashing their activities.

The discussion session was initiated by Mr. Mori as the representative of Mitsubishi Corporation, who explained the motives of the corporation for their timber trade from developing countries, which in their opinion would help the development of the country from which they would import the timber. In answer to the accusations that they were contributing to a global climate change they had started to rehabilitate tropical forests with experimental plantations in several places around the tropics and they hoped that the Bonn Symposium would provide them with further information and guidelines as how to proceed in the future. In the discussion they pointed out that the timber trade was only a marginal portion of their business and if the scientists at that meeting would convince them that they were acting unethically and/or against global ecological principles, they would drop that portion of their business altogether.

The response to the presentation by the representative of Mitsubishi Corporation came from the NGO groups, Robin Wood and Friends of the Earth, who had tried to stop Mitsubishi's logging operation in Sarawak on the ground that they were taking away the resources of the indigenous people whose interests were overruled by the government responsible for them. The case then became a political, socioeconomic and anthropological issue which was not considered the main focus of the symposium, but nevertheless needed to be discussed.

The two presentations were followed by a general discussion in which all facets of the problem with harvesting tropical forests on large scales were discussed.

The text of this report on the final discussion was collected from some written inputs but mainly from verbal presentations of participants of the symposium. These had been taped and were afterwards typed by Martina Lohmann.

H. Lieth and M. Lohmann (eds.), Restoration of Tropical Forest Ecosystems, 211–212.

The written version was edited so as to reflect the major concerns and suggestions of the participants. In this form we believe that the discussion provides a valuable statement for future work on the restoration of tropical forests.

The discussion starts with a verbal presentation of Prof. Müller given prior to the discussion which he could not attend and was closed with a final statement by Prof. Haber which is presented as a separate paper. All other contributions are embedded into the body of the general discussion which started with a keynote speech by Dr. Ashari Muktar and was subsequently chaired by Prof. Lieth.

The discussion of the topic among the participants was not finished at all during the symposium as the editors of the symposium volume found out during the editorial procedure of all papers. The senior editor had the opportunity to attend a follow-up symposium in Kuching/Sarawak where the subject of the Wakker and Hüttche paper, the governmental concern about the Penan in the Sarawak remote forests was explained by the officers concerned with the integration of the forest dwellers into a modern tropical society. This led to the request of the senior editor that the paper by Wakker and Hüttche be substantially revised. The paper had initially many unqualified accusations to the international corporation which was pointed out to us by the representatives of Mitsubishi Corporation. The paper was therefore sent to several colleagues of the other sponsoring scientific societies. As a result of these editing processes the two authors of the revised versions collated facts about the humanitarian problems related to the tropical timber exploitation. The papers, however, demonstrate also that the NGOs in their "puristic" concept forget that the human population worldwide needs to utilize the tropical forest resources, a point which was stressed by Dr. Haber in his closing statement.

It is not the task of the editors to bend the opinion of an author once he was invited to contribute to a discussion. All we demanded was the correct presentation of facts, the elimination of accusations and furthermore a statement of the reactions by the NGO they represented.

Since the editors of the volume did not present their views in a separate paper they may be allowed to voice their opinion at this point. The problems with the use of the tropical forests can be formulated as done in the following points:

1. The tropical forest must be incorporated into the management of the human society.
2. The present exploitation by international corporations and the unplanned utilization by local populations need changes.
3. The practice of clearcutting of forest areas without replanting should be abandoned.
4. The logging operations should be improved.
5. The conversion of forest areas to savannah type cultured lands should be drastically reduced.
6. The rehabilitation of already degraded areas should be accelerated with appropriate methods leading to forest type ecosystems.
7. A clear rational and international sound concept of tropical forest utilization as outlined by the ITTO should be implemented.
8. Such a concept needs to satisfy the demands of
   - global environmental protection
   - nature protection and species preservation
   - indigenous populations' rights and welfare
   - global economic development
9. The sustained forest utilization and management concepts should become the basis for legislation in tropical forest areas.
10. The international associations of biology and ecology should cooperate with the political and economic forces rather than oppose the future developments.

It is hoped that the discussion contributes to the further improvement in the implementation of the above-mentioned point.

20. Contribution to environmental action by a timber processing corporation

K. MORI
General Manager, Environmental Affairs Department, Mitsubishi Corporation, Japan

Keywords: Indigenous species, joint research project, Malaysia, Sarawak, sustainable forest management

Abstract. Recognising the scope and complexity of the world's environmental problems, changing Japanese attitudes to them and the high international expectations for appropriate action by Japan to contribute to the solution, Mitsubishi Corporation organized a Global Environmental Committee in October, 1989. In April, 1990 it established a new Environmental Affairs Department.

The main cause of tropical rainforest depletion is generally agreed to be the vicious cycle of local poverty and increasing population levels in developing countries. At the same time timber, unlike other limited natural resources such as oil and gas, is a renewable resource — if it is managed in a sustainable way. In other words, logging must be controlled at such a level that natural regeneration takes place at the same or lower rate. Therefore we are strictly abiding a policy of sustainable forest management and are studying other materials substituting tropical timber and ways of increasing value added to the local economy.

One major environmental issue is global warming, caused by excessive CO_2 emission, besides other "greenhouse gases" into the atmosphere, produced primarily in developed countries. To prevent the increase of such gases in the air, it is important to grow plants which store carbon for long time periods. Tropical hardwoods are especially to be enhanced since tropical trees are said to absorb more CO_2 than other trees.

Mitsubishi Corporation is considered as the first private company to adopt the method of using only indigenous species of tropical trees in reforestation projects when it organized an experimental research project between Universiti Pertanian Malaysia and Yokohama National University, supervised by Professor Miyawaki.

The project will last four years at an estimated cost of about 200–250 million yen (US $2 million). Though only started in April 1990, some valuable lessons have already been learnt.

Introduction

Environmental pollution has been a national issue in Japan for nearly 30 years. The original problems Japan faced were the result of our country's high rate of economic rate in the post World War II era. High growth also meant high pollution, and when the problem became severe in the late 1960s, anti-pollution movements began all over Japan.

The government reacted promptly. In 1970, 14 pollution-related laws were passed by the Japanese parliament. In 1971, a new Environmental Agency was formed to enforce these laws.

Over the last 20 years, there have been dramatic changes in Japanese attitudes about the environment. Both political and business leaders have taken part in this change. Keidanren, the largest business organization in Japan, recently announced a Global Environment Charter. This charter is a series of recommendations to corporations, urging them to adopt policies and strategies for dealing with environmental issues; its basic concept reflects the principles of the ICC "Business Charter for Sustainable Development".

Taking action on environmental affairs

Mitsubishi Corporation, in October 1989, in dealing with rising concern about global environmental issues, established an executive study committee headed by an executive vice president to determine what action the company should take.

Mitsubishi Corporation, a global trading and investment company with 60 domestic offices

H. Lieth and M. Lohmann (eds.), Restoration of Tropical Forest Ecosystems, 213–216.

and 180 offices overseas, handles more than 25 000 types of raw materials, products or services spanning to almost every industrial sector. As a truly global company, Mitsubishi's desire was to become a responsible "global corporate citizen."

The study committee recommended the establishment of an Environmental Affairs Department, which came into being in April, 1990.

The department has four basic functions:

1. To act as a corporate information centre regarding environmental issues.
2. To establish a company-wide code of environmental conduct including paying special attention to the environmental impact of our business investments.
3. To encourage an environmentally friendly way of thinking among all employees. We are organising lectures on environmental awareness for our employees, especially managers posted abroad.
4. To collect and disseminate accurate information to the general public on the company's business activities.

The department also established three basic action programmes to deal with environmental concerns:

1. Promoting the use of recycled paper for office use. I would note that the paper we use contains 83% of newsprint and we use a technology developed in Germany.
2. Encouraging the promotion of energy and natural resource conservation.
3. Organising an experimental project to restore a tropical rainforest ecosystem in Southeast Asia, where the company is involved in the logging business. This project was also the product of our internal discussions, which focused on two main issues:
   a. The threat of global warming caused by rising levels of CO_2, needed to be taken seriously. As the largest supplier of natural gas and oil to the Japanese economy, we needed to understand this issue. We expect from the involvement of a reforestation project in tropical regions a better undertanding of this problem in our own company.
   b. Based on years of experience in logging, we realized that a new emphasis on reforestation was needed to deal with both economic and ecological considerations.

The global dimensions of the tropical forest problem

Studies by the United Nations Food and Agriculture Organisation (FAO) estimate that annual depletion of rainforests is 17 million ha, 43% of which took place in Central and South America, 29% in Africa and 28% in Asia.

Total production of tropical timber is estimated at 1 800 million cubic metres, 80% of which is used locally for charcoal or fuel. 16% is processed into timber products for local use, and only 4% is exported, mostly to developed countries.

In short, the main cause of tropical rainforest depletion is the vicious cycle of local poverty and increasing population in developing countries.

Examining the record of commercial logging and reforestation

Unlike other limited natural resources, such as minerals, oil, gas and atomic fuels, timber can be a resource for future generations — if it is managed in a sustainable way. We support FAO's recent statement that "The forest is a renewable natural resource to be used by and for the benefit of people so long as it is done in a sound, sustainable manner."

In tropical timber-producing countries in Southeast Asia, each local government has forestry codes to enforce sustainable forest management rules in logging concession areas.

If operations are conducted within the rules, there will be no soil erosion, and with abundant sunlight and water, natural regeneration is assured.

While proceeding with our project, we have adopted a policy of Sustainable Forest Management, and are taking the following specific actions:

1. The major use of tropical hardwood in Japan is as panels used for construction projects. Mitsubishi Corporation has begun

joint studies with a plywood factory as well as construction companies in Japan to develop alternative materials for such panels.
2. In Malaysia, we are working with local officials and the private sector to find ways to increase the "value-added" level of their timber products, so that they can increase revenue without increasing logging.
3. We are also lowering our volume of tropical hardwood imports in support of Japanese Government guidelines.

Launching an experimental reforestation project in Malaysia

Tropical hardwoods also play a key protective role in maintaining stable climatic conditions, minimizing damage to agricultural land, and conserving soil, genetic resources and water supplies.

Commercial reforestation can be a good example of economics in harmony with the environment if it uses native trees according to the demands of the respective habitat.

Ecologists criticize the use of exotic species. Under their pressure, critics say, the natural flora and fauna of the area will suffer and may eventually perish.

To address this issue, an advisory group to the Japanese Ministry of Agriculture, Forestry and Fisheries submitted a report in May, 1990, recommending that indigenous species be used in reforestation projects.

Mitsubishi Corporation became the first Japanese private company to adopt this policy when it organised an experimental project in May 1990.

There had been some attempts at reforestation using indigenous species early this century in Southeast Asia, but unfortunately there are not enough records of such experiments.

In January, 1990, Mitsubishi Corporation contacted Professor Akira Miyawaki, Director of the Institute of Environmental Technology and Science at the Yokohama National University. Professor Miyawaki had implemented more than 200 reforestation projects in Japan in which indigenous species were successfully planted.

Professor Miyawaki suggested that an experimental project be organized in an area close to a Mitsubishi Corporation or its subsidiary office, for the sake of efficient communications, and with the cooperation of a local university. After a team from Yokohama National University (YNU) and Mitsubishi Corporation visited Bintulu, Sarawak, Malaysia — site of a new campus of the Universiti Pertanian Malaysia (UPM) — an agreement was reached organizing a cooperative project between the two universities, to be supervised by Professor Miyawaki.

In May, 1990, it was agreed that about six plots totalling 100 ha of barren land in a logged-over area on the UPM campus would be used for the experiment. After an intensive survey, it was decided to use four plots, totalling about 50 ha.

By November, 1990, thanks to the cooperation of the UPM/YNU team plus local officials and logging companies, more than 200 000 seeds were collected and then sown for germination. A new nursery and planting site also were prepared. Both were completed by June, 1991. As some seedlings were nurtured and part of the planting plot was ready earlier, the first plantings took place in late March.

The project will last four years (April 1990–March 1994), at an estimated cost of about Yen 200–250 million (US $2 million).

Although the project has just started and its success is not yet assured, we have already learned some important lessons:

1. There are opportunities to form innovative collaborative arrangements among government, academic institutions and commercial enterprises on a trilateral basis.
2. Our company has encouraged and increased information exchanges. We joined a tropical forest seminar in Malaysia in July 1991 and we also are exchanging information at this conference.
3. Truly creative solutions to environmental problems may come as companies are willing to create long-term policies.

Our investment in the Sarawak project is not expected to yield direct commercial return. It is our hope that the know-how gained in the project will lead to other attempts to restore tropical forests, not only in Southeast Asia, but in tropical regions all around the world. As a final payoff, we hope that this project will benefit not only Sarawak, but the whole world.

Finally, my thoughts and belief are that the problem of the tropical rainforest is exceedingly complex and should be studied in a global context. Our experiment is understood in Mitsubishi Corporation in that context.

International cooperation is a key element in this effort. There is no time for accusations and criticism. It is a time for action. We all have a duty to work together to preserve or to recreate a healthy environment. Mitsubishi Corporation intends to demonstrate its readiness to cooperate in this direction.

21. Mitsubishi's unsustainable timber trade: Sarawak

E. WAKKER
Friends of the Earth, The Netherlands, Damrak 26, 1012 LJ Amsterdam, The Netherlands

Keywords: Ecological and social impacts, Japan, Malaysia, timber exploitation, tropical forests

Abbreviations: cu.m. – cubic meters; EAD – Environmental Affairs Department; FoE – Friends of the Earth; PFE – Permanent Forest Estate; SAM – Sahabat Alam Malaysia/Friends of the Earth Malaysia; SMR – Sarawak Mission Report; TPA – Totally Protected Area; RWE – Round Wood Equivalents

Abstract. Japanese trading houses (*sogo sosha*), such as Mitsubishi Corporation, play a critical role in the future of tropical forests. So far, this role has been largely negative. Because of its major role in Japan's tropical timber imports, Mitsubishi has been one of the target companies in a world-wide public tropical forest campaign "Ban Japan from the Rainforest". The company undertook some steps, such as a reforestation project in Bintulu, but it continues to deny any responsibility for ecological and social disruption in the tropics. The steps undertaken so far are in fact merely 'green image' promotion activities, serving no one but Mitsubishi. This paper discusses the company's involvement in Sarawak's logging industry, which is unsustainable in respect to the tropical forest ecosystem, timber yields and people's livelihoods. It is argued that if the company really aims to reduce the burden which their business place on the environment, it would immediately stop importing timbers from ecologically sensitive environments and native customary lands under legal dispute.

Introduction

In 1990, Mitsubishi Corporation's ('Mitsubishi' hereafter) Environmental Affairs Department initiated a 50-hectare forest rehabilitation project in Bintulu, Sarawak. According to Mitsubishi, it would be 'the first major scientific attempt to test techniques for restoring open barren areas to their original, natural state' (Mitsubishi Corporation 1990). Although this project is definitely not the first attempt world-wide, it is not impossible that the Bintulu experiment will show that a small-scale forest-like ecosystem can indeed be established because the necessary inputs (time, funding and care) can be made available. This may also be observed, for example, at the grounds of the University of the Philippines in Los Baños, where a small rainforest has been established early this century. There are, however, convincing arguments, both of an ecological and social nature, that the Bintulu project will not be able to achieve its objective.

Mitsubishi's Environmental Affairs Department claims that the project may be successfully evaluated 2 to 3 years after planting. Considering the project's objective, this evaluation would have to include assessments of future originality and naturalness. After such a brief period, however, one may inventorize elementary factors such as the seedling survival rates, but there is no way it can be scientifically assessed that a 2- to 3-year-old plantation (25 tree species per 50 ha) will eventually match the primary rainforest. As a result of evolutionary processes, every hectare of natural rainforest contains large numbers of species in a composition dynamically changing over time. Plantations will never compare to this original, natural tropical rainforest because these processes cannot simply be copied. It will be argued below that there is not even satisfactory evidence that disturbed tropical forests can be fully restored.

Tropical forests are practically never re-established because the necessary socio-economic inputs are not available. Even if the techniques for

H. Lieth and M. Lohmann (eds.), Restoration of Tropical Forest Ecosystems, 217–226.

forest restoration would be available, implementation would not be feasible because of the immense cost and time required and the relatively unattractive ratio of merchantable and unmerchantable timbers.[1] Take for example the Bintulu project in which US $40 000/ha are invested in the first 3 years. When Malaysia's Federal Minister of Primary Industries, Dr. Seri Lim Keng Yaik, announced that the government plans to reforest 3 million ha of barren land in Sarawak (Houtwereld 1992), he probably disregarded the feasibility of this undertaking. Even when allowing for a 50% reduction in costs per ha, tropical forest rehabilitation in this area would require an investment of 60 billion dollars!

Considering these arguments, it only makes sense to conserve and enhance the existing forest area, rather than to destroy it and try to reproduce it elsewhere. Still, this logic is not followed up as Mitsubishi, the Malaysian governments and part of the scientific community are concerned about the technical aspects of projects which are ecologically, economically and socially unfeasible. In the meantime, Malaysia's natural forests are being depleted by Mitsubishi's subcontractors at rates faster than any reforestation can take place.

Since 1989, environmental and human rights groups in Europe, the USA, Australia, Japan, Malaysia and elsewhere have been campaigning against Mitsubishi's involvement in forest destruction. They believe that the reforestation project is merely a window dressing designed to hide its involvement in the tropical timber trade. It is therefore stressed that Mitsubishi's main concern should not be forest restoration, but phasing out further forest destruction resulting from the firm's trading practises.[2]

Sarawak's timber drain and forest conversion rates

In 1989, Japan imported 20 million cubic meters (M cu.m. RWE) of tropical forest products, of which 60% (12 M cu.m.) were logs. Out of these, more than half (6.7 M cu.m.) came from Sarawak. Japan's log imports from Sarawak tripled between 1980 and 1990, with both Sabah and Sarawak supplying about 90% of total tropical log imports in recent years (Jatan 1991). In 1990, Sarawak's timber output reached the highest level ever and added up to 18.8 M cu.m. (Forest Department Kuching 1991). Thereof, the bulk was exported to Japan (45%), Taiwan (20%), Korea (16%) and members of the European Community (Chin 1991).

These trade flows cause large-scale conversion of primary forests, mainly Hill Dipterocarp Forest. According to the Sarawak Forest Department, by 1989, ca. 3 million ha, of primary forest had been logged. Average logging rates in recent years were ca. 240 000 ha/y^{-1} in primary forest and 30 000 in forest that is being cut for the second (or more) time (quoted in WWF 1989a). Estimated conversion rates in 1990 vary between 376 000 ha to 494 700 ha (Chin 1991). Given Japan's log and sawnwood imports (7.2 M cu.m. RWE; Maskayu 1991) and assuming that 80% of these were logged in primary forest, Japan contributed to the conversion of 150 000 ha of Sarawak's primary forest in 1990 alone.[3]

Such logging rates cannot be sustained. Castilleja (1990) assessed that, at a production rate of 13 M cu.m./y and 38 cu.m./ha, *all primary* Hill Dipterocarp forest in Sarak (*including* Totally Protected Areas[4]) will be depleted by 2004. Since it takes 50 years for a residual stand to become harvestable under current practises (ITTC 1990), and taking 1975 as the year in which large-scale logging took off, Sarawak's logging industry will be phased out between 2004 and 2025.

Mitsubishi's involvement in Sarawak's forest destruction

With a total import of 1 254 000 m^3 (RWE), the combination of Mitsubishi/Meiwa Trading was Japan's second largest importer of tropical timbers in 1990[5] (Jatan 1991). Outside Malaysia, it has interests in logging firms, processing industries and plantations in Indonesia, Brazil, the Philippines, Chile, Papua New Guinea and other countries. Mitsubishi's log imports from Sarawak since 1987 are indicated in Table 1.

In addition to logs, Mitsubishi imports considerable amounts of sawn wood and plywood from

Table 1. Mitsubishi's log imports from Sarawak and the estimated area of primary forest destroyed, 1987-1991

F.Y.	Imports by Mitsubishi/Meiwa Trading (in m³)	Area of primary forest converted in Sarawak (in ha)
1987	263 000	5 537
1988	261 000	5 495
1989	298 000	6 247
1990	248 395	5 227
1991*	99 252	2 090
Total	1 169 647	24 624

Unit: cubic meters; processed timber is converted into roundwood equivalents; Extraction rate: 38 $m^3 ha^{-1}$, allowing a conservative 20% of the imports originating from second growth forest. (*) January–June 1991.

[*Sources:* ITTC 1990 (extraction rate of 38 $m^3 ha^{-1}$); Japanese Lumber Importers Association, prepared by Jatan 1991 (import statistics)]

Southeast Asia. Its sawn wood imports from Sarawak for example, increased from 4447 m^3 in 1989 to 9109 m^3 in 1990 (Jatan 1991), equal to RWEs of 8100 and 17 000 m^3, respectively. Furthermore, Mitsubishi has financial interests in two plywood plants located in Malaysia peninsula.

Japanese general trading houses (*sogo sosha*), such as Mitsubishi, are directly involved in all stages of the extraction, importing, processing and distribution of tropical timbers products (Nectous and Kuroda 1989). This was acknowledged by Malaysia's Federal Minister Moggie in 1984, who said that Sarawak timber companies are largely dependent on these trading houses for their intricate line of credit (Hong 1987). For example, Mitsubishi provided M$200 000 (30%) of the capital investment for Daiya Malaysia in 1974 (Jatan 1990).

Daiya Malaysia produces about 40 000 m^3 (Colchester 1989a) to 70 000 $m^3 y^{-1}$ of which 60% is exported to Japan (Jatan 1990). The concession is located near the Upper Tatau river and covers 90 000 ha, of which 35 000 ha, is located in production forest (Fig. 1).

Since 1984, Daiya Malaysia has been repeatedly in conflict with the Iban communities. In 1989, Ibans from Long Sangan met Mitsubishi's Chinese-Malaysian local managers and asked them to stop logging their customary lands, but their pleas were ignored. A few months later when the community set up a road blockade, negotiations finally led to Daiya Malaysia's withdrawal from the communal forest (Colchester 1989a) after which the company planned to move further upriver, thereby further encroaching in Iban territory. In April 1990, Shin Yang Corporation, a local company subcontracted by Mitsubishi, started logging in the same area, which resulted in another conflict (Jatan 1990). In 1991, protests revived and 500 men, women and children from 11 different longhouse settlements took part in anti-logging blockades in the Upper Tatau area (C. Hüttche, Robin Wood, pers. comm.).

The bulk of Mitsubishi's timber imports from Sarawak originates from areas other than Upper Tatau. Mitsubishi has never provided satisfactory evidence that this timber is cut in sustainably managed areas nor that it comes from areas undisputed by local people. Therefore it must be assumed that the general situation in Sarawak, namely that 'there is no evidence whatsoever, of attempts being made to address the major impediments to sustainability' (Chin 1991), applies to all timber traded by Mitsubishi.

The destruction of ecosystems, timber resources and livelihoods

Sustainable forest management and a sustainable timber trade can be defined in various ways (compare for example Poore 1989; ITTC 1990). We believe that an analysis from three dimensions, (a) the ecosystem, (b) timber yield and (c) peoples' livelihood is most comprehensive.

a. The collapse of untouched ecosystems

The World Bank stated: 'In practical terms, no commercial logging of tropical moist forests has proven to be sustainable from the standpoint of the forest ecosystem, and any such logging must be recognized as mining, not sustaining, the basic forest resource' (Talbot 1990). This is particulary true for logging in the Hill Dipterocarp Forests of Sarawak, which are among the richest, most diverse, and possibly one of the world's

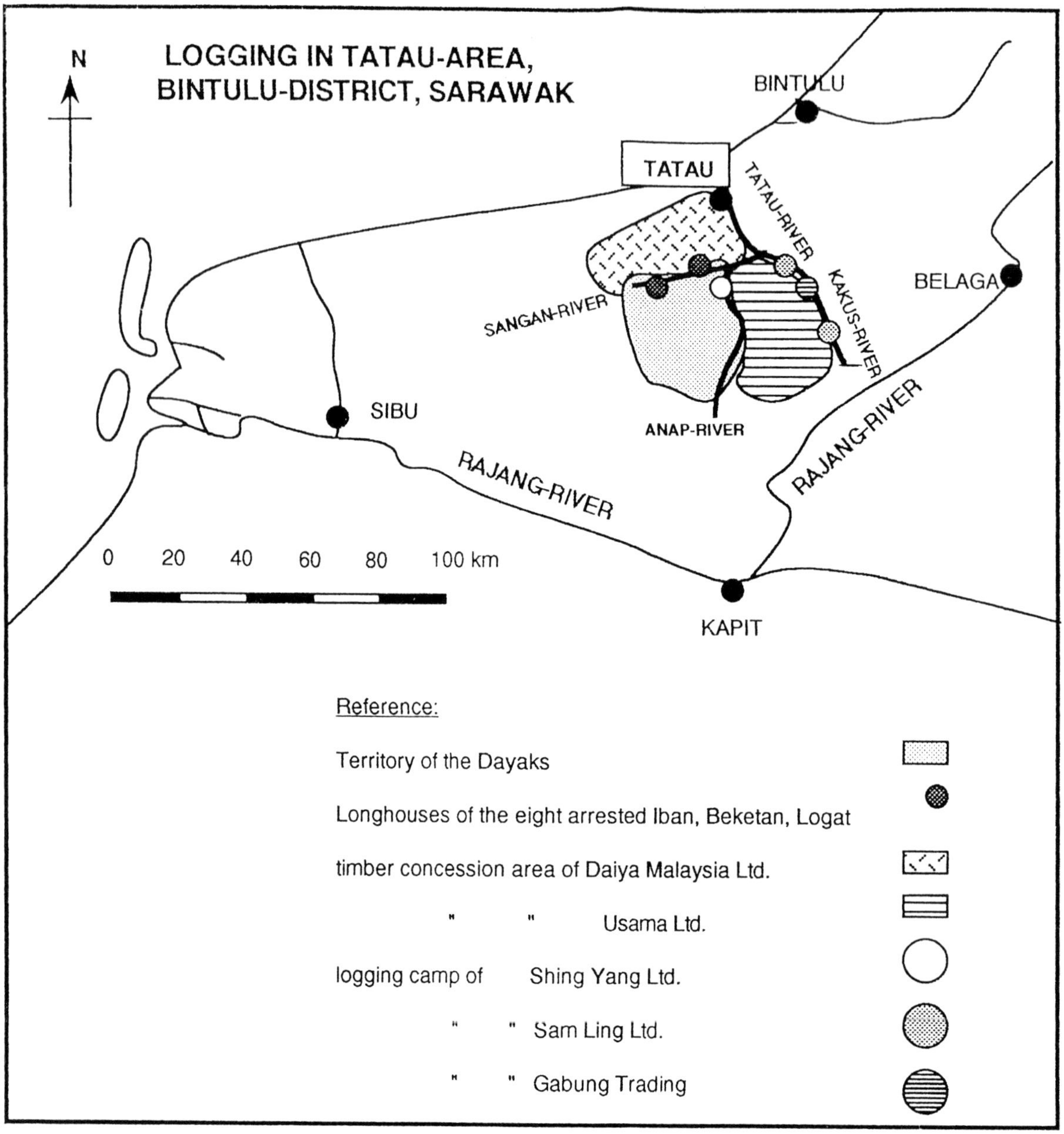

Fig. 1. Sketch of the Tatau area, Bintulu District, Sarawak.

oldest terrestial ecosystems. In Sarawak's lowland dipterocarp forest, 284 distinct tree species per 1.2 ha were found (inventory by Ronald 1990). It has a floristic diversity which is as high as that of the most prolific portions of the Amazon (Ronald 1990; also Collins et al. 1991). It is unfortunate that this ecosystem, diverse and unique as it is, has an economic value for only few dispersed timber species, because 'the more untouched, floristically heterogeneous and humid a forest is, the more difficult it is to manage it sustainably' (Budowski, quoted in WWF 1989b).

Particularly high-speed, extensive felling and road construction severely damage the rainforest ecosystem. Excessive erosion is one of the major impacts, because 'in South East Asia, roading, loading areas, and log extraction paths can leave as much as 40% of the soil exposed, and a canopy reduction of up to 60% is not uncommon' (Chin 1988). An extensive literature review by FoE-UK concluded that 'studies of fauna and flora of logged forests show that these are generally far from fully representative of the composition of the prelogging community. Disturbance may initiate a long series of reactions that spell ultimate decline and biological disintegration for many such communities (FoE-UK 1991). A study by World Conservation Monitoring Centre pointed out that 'all dipterocarp species in logging areas of Kalimantan ... can be considered directly threatened by logging because felling cycles are too short, size class regulations are not followed and areas are re-logged too quickly' (WCMC 1991).

The overall impact of logging in tropical forests is that the crucial elements of the ecosystem and their inter-relations are irreversably affected. Consequently 'the complex, primary humid tropical forest can be called a non-renewable resource, since human disturbances result in a forest that differs from the original in species and structure' (Gomez-Pompa, quoted in Lugo et al. 1985).

b. Unsustained timber yields

In Sarawak, there is no logging system practised which sustains timber yields. The Sarawak Liberation Thinning (SLT) has been 'largly abandoned as too complicated for practical application' (Poore 1989) and forestry regulations are often not lived up to. Poore's study showed that such a situation is not unique to Sarawak: 'The extent of tropical moist forests which is being deliberately managed at an operational scale for the sustainable production of timber is, on a world-scale, negligible' (0.1% of the total area of tropical forests). This is because 'if production of timber is to be genuinely sustainable, the single most important condition to be met is that nothing should be done that will *irreversibly reduce the potential of the forest to produce marketable timber*' (ibid.).

One may well state that the less intensive logging is, the less destructive it is, but 'the pressing question of to what extent a natural forest ecosystem may be modified for economic reasons without seriously impairing its ability to function and survive cannot be answered on the basis of present state of knowledge' (Lamprecht 1989). According to Budowski, 'it may be worthwhile to destroy the myth that natural forests can be successfully managed on a sustainable yield basis, because there is lack of evidence or good case studies' (quoted by WRM 1990). There is, however, an overwhelming evidence that selective logging does not sustain ecosystems nor timber yields.

Various authors have attempted to estimate a sustainable level of timber exploitation in Sarawak. The SMR (ITTC, 1990) concluded that yields should be reduced to 9.2 M cu.m/y in order to maintain sustained timber yield. Within the Hill Dipterocarp Forest, a 50% reduction from the current level of 13 M m^3 y^{-1} to 6.5 M m^3/yr is proposed. Chin (1990), however, argued that the situation in Sarawak dictates that a sustainable annunal yield for the next two to three decades, is closer to 3 M cu.m. Castilleja (1990) stated that the assumptions of the SMR are unsatisfactory and concludes that 'a reduction between 85 and 88% (depending on how much land is under legal dispute) in the current annual harvesting volume is necessary to sustain timber harvesting in the Hill Dipterocarp Forest in the short and long run.[6]

Despite these recommendations, Sarawak's timber yield was increased to 18.8 M cu.m. in 1990 (Forest Department Kuching 1991), partly by forest clearing of large patches of State Lands (Lim Keng Yaik, quoted by *New Sunday Times*, March 17, 1991). In the first four months of 1991, including the Muslin *Ramadan*, a month of low productivity, Sarawak had already exported 5.4 M m^3 (Maskayu 1991). It is therefore unlikely that, in 1991, any of the recommended reductions have been implemented as well. The increased output may be the result of logging in TPAs, State Forest, slopes steeper than 60% and early relogging.

c. Distracted lifes, cultures and livelihoods

The 220 000 people in Sarawak's interior who depend on the forest for their livelihoods are generally referred to as Dayaks. Among others, Colchester (1989b), Hong (1987) and Chin (1988, 1991) stated that the Dayak communities have sustained their livelihoods and the forest for centuries by applying sustainable shifting cultivation technologies, generally including hunting and gathering of non-timber forest products. As will be described below, logging affects this traditional lifestye, directly and indirectly, and non-agriculturalists (like the Penan) are especially hard hit.

A direct impact of logging is reduced availability of food-stuffs and other non-timber forest products hunted and gathered by the Dayaks. The degree in which the availability of non-timber forest products is reduced depends on the intensity of exploitation (De Beer & McDermott 1989). Since logging in Sarawak is very intensive indeed, communities confronted with logging operations nearby their settlements reported reduced availability of non-timber forest products, such as medicinals, wild boar, sago palm, fruits and rattans. This affects their subsistence base as well as commerce and thereby their health situation (WWF 1989a). According to Caldecott (1986), consumption of wildlife meat declined from 54 kg/person/year to only 12 kg, a decline which he largely attributed to the timber industry.

Disruption of watersheds is another direct impact of logging which affects Dayak communities. In 1986, the Federal Malaysian government noted that 59% of Sarawak's rivers are polluted (quoted in Colchester 1989a) due to excessive sediment loads and diesel spillings. Pollution of streams and rivers causes muddying of drinking and bathing waters, blockage of piped water supplies and reduces fishery stocks, thereby further improverishing the local diet. Various studies (Anderson 1979; Aspinall 1990; JIV 1990) have revealed the negative impacts of logging on health of the Penan and other cultural groups. A 1987 study by Dr. Paul Chen of the University of Malaya Medical Faculty states that malnutrition among Penan children of the Apoh, Tutoh and Ulu Baram areas was higher in recently logged areas (31.4%) than in unlogged areas (10.9%) (quoted in Aspinall 1990).

Logging affects all aspects of the ecosystem in Sarawak, and it causes deterioration of the very foundation of indigenous peoples' traditional life. 'When the loggers came to their forest, the natives suffered the shock and bitterness of seeing their forest farms, ancestral lands and sacred places destroyed and defiled before their very eyes' (Hong 1987). Consequently, the indigenous peoples have to compete with loggers for access to land and 'more often than not', says Hong, 'it was a losing battle'.

During the Penan World Tour in 1990, a Penan expressed the crisis as follows:

> Our situation now is like a child who has fallen into a fast flowing river and cannot swim. The child cries out, extending its arm for someone to help. If no one takes the hand, the child will surely drown (WCWC 1990).

The 'political ecology' of Sarawak's problems

Since 1987, Sarawak's indigenous peoples protested against the deterioration of their livelihoods, raising the pressing question of who has rightful access to the natural resources. The Forest Ordinance is a legal instrument which provides for the establishment of Communal Forest Reserves but by application of Section 6 of the Forest Ordinance, the Minister can at any time declare the existing reserves to be terminated. Since 1968, the area under Communal Forest Reserves thereby shrunk from 30 300 ha to 5 600 ha in 1984 (SAM 1990), representing only 0.05% of all forest as of 1984. The size of the average reserve per community is about 1 km^2 and even these forests are being logged since these logging concessions often overlap with Communal Reserves. According to Colchester (1989b) the legally valid claims, which extend to the primary forests (*menoa*), may encompass some 22% (2.73 million ha) of the State's territory. In contrast to these figures, 60% (5.75 million ha) of the PFE is leased out for logging. The area of concessions held by immediate past and present Chief Ministers of Sarawak, their relatives, and associates totals 2.9 million ha, or *over half of all concessions and a third of forested lands* (Colchester 1989b).[7]

Mitsubishi considers the settlement of land disputes a matter to be dealt with by the Sarawak government, but 'even governments in the process of establishing strict quotas and rules, will turn a blind eye to environmentally destructive practises for the sake of short-term currency gains and at the prospect of new land being released for landless farmers' (Nectoux and Kuroda, 1989). In addition to this, Tsuruoka (1991) found that illegal logging and tax fraud are widespread in Sabah and stated that 'similar practises are widely believed to be taking place in Sarawak, where there may allegedly be even less enforcement of forestry regulations'. According to Colchester (1989a), senior representatives of Mitsubishi acknowledged that the government does not provide them with guidelines on compensation. Therefore, a 'company code of conduct' is required which goes beyond developing government standards. However, the company should not ignore foreign policies, which was practised in Papua New Guinea, where a Commission of Inquiry, chaired by Judge Barnett, uncovered Mitsubishi's involvement in transfer pricing[8] and misdeclaration of species (Marshall 1990).

The distribution of access to and benefits from Sarawak's natural richness is highly skewed. Only about 30 000 indigenous people are directly employed in the logging industry.[9] Daiya Malaysia for example, employs only 170 people (Colchester 1989a). For the remaining 200 000 people, logging has brought little wealth and even a breakdown of average standards of living. Those involved in logging are working under extremely dangerous conditions, as for every 7 000 m^3 logged one serious injury occurs and for every 1 M m^3, one fatality occurs. The accident rate per cubic metre in Sarawak is thereby 21 times greater than in Canada (ibid.).

On the other hand, log exports from Sarawak were valued at M $2.9 billion in 1990 (Maskayu 1991), which is a great deal of money, especially since all the royalties collected go to Sarawak's treasury, contrary to oil product revenues of which 95% is syphoned off into the federal coffers (Colchester 1989b). However, the division of revenue between producer and consumer country for log exports rates 90.8% in favour of the consumer country and 9.2% in favour of the producing country (Oxford Forestry Institute 1991). When only logs exported to Japan are considered, these figures suggest that M $13.8 billion were generated in Japan, while only M $1.4 billion remained in Sarawak. This clearly shows Japan's interest in massive imports of unprocessed timber. Increasing processing within the producer country, one of Mitsubishi's planned actions, would increase the value of production. However, higher prices (for timber) do not necessarily lead to sustained yield management, 'if anyting, they led in the recent past to more rapid timber mining' (Gregerson et al. 1990).

Concluding remarks: The need for constructive action

Mitsubishi's EAD clearly does not have the intention nor the position within Mitsubishi's business conglomerate (*keiretsu*) to push for an improved forestry policy. Soon after the environmental desk was eststablished in 1990 it became clear that it had the purpose to promote the company's 'green image' rather than to clean up its act. Even if it wanted to, the EAD has not got the influence to prevent the companies within the keiretsu from carrying out environmentally destructive projects. Because the switch to sustainable forestry would financially hurt the other (timber, paper) divisions, a much cheaper public relations campaign was launched.[10]

The EAD went for a direct confrontation with the general public. It distributed comic books and leaflets in Japan, with the clear message that not Mitsubishi, but shifting cultivators are to be blamed for forest destruction.[11] Despite the overwhelming amount of evidence that Sarawak's timber industry is a total chaos, the company consistently claims, without giving any reliable evidence, that no harm is caused. Japan's number one 'environmentally aware company', was also Japan's second largest importer of tropical timber in 1990.[12] It has been shown above how these imports have contributed to the destruction of tropical forests and people's livelihoods, causing problems which cannot be resolved by the company's campaign, nor by the Bintulu reforestation project. Within the context of logging practises in Sarawak, the Bintulu project is an offense towards the people who are directly confronted with the negative impacts of the logging operations within their ancestral domain.

Mitsubishi's planned reduction of timber imports from Malaysia cannot in fact result from environmental awareness, as the company claims that Malaysia's logging practises are exemplary. It is however questionable whether this reduction will be put into practise. In a letter to WWF Sweden, dated June 28, 1991, Mitsubishi stated that 'we are gradually decreasing our volume of imports. In 1987, we imported 444 000 m^3 from Malaysia. This year it will be about 210 000 m^3 (Mitsubishi Corporation 1991). Recent data derived from the Japanese Lumber Importers Association (Jatan 1991) indicate that the firm had imported a volume of logs equal to 193 000 m^3 from Sabah and Sarawak by July 1991, and this excludes all processed timber. The reduction in log imports is likely to be substituted by processed timber, another of Mitsubishi's goals. This measure cannot be motivated on environmental grounds as there is no guarantee that forests will be managed sustainably, nor that fewer trees are cut.[13]

When reviewing the action undertaken by Mitsubishi, it is not crude to state that the company has done next to *nothing* to promote sustainable forest management and trade. Consequently, long before 2004, Mitsubishi will be out of the logging business in Sarawak as there will be no more forest to log and no more timber to trade. A company with international mobility will continue similar practises elsewhere in the world until the bulk of the natural forests are depleted. Mitsubishi has already stengthened its interests in the exploitation of the remaining timber resources in the world (notably Canada, Papua New Guinea, Indonesia, Siberia), repeating the tragedy. Such business practises are not any longer accepted by the general public.

Friends of the Earth has therefore repeatedly called on Mitsubishi to undertake constructive action. It must stop its involvement in logging operations in environmentally sensitive areas (especially primary forests) until truly applicable forest management systems are developed, which sustain both the ecosystem and timber yields. In addition to this, as recommended by the SMR (ITTC 1990), logging in disputed lands must be halted until native customary rights issues can be adequately settled (par. 30). This recommendation must also be followed up by Mitsubishi and its subcontractors, in Sarawak and elsewhere. Until such constructive action is taken, Friends of the Earth will continue its public campaign against Mitsubishi.

Notes

1. This may be one of the reasons why Mitsubishi does not expect the ¥250 million investment for the Bintulu project to produce profit. Without doubt, however, it must expect profits for its ¥750 million investment in an afforestation project in Chile. This 10 000 ha monoculture plantation is establish by using fast-growing eucalyptuses, exotic to the area. This project was adopted at the same time when Mitsubishi claimed it acknowledged the critique aimed at this kind of plantation (presumably the use of exotics and environmentally hazardous tree species). Responding to the critique, Mitsubishi initiated the Bintulu project, being fully aware that this project would not be economically feasible and continued with 'business as usual' in Chile.
2. It must be stressed that this study reviews only one company and one particular region where forest destruction is caused by Japan's timber traders.
3. The RWE conversion standard for sawn timber is 1.82 (FAO/EEC standard); logging rate per hectare: 38 m^3 (ITTC 1991).
4. Which is indeed not unrealistic. In 1990, a 5 000 ha logging concession was granted in Gunung Gading National Park (SAM, pers. comm., September 20, 1990). The park covers 5430 ha (WWF 1989a).
5. Because 1) the top 8 shareholders of Meiwa Trading stock are Mitsubishi keiretsu member companies, 2) Mitsubishi is the top shareholder (increased holding from 20.0% to 32.9% in 1991) and 3) that Mitsubishi keiretsu member companies' total stockholdings are 66.80% as of 1991, and that Meiwa Trading is listed as an affiliated company in Mitsubishi's Annual Report, it is not unreasonable to directly combine Meiwa Trading's imports statistics for tropical wood products with those of Mitsubishi Corporation (Jatan 1991).
6. Castilleja based these figures on the following assumptions: 1. the timber to be harvested conforms the demands of the current market: Class 1, dbh 60+ cm; 2. harvestable area: 2.545 M ha in the PFE; 3. a silvicultural treatment that insures the required status of the residual stand within liberation thinnings. The sustainable annual yield for Hill Dipterocarp Forest is 1.93 M m^3/yr. If the 500 000 ha estimated by SMR to be under dispute were within the area considered in this scheme, then the actual sustainable rate for the next 30 years would be 1.55 M m^3/yr.
7. These relations were publicly revealed during the 1987 State elections when competing politicians revealed lists

of the areas under timber license owned by the competitor and his relatives.

8. Transfer pricing is practised by declaring a sale price that is far below the true market value. The difference between the price reported and the actual price received can be deposited in a tax haven. Costs calculated as a percentage of export value, such as royalties and export duty, are thereby greatly reduced.
9. Out of the 57 000 people involved in the industry, 46 000 are directly employed by logging companies in 1988 (Ronald 1990, internal report; Forest Department Kuching 1991). Out of these, 30% of the labourers are Indonesian or Philippino (J. Lochhead, Singaporean and Malaysian Britisch Assoc., pers. comm.). According to the Department of Statistics, however, the forestry and forest industry Sector generates directly about 80 000 employments (Forest Department Kuching 1991).
10. Note that the total costs of the Bintulu forest regeneration project are less than 2% of the total value of the company's timber imports from Sarawak (roughly ¥4.3 billion in 1990).
11. The 216-page comic book was however withdrawn by the Japanese Ministry of Education at the time this paper was in print. The Ministry's main reason for the decision was that the comic was 'propaganda activity for one particular company'.
12. Ironically, there seems to be a relation between timber imports and environmental awareness. On September 21, 1991, Nikkei Weekly ranked the 10 'most environmentally aware companies'. Among them, Mitsubishi (1), Nissoh Iwai (2), Mitsui (3) and C. Itoh (6). These companies took account for 27% of Japan's tropical timber imports in 1989. With regard to this survey, it should be noted that the companies filled in the questionaires, not the public (Nikkei Weekly, September 21, 1991).
13. So, in terms of roundwood equivalents, the total number of trees logged will not decrease.

References

Anderson, A. J. U. 1979. Subsistence of the Penan in the Mulu Area of Sarawak. Sarawak Gazette, 105 (1461).

Aspinall, R. 1990. Medical consequences of logging on the indigenous peoples of Sarawak, Malaysia. Western Canada Wilderness Committee, Vancouver.

Beer, J. de & McDermott, M. 1989. The Economic Value of Non-Timber Forest Products in Southeast Asia. IUCN, Amsterdam.

Caldecott, J. 1986. Hunting and Wildlife Management in Sarawak. WWF (Malaysia), Kuala Lampur.

Castilleja. 1990. Reducing the annual timber harvest in Sarawak: how much? National Wildlife Federation, Washington.

Chin, S. C. 1988. Managing Malaysia's forests for the sustained production of resources. University of Malaysia.

Chin, S. C. 1990. The system must change: the promotion of sustainable forest management in Sarawak. Wallaceana 61: 1–8.

Chin, S. C. 1991. Curiouser and Curiouser: Forestry in Sarawak, a critique of the Sarawak Forestry Department Annual Report 1991. Sahabat Alam Malaysia, Penang.

Colchester, M. 1989a. Survival International at the International Tropical Timber Organisation. (Internal report of a meeting with Mitsubishi senior management delegation, November 8th, 1989).

Colchester, M. 1989b. Pirates, squatters and poachers, the political ecology of dispossion of the native peoples of Sarawak. Survival International/INSAN, London.

Collins, M., Sayer, J. & Whitmore, T. C. 1991. The Atlas of Tropical Forests: Asia and the Pacific. IUCN (World Conservation Union), Macmillan, London/Basingstoke.

FoE-UK (Friends of the Earth United Kingdom). 1991. Life After Logging? The Role of Tropical Timber Extraction in Species Extinction. London.

Forest Department Kuching. 1991. Forestry in Sarawak, Malaysia, Annual Report. Kuching, Malaysia.

Gregerson, H., Lundgren, A. & Lindell, G. 1990. Contributions of tropical forests sustainable development: the role of the industry and the trade. Paper presented to the ITTC, Bali, 19th May, 1990.

Hong, E. 1987. Natives of Sarawak: Survival in Borneo's Vanishing Forest. Institut Masyarakat, Penang.

Houtwereld. 1992. Marktverkenning Tropisch Hardhout, January 8, 1992, Schiedam.

ITTC (International Tropical Timber Council). 1990. Report submitted to the International Tropical Timber Council. The promotion of sustainable forest management: a casestudy in Sarawak, Malaysia. Yokohama.

Jatan (Japan Tropical Forest Action Network/Friends of the Earth Japan). 1990. Trading company profiles (internal report). Tokyo.

Jatan (Japan Tropical Forest Action Network/Friends of the Earth Japan. 1991. Japan's imports of tropical timber (internal report). Tokyo.

JIV (Japanese International Volunteers Centre). 1990. Hyashi, Dr. T., Medical treatment section, the destruction of the Borneo tropical rainforest and its effects on the natives of Sarawak. The Health of the Penan Tribe, June 4, 1990.

Lamprecht, H. 1989. Silviculture in the tropics. GTZ (Gezellschaft für Technische Zusammenarbeid), Germany.

Lugo A. E., Clark, J. R. & Child, R. D. 1985. Ecological Development in the Humid Tropics: Guidelines for Planners. Morrilton.

Marshall, G. 1990. The political economy of logging. The Barnett inquiry into corruption in the Papua New Guinea timber industry. The Ecologist, Vol. 20, No. 5, September/October 1990.

Maskayu. 1991. Export of timber products by Sarawak, derived from Greenpeace Italy.

Mitsubishi Corporation. October 1990. Mitsubishi and the rainforests. Tokyo.

Mitsubishi Corporation. June 28, 1991. Letter to Ulf Rasmussen. WWF-Sweden.

Nectoux, F. & Kuroda, Y. 1989. Timber from the South Seas: An Analysis of Japan's Tropical Timber Trade and its Environmental Impact. WWF, Gland.

New Sunday Times. March 17, 1991. Lim: Sarawak Logging All Right.

Nikkei Weekly. September 21, 1991. Earth-watch poll ranks companies.

Oxford Forestry Institute. 1991. Pre-project incentives study for ITTO. Oxford.

Poore, D. 1989. No timber without trees, sustainability in the tropical forest, A study for ITTO. London.

Ronald, P. 1990. Penan facts/voices from the Borneo rainforest. Mind Link! British Columbia (internal report).

SAM (Sahabat Alam Malaysia/Friends of the Earth Malaysia). 1990. Solving Sarawak's forest and native problem. Penang.

Talbot, L. 1990. A proposal for the World Bank's policy and strategy for tropical moist forests in Africa. World Bank, Washington.

Tsuruoka, D. 1991. Cutting down to size. Far Eastern Economic Review, July 4.

WCMC (World Conservation Monitoring Centre). 1991. Pre-project study on the conservation status of tropical timbers in trade. Summary Report. Cambridge.

WCWC (Western Canada Wilderness Committee). 1990. Voices from the Borneo rainforest. 1990 World Tour Report. Vancouver, British Columbia.

WRM (World Rainforest Movement). 1990. Rainforest destruction, causes, effects & false solutions. Penang.

WWF. 1989. Rainforest conservation in Sarawak: an international policy for WWF. Gland.

WWF. 1989b. Tropical forest conservation. Gland.

22. Future problems and working trends for sustainable management and restoration of tropical forests

*Podium discussion**

M. LOHMANN and H. LIETH
AG Systemforschung, University of Osnabrück, Artilleriestr. 34, D-4500 Osnabrück, Germany

Keywords: Ecology, environmental concern, political problems, socio-economic problems, tropical forestry

Introductory statements to the final discussion by Prof. Müller

The socio-political developments in the tropics during this century have led to the establishment of a large number of new states whose demands on natural resources for economic development has reached thresholds several times higher than the former economic needs of a largely subsistence-oriented indigenous population required. New guidelines for the exploitation, management and conservation of tropical ecosystems have evolved since 1950 on international, regional, national and local levels. Based on these concepts, the problems of tropical ecosystem use and conservation now require studies involving various different disciplines. Research methodologies and techniques need to be fine-tuned to answer specific questions relevant to a decision-making technocracy that is responsible for the management of tropical ecosystems. An example is the formulation of the ITTO guidelines included in this volume. Tropical ecosystem studies have thus entered a phase where they need to address local and regional issues and problems using innovative approaches that could contribute to the formulation of sustainable management strategies within a generally accepted international framework. The connection between scientific studies and decision-making processes will become increasingly indispensable in determining the success of strategies implemented for the management of tropical ecosystems.

National parks, protected areas or species conventions have become the classical means of nature conservation. This "philosophy" of conservation ignores global interlinkage among all parts as well as the fact, that there are no untouched "islands" where natural processes take place independently of the global change of land use. The new concept for the establishment of Biosphere Reserves in which nature conservation will become a part of the land use pattern of a given region may become a key to the solution of the problem. This concept tries to include the conservation strategy based on "equilibrium hypotheses" and confidence in regeneration of natural systems, into the reality of modern approaches to land use management and living standards of the population exploiting or caring for their territory.

Nature conservation in the tropics can no longer be successful if it relies only on natural regeneration capacity in remaining habitat islands. This "cheese cover concept", completely isolated and as the only valid nature conservation strategy, will lead to increased extinction. It is important to create both useful conservation and development strategies for our different types of tropical ecosystems. Wise use as "the right way of use" should not rely on species only, but should be related to the habitat, the natural ecosystems or ecosystems which have been degraded to various extents. There should be a closer link between wise use and conservation of systems.

* Chairman: Ashari Muktar
Discussion leaders: A. Miyawaki and H. Lieth

H. Lieth and M. Lohmann (eds.), Restoration of Tropical Forest Ecosystems, 227–237.

Table 1. Topics covered at the Bonn Symposium on the restoration of tropical forest ecosystems

1. Tropical forestry
Management
Harvesting
Nursery
Physiology of timber species
2. Ecology and general environmental concern
Land use planning and management
Forest succession/sustainable productivity
Species/ecosystem/biodiversity
Wildlife management
Nature preservation
Biosphere reserves concept
Global aspects of wholesale consumption of tropical forests
3. Political and socio-economic problems
Guidelines for forest utilization
Ownership of the land
Rights of the indigenous population
Compulsory restoration/replantation/revitalization of logged areas
Relations between industrialized and developing countries
Rights of concession holders for logging

There will be of course no success in any re-establishment of tropical forests if the local population and the foreign industrial logging enterprises do not understand the natural biological limits for the sustainment of a tropical forest. The present outcry of naturalists and conservationists may have come just in time to get all parties involved together to develop a working concept for the future land use in tropical forest areas. The fact, that a major company has started the discussion with this symposium and started demonstration projects in suitable areas is probably a first sign of re-thinking in the "management étages" of the industrialized northern countries.

General discussion

Ashari Muktar:

Ladies and gentlemen,

As a preamble, I shall take two minutes to show what we discussed already (Table 1). In this table we try to define particular topics for this discussion. The theme of our symposium is the restoration of tropical forest ecosystems. During the last few days we were looking at many different activities and the impacts of such activities as the vision of natural areas. We have proposed natural areas for the conservation of biodiversity, species diversity, and to certain extent our animal and natural resources (forest inhabitants) the practises of which might cause certain impacts on the forest. As you can see from Table 1 we covered aspects from managing permanent forests, mainly in Sarawak to problems the people in Sarawak have with the present exploitation practice. All these topics cause some impacts on the natural resources. If this remains uncontrolled or unmanaged it will lead to degraded ecosystems. After looking at the multitude of problems listed in Table 1 we should ask ourselves: What do we expect from this symposium? In their drive for a decent survival the people there are fighting for their consumption; their lands are often ill-managed and uncoordinated. This causes certain impacts on the natural resources of the region to the point where we experience complete depletion of the area. Several examples were presented during the last few days. Another point of forest destruction is the opencast mining of bauxite or tin or whatever. That might lead to a complete degradation of the forest to desert. This case is quite serious, because it leaves poisonous tailings on the surface.

First priority deserves every advice on restoration and rehabilitation of the tropical forest ecosystem. It might arise from forests managed by forest services or universities or from those fields being utilized or from the areas managed or left unmanaged after mining operations. In several papers we received some suggestions for the restoration and rehabilitation of such areas.

The second point which we should probably consider during this final discussion are the guidelines for best management practises. We have heard that prevention is better than cure. What are the methods of prevention that we need in a managed forest? Examples such as those we heard from Dr. Brünig are spelled out in the ITTO guidelines. How should we do without modern harvesting techniques? This needs to be improved. But it is not enough having a guideline in a sentence or two. A guideline without a proper study on how we shall do the harvesting properly, and without an explanation of proper techniques suitable in a tropical rainforest, is useless. I believe not enough has been done on these aspects in order that that forest engineers and private foresters understand the importance of their impact from an ecological point of view. Such viewpoints need to be applied in managed forests.

Third, we need to restore a near natural ecosystem in degraded areas and re-establish there, at first, a managed forest. After the intense application of silvicultural practises required to enhance the growth, we can probably include other aspects too. The forest can then be managed on an ecologically sound sustainable basis.

Fourth and last, once a large area has been reestablished we can consider the proposals on the conservation of portions of areas which our last speaker from Uganda suggested. For setting aside natural areas we should look into the guidelines what is requested, which areas should be selected and what are the important species habitats or ecosystems we can conserve. For the topic of conservation of species or whatever else we need better information in all tropical countries. Too little is known in the scientific communities in our countries. So these are the areas we should cover during the general, final discussion. I hope, Prof. Lieth, this will serve as an initiation for the task before us.

Lieth: Yes I think that is already a good start and I only may add that we abstain from discussing the so called global issues and keep the discussion on the country-size level. If something should be done there whoever is doing it should ask whether the respective country has a landscape management plan. Most countries do not have that and that is one of the basic problems in the tropics. The logging companies would have a much better judicial baseline if a landscape management plan was available as we have it in our countries here. We might invite then first some inputs to the last point of Table 1:

What kind of guidelines should be given to people who want to invest in tropical timber and trying to get natural resources developed by outside investors or by some groups who want to engage in foreign aid. May I have some questions. Is there something to be said from UNESCO side?

Clüsener Godt: ITTO guideline is a good rule for the preservation of tropical forests. I'd suggest that in the guidelines for management there should also be guidelines for non-timber products, e.g. how to build roads. It should be clear that the production of timber is one point and the establishment of a good local infrastructure is also a very important point. The following three points come to my mind:

— As ITTO prepares the first local implementation of their guidelines we, UNESCO, could for example participate and work together. Different representatives of respective regions could be involved in preparing specific guidelines for rehabilitation aspects as they have been mentioned here.
— A global action plan which might serve as a framework is needed to select and establish biosphere reserves in all critical regions and ecosystems of the tropics.
— The implementation of national biosphere reserves must consider local properties and the interests of the indigenous population. Any biosphere plan must be adapted to the region.

Lieth: Are the ITTO guidelines enough? Or is there something missing that should be incorporated?

Clüsener Godt: ITTO guidelines are the first attempt. They should include a clear part of nontimber products. They need to include ecological and economical aspects. Ecology and economy should come together and not work against each other. I believe the ITTO guidelines require improvements in this respect.

Lamb: I would like to take up the theme of economics first. What about lands with no great economic output? We have heard that one can restore degraded forests because there will be a future economic output. These is a number of ways of doing that and all you need is to find money and the political will to do it. But that leaves undone large areas of degraded tropical lands for which no immediate economic return can be expected and the costs of restoration might be high. I would like to get some feedback from the audience about the possibilities to reforest those lands, in particular about the enormous grassland areas that once had forest. The technologies that are needed are low in cost per hectare. With John Parrotta we were talking about restoring several small islands scattered through large grassland areas. From these islands the entire area may be reforested by natural dispersal mechanisms.

I can only admire the work that Prof. Miyawaki is doing. But there is a need of complete reverse approach to putting a lot of effort at one site at one time namely cover large areas with small input. For this we need to know what research strategies one might use. A cheap form of restoration is needed that can be easily applied over large areas.

Lieth: Are you far enough that you can start a pilot project? If we know what is feasible we should select an area. If we have these two things together we only need a group of people willing to work on such a project and have to get the money for them.

Do you have a larger area somewhere where a project you propose can be materialized?

Lamb: There is a large number of places where that can be done, but the problems are less ecological than sociological. Who claims to own the land or who really does own the land are often two different things. Can you get a commission to go ahead with that? If you can do that I think everything else follows really simply. What the tropics need are one or two demonstration areas in order to show what can be achieved.

Lieth: I know a very good government that has pledged to give 250 million marks for saving the tropical forest; they do not know how to spend the money.

We need such examples as you propose. The group I have in mind is very competent and they will probably listen. We can try and we have to get some politicians behind it who we can convince. Projects of this kind will only be successful if one can identify himself with such a project.

Ferraz: In many countries the typical situation is that degraded areas are owned by big companies. In other situations you have a lot of very small parts owned by peasants without any possibility to invest in this land.

You can try to promote restoration of degraded areas by showing examples in other countries. Sometimes you must show examples that it is economically more interesting for an enterprise to restore clearcut or degraded lands. The highest amount of degraded areas we have is covered with small or medium farms. How can we promote the restoration of forests there? I think we should later speak about the question of the vocabulary for this branch: restoration of degraded ecosystems is quite confusing at the moment. Some speak of restoration and some of rehabilitation. In a lot of tropical systems it is very dangerous to speak of restoration. We cannot restore one ecostystem with huge amount of species and enormous diversity. We should have a unifying concept in defining the words.

Lieth: You mean we should definine the goal we want to achieve. Some people are satisfied if we make the world green and others want to have the full species composition back which is not feasible. Somewhere between these extremes we must try to reach a certain goal. Is that satisfying? What is your opinion?

Wakker: A tropical rainforest cannot nearly be restored in the Mitsubishi project about which we heard here. It is a plantation; only more species are involved than eucalyptus. I ask Dr. Miyawaki how would he evaluate the project after two or three years for its success? Normally this process takes many many years for a forest to grow back in size and species diversity. What criteria would he use to evaluate the project?

Lieth: This question brings us back one step to the local site.

We are leaving now the rehabilitation of tropical forests in general and going to individual projects.

Ambasht: I have the following suggestions:

1. In the ITTO guidelines there should be sufficient flexibility in the general suggestions for the management of tropical forests for site-specific needs, taking into consideration the human, ecological, environmental constraints on any particular forest site.
2. Inclusion of already savannized forests in tropical forest restoration is desired. Social forestry of village woodlands may also be included: Local inhabitants can make their villages green.
3. A need of some bulletin for publishing success or failure stories or efforts or plans on important tropical forest restorations. UNESCO-Nature & Resources or INTECOL Bulletin or

UNEPs Connect or ISTE-Tropical Ecology (with some grants) can handle such scientific publications.

Maury-Lechon (to Ferraz): In Amazonia one could try to disperse early successional arboreal species like Cecropia to see whether the regeneration with climax species will come. What do you think?

Ferraz: We do not work with this method because the natural dispersion of Cecropia and the time the seed can stay in the seed bank (soil) is so long that we do not have problems with the growth of Cecropia.

Parrotta: Concerning the presentation of Dr. Ferraz, I think it is important to recognize the diversity of degradation types, the starting points for rehabilitation. Four-thousand-year-old degraded forest to last weeks harvest in Amazonia. Two major distinctions are needed for planning any experiment on rehabilitation.

1. Definition of classes of degradation types in order to think about strategies to deal with different situations.
2. Classifying different managing strategies based on history, social and economic conditions of the region. This classification would be useful for international comparisons.

Lieth: That is a good suggestion but does not withstand the possibility of starting an experiment. It should, however, be an inexpensive experiment.

Ray: Concerning Dr. Lamb's suggestions, I think this is a good idea. The islands of diversity strategy sounds very good. See to it that we have a range of different ecosystems represented in that future experiment. You can accomplish with the same amount of money more in one type of ecosystem than you can in another.

Lieth: Find a project for some guidance for particular countries. There were already projects proposed from different vegetation types. Originally called rainforest Ecosystems. I cut out the "rain". The tropical forest is nevertheless important everywhere.

Clüsener Godt: I suggest to use the American way to start an activity. You have a problem, look for people and then you can establish a project. You can have some sort of network established. We heard about different approaches during this symposium and about different areas. We know about different techniques as well. All of these must be coordinated.

Lieth: Got yourself a job. Need coordination, need integration, consider the places where it can be done, include most forest types in the world, good people and money.

Let us finally close the general level now and go to the local level.

Dr. Miyawaki is asked to answer the first question by Dr. Wakker.

Miyawaki: In Japan we have 20 years experience. We are now challenged with restoration of tropical forest in Malaysia. Living material is used to reconstruct the final vegetation. Calculate only time. After 2–3 years no more necessity for management. We can suggest natural management/selection after 3–4 years.

Wakker: But I need to repeat my earlier question, what are your criteria for success: If all trees grow up or if it has grown an ecosystem which is like a tropical forest or like another plantation. I can imagine that these dipterocarps just like grow up in height; like an ordinary pine forest in Japan. Will project aim at high diversity? Will that high diversity be a criteria for the characteristics of these ecosystems and how do you evaluate the success?

Miyawaki: This is a nice question for me. In Japan people never believed in natural restoration 30–40 years ago. In our institute we joined work with animal ecologist. We have planted for more than 20 years our main tree species. We conducted comparative studies in different natural and planted forests.

During June, July, and August the root system developed well. Growth was more than 80 cm in the first year. The first winter was difficult in Japan. After one year the growth was 1 m until 20 m. Soil fauna developed in time. As species richness index we counted after 20 years the same fauna community as indigenous. We have only one year now in the tropical rainforest project. The same height of young trees which was reached in Japan in 2 years needed only 4–5 months in Malaysia. This experience we gained in Kalimantan since 1973 in a riverside forest. How to re-establish an ecosystem with the same fauna as in a natural community? We expect that will take 40–50 years.

Examples: We have from the English colonies in 1926/27/28 were 30 dipterocarp species planted. These plantations developed well including litter and soil. We must now study how soil fauna developed with time.

Lieth: As ecologists we work with the fact that about 60–90% of the biomass is stored in 10% of the species. So we take care first of these 10% of the species and try to get them growing and assume that the rest comes by itself. This is basic experience in general forestry all over the world. Miyawaki says he expects success after 5 years for the species he plants there. This goes on and becomes in 40–50 years a near to nature forest type.

Wakker: I hope that Mitsubishi will not cut down primary rainforest and claim that they can rebuild tropical forest. Because this project exists by the grace of virgin forest.

Lieth: Yes, you are right. But the issue is to use the resources in a more logical way in the future. This is very important for a number of reasons.

We could provide a useful way to reduce the CO_2. Not store it in the deep sea but store it in the biomass from where it will be available again for use by humans and animals.

Ziegler: About time scale. This project is only on 50 ha. When can you charge its success so it could be applied on a much larger scale?

Miyawaki: This is a first experiment. We try to improve methods together with the colleagues from South America and Africa. Please help!

Lieth: I think the present experiment should not be bigger than necessary. Henry Ford started to plant rubber trees on many km^2 in cleared forest only to find out after 5 years that the project would not work. After some years he sold it to the Brazilian government for one penny. We are not as far yet we can say we can do in Malaysia what Dr. Miyawaki did in Japan. We have good experiences here in our country without these kinds of experiments. Look at the Autobahn, on both sides of the road. In ten years you see a forest growing there that requires recutting. It is not a natural forest but it is green. Try to implement good experiences in similar projects in the tropics. There are some good examples what the French did in Vietnam. They planted roadside trees, which were native trees already. From this background we can support Miyawaki's statement that if he is successful within the range of 5–10 years that he can say that this will grow into a forest. It will be a near to nature forest comparable to what we have in our forests in Germany. They are only plantations. Natural forests are not even-aged. What Dr. Miyawaki initiates is even-aged. One starts at that point and only the later development will be closer to nature ecosystem.

Wakker: But they are cutting trees before knowing enough to restore all trees in those places. We should discourage clear cutting and encourage more selective cutting.

Lieth: Dr. Brünig says he can make some management systems work (selective cutting in Sarawak); since he is the expert there I must accept his viewpoints. If you show me it is not going to work he will probably look at that. He has very good experiences. It did work in the area where he was a forester. It is not in all cases possible; it is a specially big problem in the old rainforests. The trunks there are so big around, you have to have very heavy equipment available to handle that. That is a practical problem. The question is, therefore, if they better keep their hands off such areas or develop better equipment. That requires to be analysed.

Ray: As for the legacy of the biological dispersal, consider ecological principles. What will be the impact of invaded species. Considering restoration you should have an assessment beforehand what exotic plant and animal species one is dealing with. Some ideas on the impact of exotic pests and animal species will have to be written into the regional guidelines.

Lieth: The next symposium on restoration of topical forest ecosystems should deal with consumers in general.

Ray: And producers as well as far as they can be competitors.

Lieth: Both aspects require an individual symposium. We should have some meetings on the consumer level and on the competitor's level.

Kigenyi: In the process of restoration you may find that you have invaded species. They are alien to your ecosystem. It is not typical and grows very fast for example in Uganda. What do we do with it? Are you going to clear it because it is foreign?

Lieth: This is not the only example. One hundred or so were reported from different parts of the world. This is the biggest problem to island ecosystem, e.g. broomesedge in Hawaii, aggressive grasses, mongoose in Caribbean Islands and so on. But let us go to the next point the management practices. Do you have any suggestions?

Ashari Muktar: Yes, the ITTO guidelines are rough guidelines not specific enough for implementation.

Lieth: This may be. I take the advice from many speakers: We should not go into the forest especially not in the mountain forest with heavy equipment. Equipment should be smaller in size so that they can go between the trees, and will not scratch the soil. We need better logging equipment for an ecologically minded logging operation.

Vanclay: Not only the equipment. We also need better training of the operators, in order to reduce damaging impacts of the tropical forests.

Lieth: Smaller equipment and better trained people.

Kigenyi: Minimize compaction. To go the shortest distance possible.

Vanclay: ITTO guidelines are meant to be guidelines not a specific prescription. No one knows exactly if buffer should be 10 m or 20 m. If you have a strain buffer it will be good whether it is 10 or 20 m. What we are talking about is not the finesse of having a strain buffer of certain meters; it is having a buffer at all. It is just the implementation of some basic principles. The ITTO guidelines allow flexibility in individual countries.

Lieth: I hope that the ITTO guidelines do not experience the fate of the instructions for Indian forestry. Each forest section has a thick book on how to use timber. The books are well written like all basic books of British colonial forestry. Every ten years a new book appears. The problem is what adheres to it.

Lamb: Yes, I am also concerned about theory and practise. Prof. Huss showed in one of his papers a system that should have worked ideally. He went and selected weedy species from his commercial species. Malayan foresters looked at enrichment planting. These experiences, the descriptions, only occur on experimental spots. If you look where these experiences have been implemented you look in vain. Ecologists do not know enough of sociology and economy to transform theory to practice. My recommendations: Scientists should look for partnerships with other people to implement their findings.

Lieth: This seems to be a suggestion to Mitsubishi: Please work together with a group of people parallel to the experiment in Malaysia who knows how to implement the new results.

Ziegler: I am pessimistic about the implementation of ITTO guidelines alone. Mitsubishi could implement certain ITTO and Sarawak mission guidelines on that site in Bintulu. That would be constructive apart from their experiment.

Lieth: I am not as pessimistic about implementation of ITTO guidelines because if the big companies stay together there will be a tremendous economic pressure on the countries to adhere to that. If all importing countries require to show where the tree came from, and where the lumber was felled. If this would be followed that would become a tremendous control. We now have guidelines on the table, and if they become a law it costs a few million dollars if a company does adhere to it. We need these kinds of international agreements. The NGOs are a very good instrument of enforcing that.

Ferraz: You brought in management and silvicultural plans. Maybe it is too early for management plans. First we need monitoring of all the sites. A good monitoring plan can tell you what can be included for the next plantation. This is better than to start with the management question. This is silviculture. Very local if you collect seeds, etc.

Lieth: For the experiment gives us some idea what should be monitored. A number of people are concerned about the logging operation to be done in a certain form that the forest grows by itself then. We need a management plan so that it grows back to a valuable forest. We need monitoring.

Hüttche: Think about sizes of equipment and training standard of the people. You need roads to transport logs. Roads are of major effect for the settlement. We cannot separate these issues from one another. Every management plan should contain conclusions to the effect of the built roads.

Parrotta (to Dr. Lamb): It is a good idea to superimpose experiments on timber harvesting areas. Companies may be interested in trying techniques, various operations of procedures. The scientific approach is getting the baseline data and looking for the effects many years later.

Lieth: This is the normal way things are done in practical forestry. But the time zero is passed already in most cases.

Suggest to them what they should do from the very beginning. Unfortunately it is the same in the tropics as it was in our region. When the forest is gone the scientists come.

Let us leave the management now and come to protection.

We had already said that one should make a landscape plan as we have it in our country. One category is nature preserves. Once a plan is there it is very difficult for most of us to do anything else legally with that land other than what it was specified for. This is a good instrument for nature protection. But in a country that has to feed people we cannot preserve everything. The population increase has taken care of that argument already.

Problems to consider: What do we do about the locals. In Africa in 7–10 years locals destroy areas. How much land has to be set aside. Who decides it and what kind of protection should be given.

Ray: Before you begin one should make sure that you have taxonomic experience and information. Find out species that are important to protect. This takes more time and more experience than we are normally willing to provide.

Lieth: Here I would invite Dr. Haber since here in our country we think if he preserves a particular habitat where we know that particular rare species is living we make the size of the area large enough so that we protect a number of these species simultaneously. Would you comment on that, Dr. Haber?

Haber: One of the main problems is the spatial distribution of preserves across the different land use types. Because there are specific requirements every culture for instance will always look at the areas which are very suitable for intense agriculture. And also in the long run we need intensive agriculture to feed the growing number of people. And similar question holds for silvicultural areas, where we want a timber production or other forest products and so on. You will select those sites and also those forest communities which are the most suitable forests and again there is the question where to place the forest preserves, and the same problem holds for the urban industrial areas. There you want to have open spaces for recreation of the city people but also to help the element of nature preserves even in those built-up areas. This ends up into recommendations like I have made myself 20 years ago to preserve as a general rule a minimum of 10% to nature in every type of land use but in reality this is difficult to implement. One shall end up in preserving only those sites as nature preserves which are not so suitable for agriculture. If you want to go against this trend of making usability criteria then it becomes very very expensive. You mentioned landscape planning. It is a very good thing but landscape planning in reality means land use planning and this intervenes with the constitutional rights of using property, privately owned property for any agricultural use or whatever. So I meant that planning will end up in plain economic considerations and we know that this would not work for our ecological goals. So we have to find a way between these extremes.

Lieth: Thank you very much, Dr. Haber. I see the claim for an ecologically minded land use planning is something that we need.

Lamb: Regarding the percentage used to be set aside for nature preserves, I think you can say 10% is probably good. There is a danger, however, that because you might not even get to 10%. In fact in and around some cities there is no nature reserve under permanent nature preservation. What they have is a shifting pattern of national park, for perhaps 40 years. A garden will be cut down by its owners and a new one will be set up somewhere else. So there is always the same percentage. This concept is far too simple for national parks but such different views exist in different places. This may not be ecologically the best approach but it will be socially accepted in the particular country. I do not think we should get hung up on fixed percentages of areas for national parks because different countries will have different opportunities.

Lieth: O.k. but it should not be less than 10%. The legal situation in Europe is: We do not only have the national parks. We have also the certain hectare size pieces of a preserved ecosystem or vegetation types. So we have this legal background and the owners cannot do what they want with it anymore. They have to ask the authorities whether he can even build a fence around it. So we do have a strict policy implemented, probably some of my colleagues might say it is not strict enough, but I think we have good laws. There may be better ones but I think basically we have some very good protection for these areas. What does this protection mean? The big output of NO_2 from industry changes the entire situation changes anyway. In this country we have lost practically all species sensitive to nitrogen. All rare species. They are the biggest loss here in this country. And so, there is something else that is involved which is again globally of importance and of course we cannot even say that we can preserve all plant species in our country. Let us preserve from nature as much as possible; for its own benefit and not for the benefit of men. I think the latter is anyway the wrong concept. I think the ecosystems are there for themselves, not for us. But this is of course my personal view and I am an ecologist. (There was another show of hands.)

Ziegler: Yes I propose that in certain instances apart from setting land aside for protection there should be a social assessment too, because for instance in the case of Sarawak, land was actually taken away from an indigenous tribe without their consent.

Lieth: Just a minute. That is just one step behind because we have an entire unit after that. Let us close first the scientific issue and discuss then the social issue because it is very dear to my heart.

Ray: Staying within the concept of increasing development you were talking about earlier and responding to a very important point Dr. Haber just made regarding the expense of giving areas of arable lands for the reason of preserving some species growing particularly within those habitats. We may have a chance to get such areas for nature protection. I am sure that is happening in other countries as it is happening in the U.S. as a part of land use plans. Often we can get such land through our conservation association which obtains so-called easements. They are non-governmental organisations that are compiling huge amounts of acres of private company issues with which they go again and obtain for particular unit areas such as the Virgin Islands National Park as easements. They do not purchase land but use more efficiently a final part of their money in order to gain the protection scheme implemented.

Lieth: I think that is a good idea. In Germany like in the U.S. we put much of our land back into a soil bank. They want to reduce the production surplus in the European Community and we should even leave more land alone whereas in the last thirty years after World War II they tried to make every corner of the country productive. And now we find ourselves in the situation that we have such an enormous overproduction on that little land which we have left. So you see what science can make out of the land area. I think the same will hold true in the future with the tropical areas. Under such circumstances I hope that not all of the rainforest is gone until that time. I think that land in the tropics can be made more productive than it is now. And then we would of course have a solution and the possibility to buy easements. I think they would give then more land to the nature protection and nature conservation if they could collect some money for it. We are actually trying to develop a system here whereby we suggest to certain farmers they should raise us birds and butterflies rather than grain and beans. And get paid for that. It is just a natural product. You can make a living out of that, if you do it the right way. We are still far from that. At the moment the farmers get the money for not raising more grain and not getting more milk. But I think this is an entirely wrong concept. The farmers should be paid for production and not for non-production. We are discussing that with our politicians here in our country. I think we are now practically closing this particular part of our discussion. I think Dr. Haber wants to have the last word. You can add what you think has been neglected.

Haber: I would like to comment on one point Dr. Lieth just made. Nature protection or establishment of nature reserves are very desirable and necessary, but are not enough. There is need of supervision, of monitoring, of management, and this must be taken seriously. In many cases, especially in Europe, areas declared as nature reserves or protected sites preserve successional stages of quasi-natural or semi-natural communities such as heathlands or dry meadows. Regarding such sites, the German conservationists are split into two groups. One group wants to let nature go its way, without any intervention. This means that many interesting communities and species will disappear. The other group wants careful management to hold on any succession, and to take special care of certain communities or species such as orchids, butterflies, etc., which are restricted to heathlands, dry meadows or fenlands. I suppose that similar situations occur in many tropical regions.

Such situations require very good ecological knowledge of tropical ecosystems in order to develop predictive capabilities of what will happen in such sites in response to air pollution, a threat to nature reserves as Dr. Lieth mentioned. Here in particular the import of nitrous oxides from the air needs to be monitored because these gases had a severe impact on European forests already.

I think we will never again arrive at a state of non-pollution; we will perhaps considerably reduce the present pollution levels, but there will always remain some pollution which will have to be taken into account in managing nature reserves, as well as the entire tropical forest types remaining in the future.

Lieth: Thank you very much for the closing remark of the scientific aspect. That brings us to the next part of the discussion. We want some discussion on the social aspects we started already but which was much too short. We heard the recommendation that we also should organise a symposium on the social aspects of the restoration or overutilisation of tropical forests. Turning to this question, I would like the representative of Mitsubishi Corp. to present their views.

Ozawa: Let me introduce myself: I am Mr. Ozawa. Mr. Mori is out of town and he asked me to represent Mitsubishi Corp. in this discussion. I am now in charge of the Malaysian rainforest regeneration project. I went to Sarawak twelve times since last May. Maybe I know more than you about Sarawak but if I make a mistake, our Malaysian friends Ashari Muktar and Abas bin Said may correct me. I will try to answer for Mitsubishi Corp. as far as my knowledge allows it.

Ziegler: In all scientific projects also social assessment should be made. In Japan you have very strict environmental rules. Why do we not find these assessments in countries which you exploit?

Ozawa: In domestic (Japan) we have a very strict, maybe the strictest rule in the world. Now the standard of pollution measures in Japan is one of the highest in the world. Regarding foreign investment there are many environmental impact assessments when you invest in the country and we have almost 180 branches in other countries. More than 10 or 20 investments abroad every year. Our department even proposes laws. We are recommending to take into account the standard we are giving to them.

Ziegler: What about social impact assessments?

Ozawa: Yet this also exists. We abide by the laws, the domestic laws of the country where we are operating. We are abiding by the guidance or regulations of that country.

Wakker: But in some cases you cannot rely on domestic policy of a country where you will be operating. In that case it would be good to have a company code of conduct including environmental impact assessment and social impact assessment regardless of the rules or regulations existing in that country where you are operating.

Ozawa: Yes. Other companies call the conduct that we established since 1973. We are now improving our comprehensive conduct of our company since last year. Before that we did not have much information from abroad. We got that now from more than 100 companies and we are now re-establishing our conduct code.

Wakker: Concrete rules for conduct regarding forestry later on!

Ozawa: Have you got the new chart? We are now taking into account environmental charts. All different firms (automobiles, etc.) are making their own individual conduct codes.

Lieth: If they want to consider the social aspect for any given project there is the official opinion in that particular country and then there is the concern of minority tribes which the country does not even recognize. The big corporations coming from our countries should know about that. It would be of great help if they would have an anthropologist in their advisory group for the environmental impact statements in those particular countries. Is there a possibility in the future that Mitsubishi Corp. would add an anthropologist to their staff in order to get some information whether they harm local tribes? Would you support such a demand?

Ozawa: We have 180 branches all over the world. These branches claim already to be responsible for social matters in that country. We consult with the branches responsible to that country. Primary branches might take some consultances from specialists abroad.

Lieth: No clear answer. Would you do that or not. Would you support to add an anthropologist at home?

Mita: It is not the right place to say yes or no in this kind of discussion.

Lieth: Question only whether we convinced him that there is a need.

Mita: It is clear that in the environmental problem social aspects must be respected.

Hüttche: Support the idea to put an anthropologist in your department. But also good to get a point of view from the local side that you consult local non-governmental organisations before you take logging concessions. Could you agree to this?

Lieth: I would support this idea. Get some background information which you do not get from the official of that particular country. The economists may not even know that. You would get a much better advice from an anthropologist.

Ozawa: Yes prior consultation appears to be meaningful for the future.

Lamb: If you talk to ten organisations you will get ten results. Who you talk to, what you ask, everybody will have different interests. Different viewpoints. The decision is not very simple for you.

Ziegler: One should ask indigenous NGOs.

Lamb: That would not be any different.

Lieth: If I were an anthropologist at a company I would listen to all of them and then compromise after having received the best information. Picking up the responsibility for the entire group must involve local population unit to survive from the beginning of any operation. Maybe everything becomes more expensive afterwards. The economic community in the world is learning this at the moment. This increases also the respectability for themselves. We need to improve the conciousness at the decision makers. It is, therefore, justified to put pressures on them.

We have made the major points. You, the representatives of NGOs, made a very good impression. And I would propose to invite nongovernmental organisations in the future.

Ziegler: I would like to thank that we were invited and sincerely hope that we can keep the constructive dialogue towards Mitsubishi Corp.

Lieth: Thank you all for being here. If we have succeeded in this discussion to initiate future dialogues between the major operators of large logging operations in tropical countries and also provided some help and guidance for the projects sponsored by Mitsubishi Corp., we have made a substantial contribution to the solution of problems with large-scale destructions of tropical forests.

I suggest now that Prof. Haber finishes the discussion with a closing statement.

23. Challenges in extending development to the Penan community of Sarawak

HATTA SOLHEE[1] and J. LANGUB[2]
[1] *Department of Development (State Planning Unit), Kuching, Sarawak, Malaysia;*
[2] *Majlis Adat Istiadat, Sarawak Kuching, Sarawak, Malaysia*

Keywords: Development aid, Penan, Sarawak, tropical rainforest

Abstract. A major problem for the governmental agencies involved in tropical forest management in Sarawak is the caretaking of indigenous, forest dwelling populations. Especially prominent is the necessity to work for the nomadic groups of the Penan in the Masok and Limbong districts. The state planning unit in the Department of Development has installed a state-wide scheme to assist the approximately 10 000 Penan in Sarawak to manage the transfer from a traditional subsistence in the forest to a more advanced farming system.

The problems related to this transfer, the organisation of the extension work and some results are reported. The function of a service center is discussed. Help is primarily rendered for health services and forming skills along with writing skills and family planning. The basically positive reaction of the Penan population and some reasons for complaints are mentioned.

1. Introduction

Since the mid-1980s the Penan community has attracted international attention, and the Malaysian Government, especially the State Government of Sarawak, has been accused of not caring for the community. The accusation centres around the issue of logging activities which have often been blamed, wrongly or rightly, for the fate of the Penan. However, if one were to look at the history of initiatives undertaken by the government to help the Penan, much of the concern expressed is not quite correct. In actual fact the government has, since the time of the Brooke regime, initiated various programmes aimed at assisting the Penan to adapt to the changing environment around them.

In order to appreciate the above issue and the challenges of extending development to the community, it is necessary to (1) understand the socio-cultural structure of the society; (2) examine the history of development initiatives undertaken by the Brooke (pre-1946), Colonial (1946–1963) and Malaysian (1963–) Governments; and (3) identify the constraints that impede the process of development within the society. It is hoped that this may help us understand the problems of the society in the proper perspective as well as the problems involved in extending development to them. It is also our hope that discussion on this issue in its relation to the tropical rainforest can help us to formulate better strategies to help the community.

2. Some basic facts about the Penan

The first important point to note is that when we talk about Penan we are talking of two quite distinct populations, classified by Needham (1972) as Eastern and Western Penan. The Eastern Penan are those groups living in the Tutoh, Patah, Pelutan, Apoh, Ulu Akah, Sela'an, Selungo, Ulu Limbang and Ulu Baram Rivers and those in Brunei. The Western Penan are those in the Belaga District, the Silat and at Long Beku, the last two places being in the Baram District. The Penan of the Tinjar, Suai, Niah, Jelalong, Labang and coastal Bintulu have largely migrated from the Belaga District via the Tinjar. Map 1 shows most of the locations mentioned.

Broadly, the two populations have a lot in common. For instance, they speak the same language with minor dialectal variations and they are traditionally a nomadic people and have relied on hill sago (*Eugeissona utilis*) as a staple food. However, despite these similarities there are marked differences between them, for

H. Lieth and M. Lohmann (eds.), Restoration of Tropical Forest Ecosystems, 239–250.

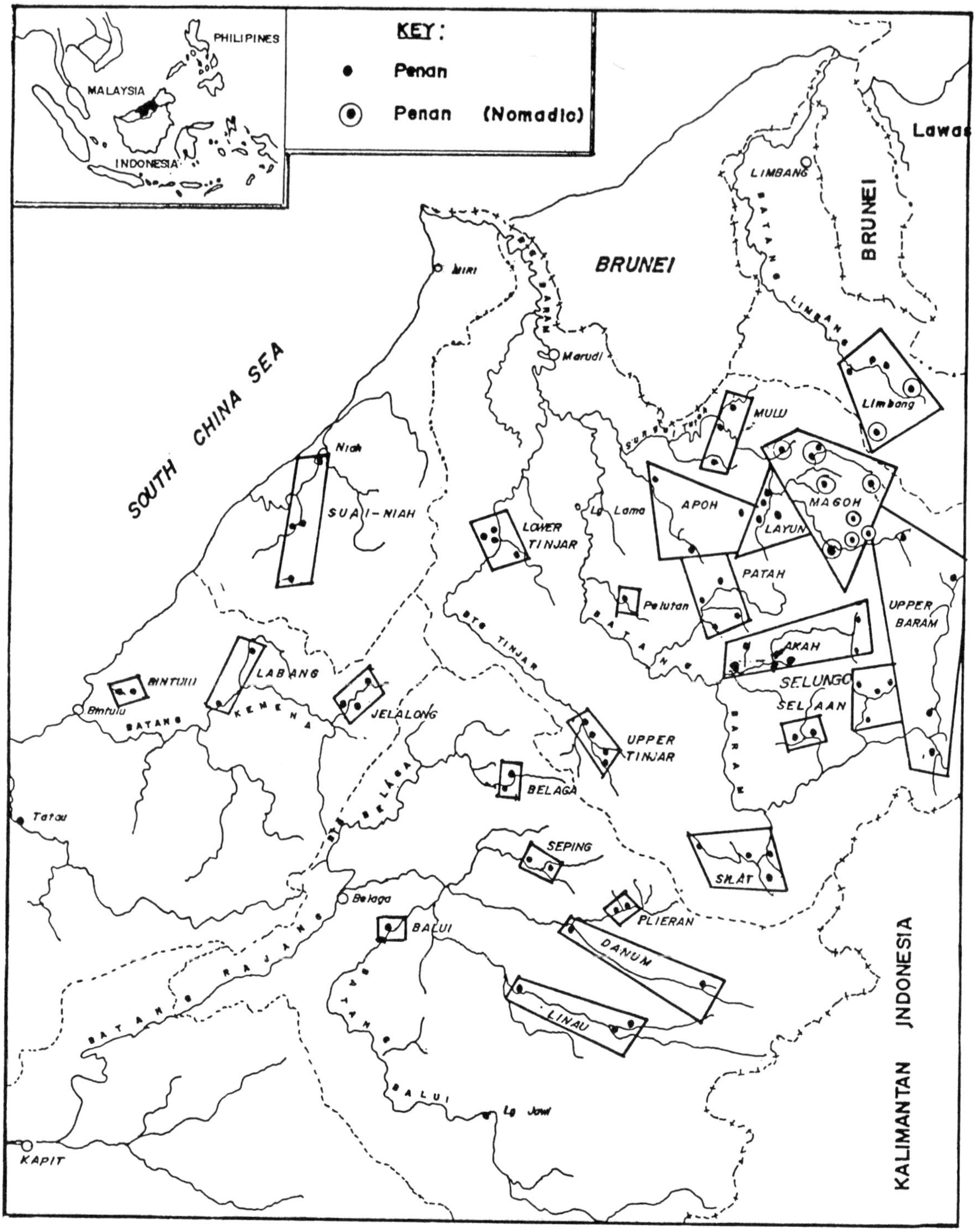

Map 1. Areas occupied by Penan.

instance in group size, average household size, settlement location and settlement system, foraging area size, hunting technique, leadership institution and genealogical knowledge (see Brosius 1992 for detailed information). Brosius (1992) points out that understanding the similarities and differences between the two groups of Penan is important because they have implications on development policy for the community.

It is also important to note that the vast majority of the approximately 10 000 Penan in Sarawak occupy some of the most rugged terrain in the most remote interior of Sarawak and far from the administrative and commercial centres downriver. Several modes of transportation in one journey, that is by outboard motor, paddling and walking on foot, are required to get to some of these remote settlements. To reach most Penan settlements are, therefore, expensive, time consuming, arduous and at times dangerous. It must also be pointed out here that the Penan population is spread out thinly over a vast area of the remote interior.

This situation limits the number of interactions between the Penan and outsiders, including development agents who are required to do extension services. It also limits visits by other government officers who are required to explain government policies and programmes, including logging activities that affect the livelihood of the interior communities.

Another important point to note is that different groups of Penan lead different lifestyles. Although traditionally a nomadic people, presently only about 400 individuals out of the 10 000 Penan, or 4% of the total population, are truly nomadic (Langub 1990b). These 400 individuals, grouped into eleven bands reside in the Tutoh and Limbang watersheds. The rest of the Penan are either settled or semi-settled.

Nomadic Penan live in groups ranging from 15 to 75 persons, with a mean of about 32 (Needham 1972, p. 178). They live in temporary camps made out of forest materials, such as saplings and leaves. Movement of a group from one camp to another is dependent on several factors, such as supply of wild sago, game, rattan and other jungle produce.

For the nomadic Penan, the forest plays a central role in their lives. It provides them their staple foods, such as the starchy pith of a type of sago palm (*Eugeissona utilis*), game, especially the bearded pig (*Sus barbatus*) and a variety of fruits and other food plants. The forest has also for centuries provided them with necessary trade items such as camphor, jelutong (a wild rubber), damar (a resin), gaharu (incense wood), bezoar stone (gallstone from a monkey to which the Chinese attribute medicinal properties) and rattan for making baskets that can be sold to merchants in the coastal towns. The forest also provides certain species of plants which are used as medicines. For Penan who have adopted the settled life, the forest provides them the materials for building houses, boats and other implements.

The semi-settled Penan, numbering 7000 people, are those in the transitional stage of becoming settled. With regard to the settlement pattern of the semi-settled groups, most maintain a two-tier settlement system (Brosius 1986, p. 176). The first component of this system is the main settlement, *lamin jau*; the second component is a series of satellite camps, *lamin tana'* in the vicinity of sago groves. The lamin jau is characteristically a permanent settlement while the lamin tana' are temporary camps occupied for a period of anywhere from a few days to six months, until the sago in the particular area is exhausted.

The main settlement of the semi-settled Penan may be formed by a cluster of several houses built on the fashion of a Malay *kampung,* or by a single longhouse, or again by two or three longhouses grouped together (Langub 1974).

The subsistence of the semi-settled Penan rotates around rice, tapioca and wild sago. They cultivate small plots of hill padi through swidden cultivation. Their method of cultivation is still very crude; no proper care is provided once the seeds start to germinate. The supply of rice from these small plots of hill padi normally lasts a family three to four months. When the supply of rice is exhausted, they resort to tapioca which they cultivate around the settlement and wild sago. Their dependence entails going back to the forest to locate and process the sago. In general, while they are learning to grow their own food crops, their dependence on the forest and forest products is still predominant.

The settlement of the permanently settled Penan are established much better. Their houses are built out of either light materials or more durable materials such as *belian* or other hardwood, with corrugated zinc coated metal sheets for roofing. The architectural style of their houses is influenced by their neighbours. Those settled Penan with Orang Ulu or Iban neighbours tend to construct their houses in the form of longhouses; those with Malay neighbours in the fashion of a Malay *kampung*.

Some of the better established communities are provided with modern amenities such as piped water, sanitary latrines, clinics and primary schools.

The permanently settled Penan cultivate much larger plots of hill padi than those of the semi-settled Penan. This is to make sure of sufficient production to last them until the next season. A few settled Penan settlements in the Ulu Baram have started the cultivation of wet padi, having learned this from their Kelabit neighbours. For the settled Penan, their staple food is rice.

Some of the settled Penan, especially those in Suai and Niah have long practised the cultivation of cash crops such as rubber, pepper and cocoa. The standard of farming and cultivating of cash crops is on par with the Chinese and other farmers.

One of the most important aspects of Penan way of life, especially the nomadic and semi-settled Penan is their relationship with the forest. The relationship between the Penan and the forest, and their management of its resources has been extensively discussed elsewhere (Brosius 1986, 1989, 1990; Langub 1988, 1989, 1990a) and here we need only summarize it.

The Penan culture has its orientation in the forest environment. Their history and ecological knowledge are encoded in the river systems, peculiar landscapes, types of vegetation, etc. To the Penan, the forest is like a 'supermarket', where they get the food and trade items. They consider the forest as something very precious which they want to look after forever: they make use of it for subsistence and when they die, they leave it for the next generation to look after. Ever conscious of the fact that their children and children's children will one day walk in their footsteps, the Penan see to it that they hand to the next generation a forest area capable of providing the needs of succeeding generations. There is then a sense of stewardship over the forest and land in which they live.

Perhaps the most important aspect of Penan resource management is the practice of *molong*. The word molong can be roughly explained as preservation of forest resources for the future. Penan molong all sorts of trees: fruit trees, trees for constructing house and boats and for making blowpipes, *tajem* trees for their poison darts, sago and rattan stands. Some fruit trees were molong many generations before and serve as historical reminders of ancestors long gone. Trees, fruit trees, sago and rattan can be molong communally or individually, but the basic idea is the same: to preserve these resources for the future.

Penan harvesting strategy is based on the principle of sustained yield. When harvesting sago, they cut down only one or two of several trunks, leaving the palm to resprout. They never cut down the entire plants at the root clump as this would kill it. Similarly, mature rattan is selectively harvested, allowing the young rattan plants to mature for future collection. Fruits are plucked from the tree or allowed to fall for collection rather than cutting down the whole tree. Traditional Penan hunting techniques of using dogs and spears and blowpipes are such that the stability of animal populations are not threatened.

In their overall harvesting strategy, the Penan hunter-gatherers migrate from one area that has been harvested to another where resource replenishment has taken place. As for the semi-settled Penan, they rotate their harvest of forest resources from one place to another allowing harvested areas to regenerate.

3. The Penan and the outside world

Contact between the Penan and the outside world has always been through their neighbours, primarily Kayan, Kenyah, Berawan and Kelabit, and civil servants.

The contact between the Penan and their non-Penan neighbours (*lebu*) has always been through trade. Barter trade with the lebu is an important aspect of the Penan economy (Langub

1984). In barter trading the Penan collect a variety of jungle produces which they exchange for essential goods such as clothes, cooking utensils, salt and metal. Nowadays most of these trades involve the sale of handicrafts such as rattan mats and baskets and *parang* for cash and other basic necessities. During the occasion, too, the Penan obtain information about the outside world and new ideas from their trading partners.

Visits by Penan to their neighbours' longhouse also occurs; and in some instances, these neighbours would assist the Penan to start small padi farms.

Through these relationships the Penan began to appreciate the settled way of life. Impressed with the immense wealth of their labu neighbours, as seen from the Penan perspective, they yearn to acquire such possessions and wealth. The Penan look at development of their society from the perspective of their lebu neighbours. Their lebu neighbours, therefore, become a focal point of reference worthy of emulation.

Despite all this, the Penan have great mistrust of the lebu. The Penan are always suspicious that the lebu will cheat them in trade transactions. Indeed, during the Brooke and colonial days, barter trade was supervised closely by government officials to prevent the Penan from being cheated by their trading partners (Langub 1984). The Penan also dislike the lebu for their superior attitudes, and Penan are aware that they are looked down upon by their neighbours.

As the Penan begin to adopt a sedentary lifestyle, they are confronted with a new set of problems, especially land dispute with their lebu neighbours and their lack of farming skills. In such a situation, the Penan always find themselves at the losing end of the stick, and these are factors which often discourage Penan from becoming full-fledged sedentary farmers.

Another group of people the Penan meet quite frequently are the civil servants: District Officers (DOs) and Sarawak Administrative Officers (SAOs), medical personnel, agricultural extension workers and school teachers.

The Penan consider civil servants as providers of minor rural projects, essential services, such as medical and health services, agricultural subsidies and education. Increasingly, the Penan look towards the government as provider of hand-out projects and when these are not met they blame government officials, especially the DOs and SAOs, for not being able to deliver these handouts. This is partly the reason why DOs and SAOs are nowadays held by them in low esteem.

The Penan are fond of comparing the role played by present-day District Officers and those played by District Officers during the Colonial period. During the Colonial period DOs and SAOs were considered by the Penan as their protectors because they supervised barter trading closely and made sure that Penan were not cheated. Colonial DOs, according to Penan, often visited them and had more time to listen to their problems and to discuss matters that affect the community. DOs then were held in high esteem.

In recent years Penan views of DOs and SAOs have changed somewhat. This is partly due to the fact DOs have assumed many other functions and thus have less time to visit the Penan regularly.

DOs are an important link between the Penan and development. To maintain this important link DOs must visit the Penan often and hold meaningful dialogues with them.

In recent years with the advance of logging activities in the interior of the state, Penan are increasingly in constant contact with timber operators. Although timber companies have provided Penan with various kinds of assistance, such as building materials, housing and transportation to various places such as to hospital and jobs, Penan attitude to timber companies is one of suspicion. They regard timber companies as the source of their many problems, especially the depletion of the food supply such as sago and game, and trade items such as rattan, gaharu and other jungle products.

A lot of the mistrust Penan harbour towards timber companies is due to misunderstanding and lack of communication between members of the communities and timber companies. Again, DOs could play a useful role here by organizing regular dialogues between the Penan and timber companies.

Today, the Penan find themselves in a world that is fast changing. They are not in control of the situation and neither do they understand these changes. They equate changes to their environment with problems to them — the greater the changes, the more complex the problems.

Development strategies and programmes

4. Historical perspective

The earliest government involvement in Penan Welfare was in providing health and medical care to members of the community through the travelling dispensary and the supervision of barter trades. These activities, however, were undertaken only occasionally and were never meant to get them to settle down. But some serious attempts to get the Penan to adopt a settled way of life began during the Colonial period. The Colonial administration, finding out that the community was not easily accessible and thus difficult to administer, tried to persuade them to move downriver and establish permanent settlements. Simple farming projects, such as starting small plots of hill padi and vegetable gardens were started, but most of these early attempts failed primarily because of the lack of follow-up visits by officials. Several groups did take the advice and moved downriver to establish temporary settlements. The groups that moved met with many problems, particularly with regards to food supplies; thus the majority have not fully settled.

One very popular means adopted by the Colonial administration in maintaining contact with the nomadic Penan was *'tamu'* or official barter trading. In most cases it was held once a year in specific areas earlier agreed upon by all parties. The Penan, Kayan, Kenyah and other native groups who had anything to trade would assemble at the agreed location. The officials, apart from overseeing the barter trading, also used the occasion to hold dialogue sessions with the various groups that were present and provided simple medical treatment to those who required them. It was during the occasion, too, that government officals explained policies and programs that had implications on the Penan and other Ulu people. The tamu also facilitated the exchange of views and experiences among the participants.

Colonial administrators also used the malaria eradication programme to maintain constant contacts and have dialogues with the nomadic group. This programme was particularly important, not only in its efforts to prevent the spread of malaria in the remote areas, but also in providing opportunities to officials to have frequent dialogues with the Penan and other Ulu people. Attempts were also started to have Penan children to attend schools; of course, numerous problems were encountered in the endeavour.

The independent government continued the programmes started by the Colonial administration in trying to assist the Penan to settle down permanently in the areas of their own choice. The tamu was continued until 1976. Attempt had been made to revive the tamu but the meetings are irregular and ad hoc. Thus it is not as effective as it used to be. A more comprehensive medical and health programme, however, has been carried out to improve the health condition of the commuity. The programme includes the Flying Doctor Service, Mobile Health Team, the Village Health Promoter and Environmental Sanitation Programme (include the gravity feed water system). The Education Department and the Sarawak Foundation also intensified their efforts to get more Penan children in schools, boarding facilities and free school uniforms and books were parts of the programme. Furthermore, the State Government, through the District Office and the Agriculture Department (DOA), supplied the communities with building materials (zinc sheets, timber, nails, etc.) and agriculture subsidies respectively. Due to transportation problems, a larger proportion of the materials are transported by helicopter at a very high cost and follow-up visits are infrequent.

The programmes implemented by the Medical Department were and still are well accepted by the Penan communities. The main factors accounting for this is the regularity of the visits by the medical personnel through the Flying Doctor and Mobile Health Services, as well as the participatory approach adopted in the implementation of the programmes. The presence of the doctors and other medical personnel normally bring joy and delight to the members of the communities. The sick are given treatment, and medicine are dispensed to those who need them free of charge.

The programmes of DOA and District Office,

however, met with numerous problems, mainly because of infrequent and irregular follow-up visits by officials to supervise and assist in project implementation, the nature of the projects themselves, and problems of land ownership (in the case of agricultural projects). There are many instances where the building materials are not put to proper use for the simple reason that the Penan do not have the skill to construct the longhouse or they did not see the real needs for a permanent longhouse just yet. In general, the programmes of the DOA and District Office were ad hoc and physical hand-outs in nature with no proper budget for follow-up actions to see that they were implemented in the desired manner.

Nevertheless, there are some success stories. Towards the end of the sixties and early seventies more and more Penan in the Belaga District began to settle down in permanent longhouse settlements and cultivating hill padi. Assistance from the government was minimal; instead the Penan learned most of their skills in constructing longhouses and farming from their Kayan neighbours.

At Lusong Laku, Belaga District a primary school was established in the late sixties but because of poor enrollment the school was closed down a few years later. It was not until the mid-1980s that another school again reopened at Lusong Laku. This school was more successful and the success was mainly attributed to the dedication of its headmaster. In 1991 it was judged as one of the four best schools in Malaysia in terms of academic performance as well as in extracurricular activities. Today, there are two primary schools for the Penan in Belaga District to service Penan children who come from nearby villages, mainly semi-settled groups.

In the Upper Baram River, a few groups of Penan have successfully adopted the settled life (Murang 1990). One good example is the group at Long Lamai. In 1961, the District Officer, Baram described the village of Long Lamai thus, "The settlement is the most impressive ... I have never seen a group of Penan looking so fit and clean ... settled or nomadic" (District Officer, Baram 1961).

Long Lamai owed its success in adopting the settled life to the Borneo Evangelical Mission which opened adult education classes and a small clinic for the Penan around the area. The Sa'ban Pastor who looked after the small congregation also taught them to cultivate rice, tapioca and other food crops.

From a humble community comprising a few families living in scattered huts in the late 1950s, Long Lamai has developed into a well-organized community of 68 families with a population of 329 people. It is one of the few settled Penan villages that is self-suffficient in rice. They not only cultivate hill padi but also wet padi, which is a more efficient way of cultivating the land.

Long Lamai has a good primary school and many young Penan from around the area have gone to secondary school. A few fortunate ones have found employment in the civil service, and indeed, the first Penan to gain employment as a Sarawak Administrative Officer is from Long Lamai.

Besides the Borneo Evangelical Mission, the local district administration also took a keen interest in the progress of the Penan settlement at Long Lamai from the beginning of its establishment as a settlement. One Sarawak Administrative Officer based at Marudi was specifically assigned to look after the progress of newly established Penan villages in the Baram, especially the one at Long Lamai. His job was simply to give advice and encouragement to the newly established Penan villages.

Baram District is large and the Penan population is scattered thinly in the interior of the District. With only one SAO looking after the welfare of all the Penan in the District, he could only concentrate on a few villages. Many Penan groups were, therefore, not adequately covered by the district administration.

At any rate, several primary schools were established in various Penan settlements in the Baram. Although most of these schools did not function well because of the high rate of absenteeism, they have produced a good number of young Penan who are able to read and write. Presently there are five primary schools in the Baram District, solely to cater to the needs of Penan children.

Other development assistances that Penan in the Baram and Belaga Districts received from the government include water supply projects, footpaths, bridges, housing and building materials. Again, because of the scattered nature of the population, not all Penan groups benefited from the above development projects.

5. Development programme since 1987

In 1987, efforts to develop the Penan community took a new turn when the State Government form a Special Cabinet Committee on Penan Affairs to formulate strategies and programmes for Penan development. The specific mission of the Committee is *to bring the Penan community into the mainstream of the Malaysian society, so that they can participate meaningfully in the development process of the country.* The State Planning Unit (SPU) serves as the Secretariat of this Committee.

Similar committees have also been set up at the Divisional and District levels. Their main functions are to identify programmes/projects for the Penan in their Division or District, to coordinate implementation of programmes/ projects and to monitor progress of implementation.

Two broad categories of development programmes have been adopted by the Cabinet Committee on Penan Affairs: (1) short-term development programme and (2) long-term development programme.

Short-term development programme

Essentially this takes the form of minor rural development programmes which provide projects that can give immediate benefit to the community. These include projects such as piped water supply, wooden bridges, footpaths, building materials and rest houses for Penan to stay when they visit a rural clinic in another village. Similar projects had been implemented prior to 1987, but with additional funding, they are being intensified to reach more Penan groups.

Long-term development programme

The government has introduced two long-term development programmes for the Penan. These are the Penan Volunteer Corps and the Service Centre programme.

Penan Volunteer Corps programme

Broadly, the objectives of the Penan Volunteer Corps programme are: to act as a close liaison between the Penan community and representative of the government; to provide them with basic services in agriculture, personal and community health care and functional education and to instill in them the spirit of self-reliance and a sense of responsibility to their community through participating in community works.

The volunteers are assigned to carry out the following tasks:

- To explain to the Penan the roles of various government departments and the kinds of services offerred by these agencies for the benefit of the public.
- To lead and work with the Penan in agriculture activities.
- To provide simple treatment to common ailments (the volunteers were given first aid training) and to lead the community in cleaning the village compound and also the longhouse.
- To conduct functional education classes for the adult population.
- To assist the community in implementing village development project, e.g. construction of piped water supply systems, bridges, footpaths, etc.
- To encourage and coordinate traditional activities, such as the production of local arts and handicrafts so that these skills and talents are not lost.
- To get the feedback from the Penan with respect to their aspirations and problems.
- With respect to the six Penan volunteer carpenters, the specific job is to provide technical know-how to Penan in basic carpentry and building simple houses.

The programme took off in November 1989 after the volunteers were given a 3-month course in agriculture, medical and health services and adult education. Every volunteer must complete, at least, 9 years of schooling. Today a total of 26 Penan youths (20 boys and 6 girls) have been recruited as volunteers. Of these 26 youths 15 have been assigned to work in 10 Penan settlements in Baram District, 4 in 3 Penan settlements in Belaga District and 1 in the Limbang District. The 6 carpenters are based at Marudi and assigned to build longhouses in different parts of the Baram District whenever their services are required.

So far there has been no evaluation of the programme, but from general observation the following comments may be made about the programme.

On the whole the Penan response to the programme has been encouraging. The health programme has been much appreciated; there has been a great deal of improvement in village cleanliness since the volunteer started the programme. However, the volunteers face a number of problems on the ground. This is expected from such a pilot project. For example, a few volunteers complained that they normally run short of medicine every third month before they get fresh supply when they come down to the District Headquarters to collect their allowance and give the quarterly briefing to the District Committee on Penan Affairs.

With regard to the adult education, the level of attendance in most villages is quite high, indicating good response from the Penan. However, one problem faced by the volunteers is trying to sustain the interest of the old people in the programme, and the lack of teaching aids, including electricity, for holding classes at night when the older people are free to attend such classes.

The volunteer programme is a good one, though its effectiveness will depend greatly on the type of supervision and support provided by professional change agents to the young volunteers. There are indications that such supervision is not at the level desired.

The programme that has not shown much result is the agricultural programme. One of the reasons might be that the volunteers have not shown sufficient confidence in doing farming themselves; and as the Penan is still mobile, their farms are not properly attended to. In many cases, survival rate of the crops is very low.

The Service Centre programme

At the outset, it should be made clear that the State government of Sarawak does not have a policy to resettle Penan. It is aware that there are numerous problems involved in resettling them. First there is the problem of drawing land boundary between a resettled Penan community and their neighbours. The Penan neighbour might lay customary rights to land where the Penan are resettled. Second, relocating the population from one area to another is expensive in terms of transporting people and goods. It also puts unnecessary social pressure on the people to be relocated.

Instead of relocating the Penan, the Government is reaching out to them in their existing settlements by providing basic services through the establishment of service centres at various strategic Penan settlement. The service centres serve the settlements where they are located and those nearby. Each of these service centres is provided with basic facilities and services, namely, a primary school, a rural clinic and an agriculture station. Other services will be added on to the center as the needs arise. The settlements covered by one service centre is exemplified in Map 2.

The primary school is staffed with trained teachers from the education department, the rural clinic staffed by a hospital assistant and a community nurse from the medical and health department and the agriculture station manned by an agricultural extension worker and a home demonstrator from the agriculture department. Besides manning the centre, these officials are expected to pay regular visits to nearby villages and undertake community development works in these villages. They are also expected to supervise the works of the volunteers in nearby villages.

One cash generating project that the agriculture department should incorporate in its agricultural programme, especially among the Western Penan is the cottage industry. Western Penan are highly skilled blacksmithers and weavers of rattan mats and baskets. Indeed, most of their cash income is derived from these two sources. The blacksmith industry could be further improved to increase *parang* production.

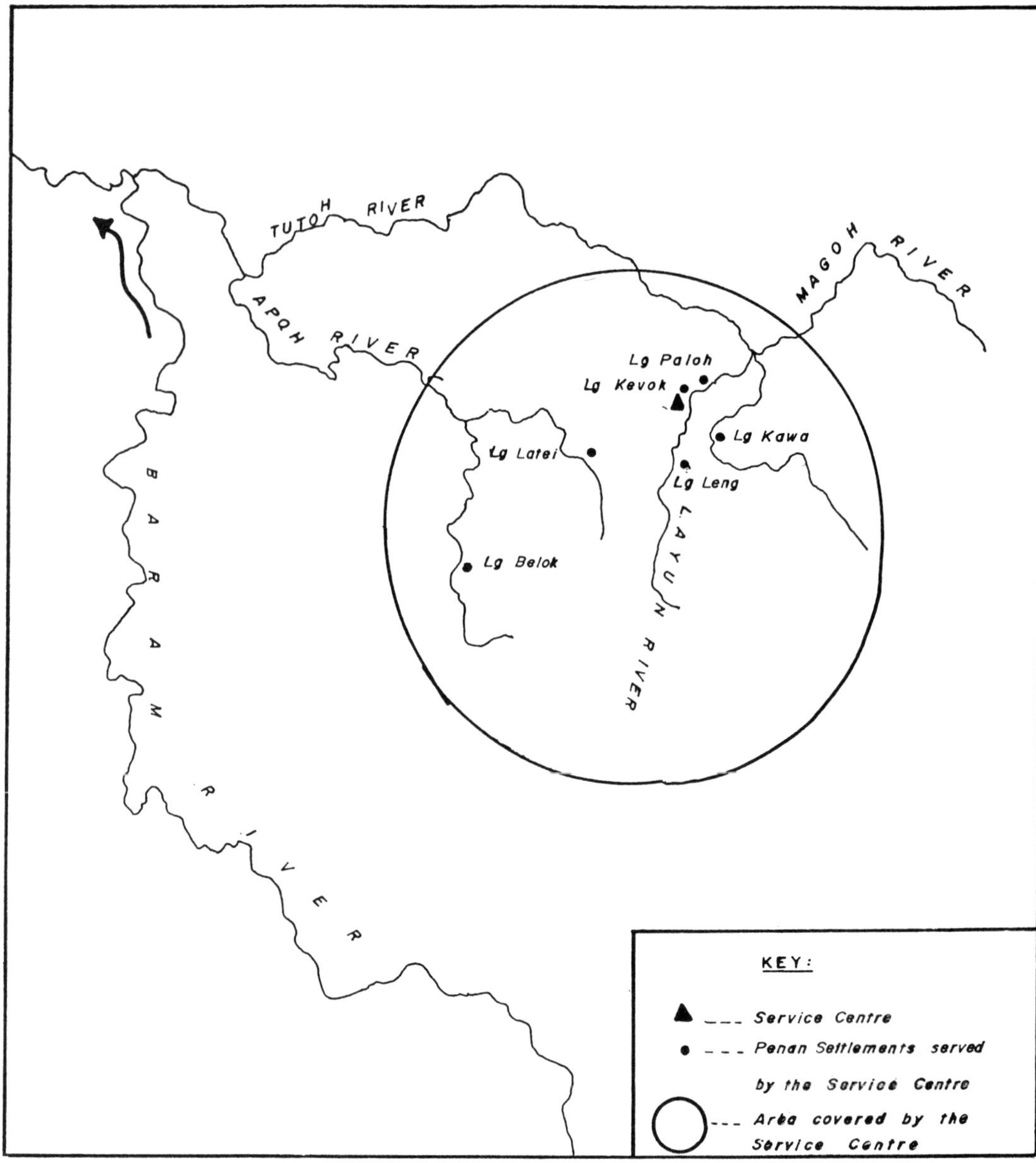

Map 2. Penan settlements covered by a service centre.

With regard to the production of rattan mats and baskets, the main problem faced by the Penan is the depleting supply of rattan. There are two possible solutions to the problem: one is to introduce the planting of rattan to the Penan. This is being done in a pilot project. The other is to identify rattan stands in the forest and reserve the area for the Penan to manage through the concept of *molong*. The first is a radical proposal for the Penan to accept, but the second option will be well received by them.

At present only two of the service centres are in full operation. One in Lusong Laku in the Linau River, Belaga District serving 351 people of the three settlements of Lusong Laku, Long Kajang and Long Tanyit. The other is at Long Kevok, Layun River, Baram District serving 678 people of the settlements of Long Kevok, Long

Leng, Long Palo, Long Latei, Long Belok and Long Kawa.

Four more service centres have been identified for implementation soon; three in Baram District and one for Belaga District. The three new service centres proposed in the Baram District will be located at Batu Bungan, near Mulu National Park, Long Sait, Selungo River in the Upper Baram and Long Jakitan, Silat River also in the Upper Baram.

At the moment it is still too early to give an assessment of the Service Centre programme. Judging from the success of the primary school at Lusong Laku which was judged as one of the four best schools in Malaysia in 1991 in terms of academic performance and extracurricular activities it can be safely expected that the programme will bring the desired benefits to the Penan. The biggest challenge in providing essential services to the Penan through the service centres is not finding the money to construct the centres but finding dedicated teachers, hospital assistants, nurses, agricultural extension workers and home demonstrators who are willing to work in very trying circumstances in the remote interior.

6. Rationale for special assistance to the Penan

As we have discussed earlier, the majority of the Penan in Sarawak are semi-nomadic — they are in the transitional stage of becoming settled community. This is a stage between two extreme modes of existence — from one being totally dependent on the forest for foods and other necessities to one in which the forest no longer plays an important role. This is a very difficult stage for any nomadic group, as it requires members of the community to acquire new skills and habits to enable them to produce foods and other items of basic needs, to learn to live with neighbours of different groups, and more importantly to trade in the open market. Without proper guidance and assistance, many of the nomadic groups will not be able to progress towards the settled mode of living.

The case of the Penan is a typical one — having been persuaded to take up a settled lifestyle many years ago and finding that the path is full of obstacles and constraints, many are not able to proceed further. Thus they remain in a transition stage; a stage where they are still learning to adapt to the settled lifestyle, while still relying on the forest for their daily necessities. Some people may say that the semi-nomadic Penan are enjoying the best of both worlds, and it is for this reason that they are not willing to completely settle down. This is not true — in reality many are suffering psychologically from embarrassment and a feeling of inferiority complex when facing their more advanced neighbours and also from a feeling of inadequacy to face the many demands of settled lifestyle. They still need the forest, because (1) the forest guarantees their security for it provides them with foods and other necessities, and (2) they are still not sure that the settled mode of life can guarantee their survival.

Penan want development; they want to live a lifestyle similar to that of their more settled neighbours. They often express their desire to own modern household items and have easy access to modern services and facilities especially clinics and schools. Every time government officials, especially the high ranking ones, visit their settlements, they never fail to give the former shopping lists containing a variety of modern items, such as outboard engines, electric generators, building materials, rice mills, footpath and bridges, clinics and schools within the immediate vicinity of their settlements. Whether what they ask for are in line with their real needs is not the point — the fact that they are asking for these modern items reflect a desire to own or have easy access to them, just like their more settled neighbours downstream. Dr. Peter Brosius of University of Georgia, U.S.A. who studied the Penan community in the Belaga District (for about three years) confirmed that the Penan want development and to live a more settled life! They want the nice things which they normally associate with settled mode of living.

The issue here is not whether we should leave the Penan in the jungle as hunters and gatherers or bring them into the mainstream of development. The crucial questions are (1) what are the best or most effective strategy(ies) or approach(es) to develop the community; and (2) what kinds of development programmes are needed and suitable for this community? In view of their stage of development and the difficult access to their settlements, the Penan need special assistance and programmes to enable them to become full-fledged sedentary farmers. While this is appreciated, numerous challenges (and obstacles) would face the development workers in trying to bring about development to the Penan. These challenges which take the form of obstacles and constraints, include difficult access and remoteness of the settlements giving rise to lack of communication between official change agents and the community, their dependence on the forest vis-à-vis of logging, rights over farming land, lack of farming skills and their distrust of their neighbours and outsiders generally.

7. Concluding remarks

Development is a slow process; it will be slower still if your target group are bands of hunters and gatherers. Our aim is at the total transformation of the people themselves, not merely changing the physical condition of their villages. A total transformation would include a change in their world view, habits and skills so as to enable them to be self-sufficient and function effectively as members of the Malaysian society. These are the real challenges facing development workers in Sarawak. It calls for special strategies and programmes including a strong communication component to win the trust of the target groups — the semi-nomadic and nomadic Penan. More importantly, these strategies require a group of dedicated and committed development workers who are willing to live and work with the Penan in the difficult environment. We believe officials at the Service Center and Penan volunteers, given the right motivation and support, can form the nucleus of such development workers.

Acknowledgement

This article was originally presented as a paper at the International Symposium on Rehabilitation of Tropical Rainforest Ecosystems: Research and Development Priorities, held in Kuching, Sarawak, Malaysia, on 2–4 September 1992.

References

Brosius, J. P. 1986. River, forest and mountain: the Penan gang landscape. Sarawak Museum Journal XXXV (57) (New Series): 173–184.

Brosius, J. P. 1989. Landscape and society in Borneo: Penan hunter-gatherers of Sarawak. A paper presented at a symposium on Rainforest Life in a Canopied World, the University Museum of Archaeology and Anthropology, University of Pennsylvania, February 4, 1989.

Brosius, J. P. 1990. Penan hunter-gatherers of Sarawak, East Malaysia. Anthroquest 42: Summer 1990, 1–7.

Brosius, J. P. 1992. Perspective on Penan development in Sarawak. Sarawak Gazette CXIX (1519): 5–22.

District Officer, Baram. 1961. Penan affairs. An official memorandum from the District Officer, Baram to the resident, 4th Division, Miri dated 21 January 1961. District Office, Baram File 81/1008.

Langub, J. 1974. Adaptation to the settled life by the Penan of Belaga sub-district. Sarawak Museum Journal XXII (43) (New Series): 295–301.

Langub, J. 1984. Tamu: barter trade between the Penan and their neighbours. Sarawak Gazette CX (1485): 11–15.

Langub, J. 1988. The Penan strategy. In: Denslow, J. & Padoch, C. (Eds.), People of the Tropical Rain Forest. Berkeley: California University Press.

Langub, J. 1989. Some aspects of life of the Penan. Sarawak Museum Journal XL (61) (New Series): Special Issue No. 4, Part III, 169–184.

Langub, J. 1990a. Hunting and gathering: a view from within. A paper presented at the First Extraordinary Session, The Borneo Research Council, Kuching, Sarawak, Malaysia, August 4–9, 1990.

Langub, J. 1990b. A journey through the nomadic Penan country. Sarawak Gazette CXVII (1514): 5–27.

Murang, O. 1990. Kelabits, Sa'bans, Kenyah and their Penan neighbours: partners in development. A paper presented at the First Extraordinary Session, The Borneo Research Council, Kuching, Sarawak, Malaysia, August 4–9, 1990.

Needham, R. 1972. Punan-Penan. In: LeBar, F. (Ed.), Ethnic Groups In Insular Southeast Asia. Vol. I: Indonesia, Adman Islands, and Madagascar. New Haven: Hraf.

Urguhart, I. A. N. 1950. Tribulations and pleasures of travelling to Long Kajang. Sarawak Gazette LXXVI (1103) (7 February 1950): 33–35 and LXXVI (1104) (7 March 1950): 67–69.

24. Concluding remarks

W. HABER
Lehrstuhl f. Landschaftsökologie der TU München, 8050 Freising 12, Germany

I am pleased to take up Dr. Lieth's proposal to make some concluding remarks on this symposium. It was, I believe, very rewarding as seen from the viewpoint of ecological science. Many ecologists, biologists, foresters and other scientists coming from different parts of the world have been able to exchange their views and to present them to representatives of the "land users", in particular to staff members of Mitsubishi Corporation who sponsored this symposium.

But I have also the feeling that the symposium, seen from a social and political context, has not fulfilled all expectations, leaving a certain disappointment. The reason seems to me very clear and should be addressed frankly: On one side there is the millennia-long history of human exploitation of the biosphere, which started with the invention of agriculture some 7000 years ago. This exploitation has been accompanied in many regions, and especially those imprinted by so-called western civilization, with a strong tradition of carelessness in treating the biosphere and its resources. Only today, after this carelessness has increased excessively, having turned even to ruthlessness, we begin to learn our lessons from it. We are learning from ecological insights and research, but — and this is the other side of my reflection — ecology is a very young science, ecosystem science even younger, and both are still not strong enough to really thoroughly change the old exploitative traditions. Silviculture is rather young too, its origins in Central Europe dating back to 200 years only. And most silvicultural knowledge is based on — predominantly empirical — knowledge and experiences gathered in temperate zone forests, being of only limited value for tropical silviculture which is so urgently needed today. However, there were some very interesting and encouraging contributions in this symposium opening up guidelines of silviculture for humid tropical forests.

Thus we are confronted with a rather young tradition of silviculture and ecology opposing, and correcting, an old "western" tradition of exploitation and carelessness which is being driven by very strong political and economic forces all over the world. It is the outgrowth of what we call the western culture and civilization, and one of its main economic components are the big multinational corporations which have become the spearheads of this tradition of exploitation. I think it is both unavoidable and necessary to firmly attack and criticize that carelessness which has invaded the attitude of the big corporations.

In my opinion as an ecologist, and even on behalf of the international community of ecologists, groups such as the Robin Wood Group who took part in this symposium deserve our support in their efforts to change general thinking and the attitude of the resource exploiters, even if we do not always agree with their kind of argumentation. As ecologists or scientists we know or understand what is right or wrong in land use, but we depend upon the support of social forces to exert pressure on those who are responsible for economic or political actions, in order to change their directions, policies and operations in using our earth's resources.

In certain regards, we, as ecologists, may also be blamed for some misunderstandings. For quite a while we have advocated the notion of equilibrium of nature, at least for climax ecosystems. Some of us have interpreted this using a picture of nature in harmony. But in reality there is no harmony, neither within nature nor be-

H. Lieth and M. Lohmann (eds.), Restoration of Tropical Forest Ecosystems, 251–253.

tween humans and nature. On the other hand, in applied ecology it is our goal to approach at least a kind of balance between resource exploitation and regeneration, and there is always a conflict between this approach to a pretended harmony and reality. In this context I would like to remind you of a book which I consider to be one of the most illuminating publications of the second half of this century, Edward O. Wilson's book *On Human Nature.* He has made very clear that this nature is an exploitative one; and of course, as components of ecosystems, humans belong to the group of heterotrophs — they are consumers, they are predators and very strong and skillful competitors. This is our inherited natural role, and the majority of humans are more or less unconsciously following the innate drive of the creatures they are.

Only a minority — hopefully a growing one — is recognizing that humans should not, or not exclusively, follow their innate drive of consumerism or careless competitiveness. Instead they should assume a new role and a new responsibility, i.e. that of prudent and balanced control of resource use and exploitation. It is the role natural ecosystems play in regulating individual and population behaviour. Of course I am simplifying the problem, but as a matter of fact, all living beings have to exploit their surroundings or their environment as much as possible to make a living, and it is the powers of the ecosystem, or the higher organization level, which regulate and balance exploitation. Ecosystem research has just begun to identify and to understand these regulating agents keeping the system durable and well functioning.

Transferring this new knowledge into social and political actions is the main task of applied ecology, and of urgent necessity. We have to learn to cooperate with natural regulating forces and agents — to cooperate in such ways that our own intellectual abilities be combined with systems' regulation in achieving a smoother and more harmonious function of human–nature relationships.

However, we are still suffering from a critical default inherent in western culture and tradition — the lack of respect for nature. This is a tragic default as we all are either members of western culture or aspiring to adapt to it. For the next decades it is crucial that we learn our main lesson from the wealth of ecological insights we have gathered: that we cannot sustain the biosphere and our society on the basis of exploitation alone. But for transferring ecological knowledge into economic and political actions for achieving sustainability, we also need social and political stability. This holds even for single sections of our system – the maintenance and sustainability of forests require an effective and reliable forest service supervising and controlling all forest operations including protection. This has been mentioned in several contributions to this symposium, but a forest service cannot function in a politically unstable society.

Political stability, however, is as desirable as it is often lacking or uncertain, and we do not need to go very far from here to testify to it. Look what happens during these weeks in Yugoslavia. We all are preaching for priority of solving environmental or ecological problems which are fundamental for human survival. But what priority has been set, and is being followed, in Yugoslavia, not far from the centre of Europe — a continent which is regarded as a model for Third World countries? From a field trip in the mid-eighties I remember the large afforestations in the Dubrovnik area where big efforts are done to restore forest on the Carstic Mountains deforested by the Romans some 2000 years ago. Now all this is set on fire by the civil war which is raging in this country.

These are problems ecologists cannot overlook in struggling for an environmentally balanced land use or for the protection of forests. Here is the interface between ecological knowledge or science and political and social forces, which have command over everyday's practice. This is for me as ecologist and as present president of INTECOL a very difficult interface. I can clearly recognize, and evaluate, what happens to the Penan in Sarawak; but as an ecological scientist I do not have any competence for handling this conflict. Of course I strongly recommend the protection of a large tract of Sarawak forests including the habitat of the Penan; but these people have to take their own decision on their future life-style. We cannot expect them to remain "ecological forest dwellers" and to resist every external cultural influence — only because we are "ecologically pleased" with their present life-style.

Behind the Penan tragedy there is a more fundamental problem. The "developed countries" are being rightly accused of imposing their political and economic system upon people living in quite different cultural traditions and conditions. Within the developed countries, there is also growing criticism from environmentalists who are fighting for a wise use of the natural resources — including primeval forests — of "less developed countries", and for a comprehensive nature preservation policy in tropical regions. Most participants of this symposium are representing this attitude. But we overlook the fact that our views and goals of nature conservation might appear as alien and as unintelligible to many tropical people as are our political and economic efforts. Personally, I often have mixed feelings when travelling through tropical forest areas and trying to apply our western views of forest conservation or regeneration. We here in Central Europe did not treat our forests so much better in the past, and even today's forest policy is by no means exempted from exploitative thinking.

Sustainability or sustainable use of resources is the leading idea of the last decade of this century. To achieve it we can only proceed by many small but interconnected steps influencing land use, respecting land suitability and ecological constraints. There is no quick or single solution for these problems. This symposium has been one of these steps — *maybe a bigger one* — in emphasizing the common responsibility of ecologists, environmental action groups, foresters, and resource exploiters for a sustainable use of natural resources. As non-renewable resources are becoming more scarce, the wise use of renewable resources is gaining more importance. One of the most essential renewable resources is forest, and especially tropical forest. This symposium has demonstrated that a sound compromise between protection and exploitation of the tropical forest resource has to be developed and implemented. The message goes to the sponsor of this symposium, to Mitsubishi Corporation, and I think they will have well understood this message. We shall look very watchfully upon what will happen in the next years in the Sarawak forests to find out if the lessons have been learned. Thank you very much.

Index of plant and animal names

Index of scientific names

General index